Plant-Bacteria Interactions

Edited by
Iqbal Ahmad, John Pichtel and
Shamsul Hayat

Related Titles

Roberts, K. (ed.)

Handbook of Plant Science

2 Volume Set

2007
ISBN: 978-0-470-05723-0

Kahl, G., Meksem, K. (eds.)

The Handbook of Plant Functional Genomics

Concepts and Protocols

2008
ISBN: 978-3-527-31885-8

Meksem, K., Kahl, G. (eds.)

The Handbook of Plant Genome Mapping

Genetic and Physical Mapping

2005
ISBN: 978-3-527-31116-3

Cullis, C. A.

Plant Genomics and Proteomics

2004
ISBN: 978-0-471-37314-8

Kayser, O., Quax, W. J. (eds.)

Medicinal Plant Biotechnology

From Basic Research to Industrial Applications

2007
ISBN: 978-3-527-31443-0

Dolezel, J., Greilhuber, J., Suda, J. (eds.)

Flow Cytometry with Plant Cells

Analysis of Genes, Chromosomes and Genomes

2007
ISBN: 978-3-527-31487-4

Plant-Bacteria Interactions

Strategies and Techniques to Promote Plant Growth

Edited by
Iqbal Ahmad, John Pichtel, and Shamsul Hayat

WILEY-VCH Verlag GmbH & Co. KGaA

The Editors

Dr. Iqbal Ahmad
Aligarh Muslim University
Department of Agricultural Microbiology
Aligarh 202002
India

Prof. Dr. John Pichtel
Ball State University
Department of Natural Resources and
Environmental Management
WQ 103
Muncie, IN 47306
USA

Dr. Shamsul Hayat
Aligarh Muslim University
Department of Botany
Aligarh 202002
India

Library of Congress Card No.: applied for

British Library Cataloguing-in-Publication Data
A catalogue record for this book is available from the British Library.

Bibliographic information published by the Deutsche Nationalbibliothek
Die Deutsche Nationalbibliothek lists this publication in the Deutsche Nationalbibliografie; detailed bibliographic data are available on the Internet at <http://dnb.d-nb.de>.

© 2008 WILEY-VCH Verlag GmbH & Co. KGaA, Weinheim

Composition   Thomson Digital, Noida, India
Printing   Betz-Druck GmbH, Darmstadt
Bookbinding   Litges & Dopf GmbH, Heppenheim
Cover Design   Adam Design, Weinheim

Printed in the Federal Republic of Germany
Printed on acid-free paper

ISBN: 978-3-527-31901-5

Contents

Plant-Bacteria Interactions. Strategies and Techniques to Promote Plant Growth
Edited by Iqbal Ahmad, John Pichtel, and Shamsul Hayat
Copyright © 2008 WILEY-VCH Verlag GmbH & Co. KGaA, Weinheim
ISBN: 978-3-527-31901-5

List of Contributors

Farah Ahmad
Department of Agricultural
Microbiology
Faculty of Agricultural Sciences
Aligarh Muslim University
Aligarh 202002
India

Iqbal Ahmad
Department of Agricultural
Microbiology
Faculty of Agricultural Sciences
Aligarh Muslim University
Aligarh 202002
India

Javier Andrés
Laboratorio de Interacción
Microorganismo–Planta
Facultad de Ciencias Exactas
Físico-Químicas y Naturales
Universidad Nacional de Río Cuarto
Campus Universitario Ruta 36
Km 601 (5800) Río Cuarto
Córdoba
Argentina

José Antonio Lucas García
Dpto. CC. Ambientales y Recursos
Naturales
Facultad de Farmacia
Universidad San Pablo CEU
Urb. Montepríncipe
Ctra. Boadilla del Monte Km 5.3
28668 Boadilla del Monte
Madrid
Spain

Farrukh Aqil
Department of Agricultural
Microbiology
Faculty of Agricultural Sciences
Aligarh Muslim University
Aligarh 202002
India

Germán Avanzini
Laboratorio de Interacción
Microorganismo–Planta
Facultad de Ciencias Exactas
Físico-Químicas y Naturales
Universidad Nacional de Río Cuarto
Campus Universitario Ruta 36
Km 601 (5800) Río Cuarto
Córdoba
Argentina

Plant-Bacteria Interactions. Strategies and Techniques to Promote Plant Growth
Edited by Iqbal Ahmad, John Pichtel, and Shamsul Hayat
Copyright © 2008 WILEY-VCH Verlag GmbH & Co. KGaA, Weinheim
ISBN: 978-3-527-31901-5

Jorge Barriuso
Dpto. CC. Ambientales y Recursos
Naturales
Facultad de Farmacia
Universidad San Pablo CEU
Urb. Montepríncipe
Ctra. Boadilla del Monte Km 5.3
28668 Boadilla del Monte
Madrid
Spain

Evelin Carlier
Laboratorio de Interacción
Microorganismo–Planta
Facultad de Ciencias Exactas
Físico-Químicas y Naturales
Universidad Nacional de Río Cuarto
Campus Universitario Ruta 36
Km 601 (5800) Río Cuarto
Córdoba
Argentina

Alex T. Chow
College of Environmental Science and
Engineering
South China University of Technology
Guangzhou
China

Frank B. Dazzo
Department of Microbiology and
Molecular Genetics
Michigan State University
East Lansing, MI 48824
USA

Dilfuza Egamberdiyeva
Department of Biotechnology
Faculty of Biology
National University of Uzbekistan
100174 Tashkent
Uzbekistan

Ana García-Villaraco Velasco
Dpto. CC. Ambientales y Recursos
Naturales
Facultad de Farmacia
Universidad San Pablo CEU
Urb. Montepríncipe
Ctra. Boadilla del Monte Km 5.3
28668 Boadilla del Monte
Madrid
Spain

Stephen R. Giddens
Department of Plant Sciences
University of Oxford
South Parks Road
Oxford OX1 3RB
UK

Reeta Goel
Department of Microbiology
G.B. Pant University
of Agriculture & Technology
Pantnagar, Distt. U.S. Nagar
Uttarakhand 263145
India

J. González-López
Institute of Water Research
University of Granada
Calle Ramón y Cajal 4
180071 Granada
Spain

F.J. Gutiérrez Mañero
Dpto. CC. Ambientales y Recursos
Naturales
Facultad de Farmacia
Universidad San Pablo CEU
Urb. Montepríncipe
Ctra. Boadilla del Monte Km 5.3
28668 Boadilla del Monte
Madrid
Spain

S. Hayat
Department of Botany
Aligarh Muslim University
Aligarh 202002
India

Khandakar R. Islam
Crop, Soil and Water Resources
The Ohio State University South
Centers
1864 Shyville Road
Piketon, OH 45661
USA

Robert W. Jackson
School of Biological Sciences
University of Reading
Whiteknights
Reading RG6 6AJ
UK

Hemant K. Jaiswal
Department of Genetics and
Plant Breeding
Institute of Agricultural Sciences
Banaras Hindu University
Varanasi 221005
India

Alexander A. Kamnev
Institute of Biochemistry and
Physiology of Plants and
Microorganisms
Russian Academy of Sciences
410049 Saratov
Russia

Abdul G. Khan
14, Clarissa Place
Ambarvale NSW 2560
Australia

Mahejibin Khan
Department of Biotechnology
Kumaun University
Nainital
India

M.S. Khan
Department of Agricultural
Microbiology
Faculty of Agricultural Sciences
Aligarh Muslim University
Aligarh 202002
India

Sarita Kumari
Department of Microbiology
G.B. Pant University of Agriculture
& Technology
Pantnagar, Distt. U.S. Nagar
Uttarakhand-263145
India

M.V. Martínez-Toledo
Institute of Water Research
University of Granada
Calle Ramón y Cajal 4
180071 Granada
Spain

Ravi P.N. Mishra
Department of Genetics and Plant
Breeding
Institute of Agricultural Sciences
Banaras Hindu University
Varanasi 221005
India

Christina D. Moon
AgResearch Ltd
Grasslands Research Centre
Private Bag 11008
Palmerston North
New Zealand

Javed Musarrat
Department of Agricultural
Microbiology
Faculty of Agricultural Sciences
Aligarh Muslim University
Aligarh 202002
India

Carolina Pasluosta
Laboratorio de Interacción
Microorganismo–Planta
Facultad de Ciencias Exactas
Físico-Químicas y Naturales
Universidad Nacional de Río Cuarto
Campus Universitario Ruta 36
Km 601 (5800) Río Cuarto
Córdoba
Argentina

C. Pozo Clemente
Institute of Water Research
University of Granada
Calle Ramón y Cajal 4
180071 Granada
Spain

Agustín Probanza Lobo
Dpto. CC. Ambientales y Recursos
Naturales
Facultad de Farmacia
Universidad San Pablo CEU
Urb. Montepríncipe
Ctra. Boadilla del Monte Km 5.3
28668 Boadilla del Monte
Madrid
Spain

Beatriz Ramos Solano
Dpto. CC. Ambientales y Recursos
Naturales
Facultad de Farmacia
Universidad San Pablo CEU
Urb. Montepríncipe
Ctra. Boadilla del Monte Km 5.3
28668 Boadilla del Monte
Madrid
Spain

B. Rodelas González
Dpto. de Microbiología
Facultad de Farmacia
University of Granada
Calle Ramón y Cajal 4
180071 Granada
Spain

M. Rodríguez-Díaz
Institute of Water Research
University of Granada
Calle Ramón y Cajal 4
180071 Granada
Spain

Susana Rosas
Laboratorio de Interacción
Microorganismo–Planta
Facultad de Ciencias Exactas
Físico-Químicas y Naturales
Universidad Nacional de Río Cuarto
Campus Universitario Ruta 36
Km 601 (5800) Río Cuarto
Córdoba
Argentina

Marisa Rovera
Departamento de Microbiología e
lmmunologia
Universidad Nacional de Río Cuarto
Ruta 36, Km 601
5800 Río Cuarto
Córdoba
Argentina

Alok Sharma
Department of Structural Biology
Helmholtz Centre for Infection
Research
Inhoffenstraße 7
D-38124 Braunschweig
Germany

Yogesh S. Shouche
National Center for Cell Sciences
Pune University Campus
Gomeskhind
Pune 411007
India

Manoj K. Singh
Department of Genetics and Plant
Breeding
Institute of Agricultural Sciences
Banaras Hindu University
Varanasi 221005
India

Ramesh K. Singh
Department of Genetics and Plant
Breeding
Institute of Agricultural Sciences
Banaras Hindu University
Varanasi 221005
India

Lai M. So
Department of Biology
The Chinese University of Hong Kong
Shatin, NT
Hong Kong SAR
China

Ravindra Soni
Department of Microbiology
G.B. Pant University of Agriculture
& Technology
Pantnagar, Distt. U.S. Nagar
Uttarakhand 263145
India

Kin H. Wong
Department of Biology
The Chinese University of Hong Kong
Shatin, NT
Hong Kong SAR
China

Po K. Wong
Department of Biology
The Chinese University of Hong Kong
Shatin, NT
Hong Kong SAR
China

Youssef G. Yanni
Sakha Agricultural Research Station
Dept. of Soil Microbiology
Kafr El-Sheikh 33717
Egypt

Maryam Zahin
Department of Agricultural
Microbiology
Faculty of Agricultural Sciences
Aligarh Muslim University
Aligarh 202002
India

Mohd G.H. Zaidi
Department of Chemistry
G.B. Pant University of Agriculture
& Technology
Pantnagar, Distt. U.S. Nagar
Uttarakhand 263145
India

Xue-Xian Zhang
School of Biological Sciences
University of Auckland
Private Bag 92019
Auckland
New Zealand

1
Ecology, Genetic Diversity and Screening Strategies of Plant Growth Promoting Rhizobacteria (PGPR)

Jorge Barriuso, Beatriz Ramos Solano, José A. Lucas, Agustín Probanza Lobo, Ana García-Villaraco, and Francisco J. Gutiérrez Mañero

1.1
Introduction

1.1.1
Rhizosphere Microbial Ecology

The German agronomist Hiltner first defined the rhizosphere, in 1904, as the 'effect' of the roots of legumes on the surrounding soil, in terms of higher microbial activity because of the organic matter released by the roots.

Until the end of the twentieth century, this 'effect' was not considered to be an ecosystem. It is interesting to make some brief observations about the size, in terms of energy and extension, of this ecosystem to determine its impact on how the biosphere functions. First, in extension, the rhizosphere is the largest ecosystem on earth. Second, the energy flux in this system is enormous. Some authors estimate that plants release between 20 and 50% of their photosynthates through their roots [1,2]. Thus, rhizosphere's impact on how the biosphere functions is fundamental.

A large number of macroscopic organisms and microorganisms such as bacteria, fungi, protozoa and algae coexist in the rhizosphere. Bacteria are the most abundant among them. Plants select those bacteria contributing most to their fitness by releasing organic compounds through exudates [3], creating a very selective environment where diversity is low [4,5]. A complex web of interactions takes place among them, and this may affect plant growth, directly or indirectly. Since bacteria are the most abundant microorganisms in the rhizosphere, it is highly probable that they influence the plant's physiology to a greater extent, especially considering their competitiveness in root colonization [6].

Bacterial diversity can be defined in terms of taxonomic, genetic and functional diversity [7]. In the rhizosphere, the metabolic versatility of a bacterial population (functional diversity) is based on its genetic variability and on possible interactions with other prokaryotic and eukaryotic organisms such as plants.

Plant-Bacteria Interactions. Strategies and Techniques to Promote Plant Growth
Edited by Iqbal Ahmad, John Pichtel, and Shamsul Hayat
Copyright © 2008 WILEY-VCH Verlag GmbH & Co. KGaA, Weinheim
ISBN: 978-3-527-31901-5

However, a question still to be answered regarding microbial communities in the rhizosphere is the relationship between the ecological function of communities and soil biodiversity. In spite of the lack of information about the importance of the diversity and the richness of species related to their ecological function [8,9], soil organisms have been classified several times in functional groups [10].

This lack of knowledge about bacterial diversity is partly owing to the high number of species present, as well as to the fact that most bacteria are viable but not culturable.

The biological diversity of soil microorganisms has been expressed using a variety of indexes [11,12] and mathematical models [13], but there is no accepted general model to describe the relationship among abundance, species' richness and dominancy. It is, therefore, reasonable that the components of diversity are studied separately to quantify them [14].

Bacterial diversity studies are more complex at taxonomic, functional and genetic levels than are similar studies on eukaryotic organisms owing to the minute working scale and the large number of different bacterial species present in the environment. Torsvik and coworkers [15] identified more than 7000 species in an organic forest soil.

The variations in populations through space and time and their specialization in ecological niches are two important factors in the rhizosphere that must be considered in studying how species' richness influences the functioning of the system. The functioning of soil microbial communities is based on the fact that there is appropriate species diversity for the resources to be used efficiently and that this can be maintained under changing conditions [14].

In the rhizosphere, as in other well-formed ecosystems with an appropriate structure, changes in some of the components can affect entire or part of the system. The degree of impact will depend on features of the system such as its resistance or resilience. The state of this system changes depending on variables such as the age of the plant, root area, light availability, humidity, temperature and plant nutrition [16,17]. Under stressful conditions, the plant exerts a stronger control on release of root exudates [18,19]. From this viewpoint, it is reasonable to assume that the changes that occur in the plant will change the root exudation patterns and, thereby, the rhizosphere microbial communities. There have been many studies that relate the quality and quantity of the exudates with changes in the structure of rhizosphere microbial communities [20].

In 1980, Torsvik [21] published the first protocol for the extraction and isolation of microbial DNA from soil. Since then, there have been many studies directed at the development of new methods and molecular tools for the analysis of soil microbial communities. However, molecular genetics is not the only tool used in solving the difficulties in analyzing soil microbial communities. A multimethodological approach using conventional techniques such as bacterial isolation and physiological studies, together with molecular genetics, will be necessary to fully develop the study of microbial ecology [22,23].

The bacterial community can be studied using several approaches: first, a structural approach, attempting to study the entire soil bacterial community; second, the relationships between populations and the processes that regulate the system; and finally, a functional approach.

Recent research has shown that, within a bacterial population, cells are not isolated from each other but communicate to coordinate certain activities. This communication is key to their survival since microbial success depends on the ability to perceive and respond rapidly to changes in the environment [24]. Bacteria have developed complex communication mechanisms to control the expression of certain functions in a cell density-dependent manner, a phenomenon termed as *quorum sensing* (QS).

Quorum sensing confers an enormous competitive advantage on bacteria, improving their chances to survive as they can explore more complex niches. This mechanism is also involved in the infection ability of some plant bacterial pathogens (such as *Xanthomonas campestris* and *Pseudomonas syringae*) [25].

Bacterial communication by quorum sensing is based on the production and release of signal molecules into the medium, termed autoinducers, concentration being proportional to cell density. When bacteria detect the signal molecule at a given concentration, the transcription of certain genes regulated by this mechanism is induced or repressed. There are many microbial processes regulated by quorum sensing, including DNA transference by conjugation, siderophore production, bioluminescence, biofilm formation and the ability of some bacteria to move, called swarming [26,27].

Recent studies have shown the importance of this type of regulation mechanism in putative beneficial bacterial traits for the plant, such as plant growth promotion, protection against pathogens or saline stress protection [28,29]. In addition, coevolution studies of plants and bacteria have determined that some plants release molecules, which mimic acyl homoserine lactones (AHLs) and even enzymes that are able to degrade the AHL molecule in root exudates. Somehow, plants have 'learned' the language of bacteria and use it for their own benefit. Some studies have discovered that this behavior leads to defense against plant bacterial pathogens, altering or blocking communication among bacteria, thus dramatically reducing their infection efficiency.

1.1.2
Plant Growth Promoting Rhizobacteria (PGPR)

Bacteria inhabiting the rhizosphere and beneficial to plants are termed PGPR [30]. Thus, the rhizosphere of wild plant species appears to be the best source from which to isolate plant growth promoting rhizobacteria [4,31].

A putative PGPR qualifies as PGPR when it is able to produce a positive effect on the plant upon inoculation, hence demonstrating good competitive skills over the existing rhizosphere communities. Generally, about 2–5% of rhizosphere bacteria are PGPR [32].

Some PGPR have been produced commercially as inoculants for agriculture, but it must be borne in mind that the inoculation of these bacteria in soil may affect the composition and structure of microbial communities, and these changes must be studied since they have, at times, been related to the inefficiency of biofertilizers when applied to plant roots [33,34]. On the contrary, many studies [35] have tested the efficiency of PGPR in various conditions, observing that PGPR are efficient under determined conditions only [36]. Knowledge of the structure of rhizosphere microbial communities and their diversity, as related to other essential processes within the system such as complexity, natural selection, interpopulational relations (symbiosis, parasitism, mutualism or competence), succession or the effect of disturbances, is the key to a better understanding of the system and for the correct utilization of PGPR in biotechnology.

Taking all of the above into consideration, it appears that quorum sensing can be a very useful tool in agriculture, with the potential to prevent bacterial pathogen attack and improve PGPR performance. There already exist transgenic plants that have been engineered to produce high levels of AHLs or an enzyme capable of degrading AHLs and that have demonstrated considerable capacity in blocking pathogen infection or altering PGPR performance [24].

1.2
Rhizosphere Microbial Structure

1.2.1
Methods to Study the Microbial Structure in the Rhizosphere

As mentioned above, the bacterial community can be studied through two approaches: structural and functional. To understand the structural approach, we must know the groups of individuals, their species and abundance. Traditionally, this has been done by extracting microorganisms from the system, culturing them in the laboratory and performing many morphological, biochemical and genetic tests. Bacteria extraction methods require a dispersing agent to disintegrate the links among cells and need to be performed using either physical or chemical agents or a combination of both.

When handling bulk soil, rhizosphere soil and plant roots, dispersion methods need to be used owing to the intimate relationship between bacteria and the substrate. The efficiency of these methods is evaluated by comparing the microbial biomass of the original substrate before and after extraction. However, microbial biomass is difficult to calculate. There are several ways to approach these parameters including direct counting under a microscope (e.g. by using acridine orange dye) [37], microbial respiration (i.e. substrate induced respiration, SIR [38]), ATP level assay [39], counting viable cells with the most probable number (MPN) [40], using biomarkers such as lipids [41] and soil fumigation with chloroform [42].

After extracting bacteria, several simple methods can be applied to isolate and count soil bacteria, such as growing them in a nonselective medium to obtain the total viable count (TVC). The data obtained with this method are expressed as colony forming units (CFUs).

These studies, in which bacteria are grown on plates, are used to calculate the soil bacterial diversity, by observing the number and abundance of each species. Diversity indexes, such as the Shannon index (H), the Simpson index and the equitability index (J), have all been used to describe the structure of communities from a mathematical viewpoint [43].

The percentage of culturable microorganisms in soil is very low; however, some researchers estimate this at only 10% [44], while others suggest 1% [43] or even lower (between 0.2 and 0.8%) [45]. Because of the limitation of some methods, techniques in which it is not necessary to culture microorganisms on plates are required. One such technique is the phospholipid fatty acid analysis (PLFA) [33,34,46–48]. Phospholipids are integrated in the bacterial cell membranes [49]. Different groups of microorganisms possess different fatty acid patterns. It is not usually possible to detect specific strains or species, but changes in the concentration of specific fatty acids can be correlated to changes in specific groups of microorganisms.

Another approach to nonculturable diversity is through techniques of molecular genetics, which, in the past 20 years, has revealed new information about soil microbial communities [50]. Techniques include DNA and/or RNA hybridization [51], polymerase chain reaction (PCR), ribosomal RNA sequencing [52], $G + C$ percentages [53] and DNA reassociation between bacteria in the community [53,54].

At present, the most notable techniques are temperature gradient gel electrophoresis (TGGE) and denaturing gradient gel electrophoresis (DGGE), both based on the direct extraction of DNA or RNA from soil; the amplification of this DNA (by means of PCR), followed by electrophoretic separation in a temperature gradient for the former or by using chemical denaturing substances for the latter. These techniques allow the separation of DNA fragments of exactly the same length but with different sequences, based on their melting properties [54–56]. Other techniques include restriction fragment length polymorphism (RFLP) [57,58], techniques related to the analysis and cutting of different restriction enzymes (amplified ribosomal DNA restriction analysis, ARDRA) [59] or cloning the rDNA 16S and then sequencing [5]. The use of microarrays [22] is also an emerging technique with a promising future, which permits the identification of specific genes [60].

Each of the methods described above possesses its own distinctive advantages and disadvantages. Generally, the more selective the method, the less able it is to detect global changes in communities and vice versa. Using these tools can provide an estimate of the microbial diversity in the soil.

1.2.2
Ecology and Biodiversity of PGPR Living in the Rhizosphere

In the last few years, the number of PGPR that have been identified has seen a great increase, mainly because the role of the rhizosphere as an ecosystem has gained importance in the functioning of the biosphere and also because mechanisms of action of PGPR have been deeply studied.

Currently, there are many bacterial genera that include PGPR among them, revealing a high diversity in this group. A discussion of some of the most abundant genera of PGPR follows to describe the genetic diversity and ecology of PGPR.

1.2.2.1 Diazotrophic PGPR

Free nitrogen-fixing bacteria were probably the first rhizobacteria used to promote plant growth. *Azospirillum* strains have been isolated and used ever since the 1970s when it was first used [61]. This genus has been studied widely, the study by Bashan *et al.* [62] being the most recent one reporting the latest advances in physiology, molecular characteristics and agricultural applications of this genus.

Other bacterial genera capable of nitrogen fixation that is probably responsible for growth promotion effect, are *Azoarcus* sp., *Burkholderia* sp., *Gluconacetobacter diazotrophicus*, *Herbaspirillum* sp., *Azotobacter* sp and *Paenibacillus* (*Bacillus*) *polymyxa* [63]. These strains have been isolated from a number of plant species such as rice, sugarcane, corn, sorghum, other cereals, pineapple and coffee bean.

Azoarcus has recently gained attention due to its great genetic and metabolic diversity. It has been split into three different genera (*Azovibrio*, *Azospira* and *Azonexus*) [64]. The most distinctive characteristic of these genera, which particularly differentiates them from other species, is their ability to grow in carboxylic acids or ethanol instead of sugars, with their optimum growth temperature ranging between 37 and 42 °C. *Azoarcus* is an endophyte of rice and is currently considered the model of nitrogen-fixing endophytes [65].

1.2.2.2 *Bacillus*

Ninety-five percent of Gram-positive soil bacilli belong to the genus *Bacillus*. The remaining 5% are confirmed to be *Arthrobacter* and *Frankia* [66]. Members of *Bacillus* species are able to form endospores and hence survive under adverse conditions; some species are diazotrophs such as *Bacillus subtilis* [67], whereas others have different PGPR capacities, as many reports on their growth promoting activity reveal [33,68,69].

1.2.2.3 *Pseudomonas*

Among Gram-negative soil bacteria, *Pseudomonas* is the most abundant genus in the rhizosphere, and the PGPR activity of some of these strains has been known for many years, resulting in a broad knowledge of the mechanisms involved [33,70,71].

The ecological diversity of this genus is enormous, since individual species have been isolated from a number of plant species in different soils throughout the world. *Pseudomonas* strains show high versatility in their metabolic capacity. Antibiotics, siderophores or hydrogen cyanide are among the metabolites generally released by these strains [72]. These metabolites strongly affect the environment, both because they inhibit growth of other deleterious microorganisms and because they increase nutrient availability for the plant.

1.2.2.4 Rhizobia

Among the groups that inhabit the rhizosphere are rhizobia. Strains from this genus may behave as PGPR when they colonize roots from nonlegume plant species in a nonspecific relationship. It is well known that a number of individual species may release plant growth regulators, siderophores and hydrogen cyanide or may increase phosphate availability, thereby improving plant nutrition [73]. An increase in rhizosphere populations has been reported after crop rotation with nonlegumes [74], with this abundance benefiting subsequent crops [75].

1.3
Microbial Activity and Functional Diversity in the Rhizosphere

1.3.1
Methods to Study Activity and Functional Diversity in the Rhizosphere

The classical approach to determining functional diversity is to use culturable bacteria grown on a plate and subject them to selected biochemical tests. Another method involves analyzing bacterial growth rate on a plate, which is considered as an indicator of the physiological state of the bacteria in the environment, the availability of nutrients and the adaptation strategy [76]. It is known that culturable bacteria are scarce in soil but are considered responsible for the most important chemical and biochemical processes. This is based on the fact that nonculturable bacteria are mostly 'dwarfs', measuring less than 0.4 μm in diameter and are considered as dying forms with almost no activity [77]. Bååth [37] studied the incorporation of radioactive precursors of DNA ([H^3]-thymidine, to assess population growth), and proteins (L-[C^{14}]-leucine, to assess population activity) in various fractions of soil filtrates. His research revealed that the culturable bacteria fraction (the larger size) is responsible for most of the growth and activity of the soil communities, whereas the fraction of cells less than 0.4 μm, considered nonculturable, had little importance in the metabolism and soil activity. Finally, using the PLFA technique, it has been demonstrated that there are no significant differences between the phospholipid fatty acids of bacteria in soil and bacteria culturable from this soil.

In contrast, other authors state that in rhizospheric communities, there are some difficulties in culturing groups of bacteria present in low densities that are metabolically very active; they can synthesize high amounts of proteins, use different substrates [78] and are believed to be important in fundamental processes in the soil. These bacteria are called keystone species, some of which include *Nitrosomonas* and *Nitrobacter*, playing a very important role in the nitrogen cycle [79].

At present, enzymatic activity measurement is one of the more widely used techniques to determine microbial diversity, in which it is possible to perform studies with a specific enzyme. An other approach is to use Biolog plates, which permit microbial communities to be characterized according to their physiological profile (community-level physiological profile, CLPP [47,80]) calculated from the different utilization patterns of many carbon and nitrogen sources, determined by a redox reaction that changes color after inoculation and incubation of the microbial communities [47,81].

New approaches such as the search for new catabolic, biosynthetic or antibiotic functions in soil samples [82] are required to identify new, potentially nonculturable genotypes. The cloning and sequencing of large DNA fragments (BAC library) will provide researchers with information about the metabolic diversity of nonculturable and culturable strains in the future and also provide important information on ecological laws and the operation of the soil ecosystem [22]. Undoubtedly, future studies on soil communities will involve microarray techniques [22] that will permit the study of differences in the structure of communities, identifying groups that are active or inactive during a specific treatment [60] leading to the identification of strains isolated from different environments and explaining differences or similarities in the operation of niches [83]. These techniques are complemented with transcriptomic techniques, based on the description of the activity of a gene by its expressed mRNA, and the proteomic approximation [22,82].

1.3.2
Activity and Effect of PGPR in the Rhizosphere

Some researchers approach the study of biochemical diversity in soil by identifying biochemical activities related to putative physiological PGPR traits in bacteria isolated from the rhizosphere (Table 1.1) [31].

Microbial activity in the rhizosphere indicates how metabolically active the microbial communities are. Using PGPR as inoculants in soil, besides altering the structure of the communities, will also influence microbial activity, and this could be related to the survival of the PGPR in the environment [34]. Some of the factors influencing the survival and activity of bacteria in the rhizosphere are physical (texture, temperature and humidity), while others are chemical, such as pH, nutrient

Table 1.1 Frequency of physiological PGPR traits in the mycorrhizosphere of *P. pinaster* and *P. pinea* and the associated mycosphere of *L. deliciosus* [31].

	P. pinaster		P. pinea	
PGPR trait	Mycorrhizosphere	Mycosphere	Mycorrhizosphere	Mycosphere
Aux (%)	14	0	50	42
Aux + PDYA (%)	0	0	0	2
Aux + CAS (%)	0	3	11	2
Aux + ACC (%)	0	0	7	0
Aux + CAS + PDYA (%)	0	3	0	0
PDYA (%)	47	35	11	32
PDYA + ACC (%)	3	0	0	0
CAS (%)	36	40	14	11
CAS + PDYA (%)	0	3	0	0
CAS + PDYA + ACC (%)	0	3	0	0
ACC (%)	0	13	7	11

Aux, auxin production; PDYA, phosphate solubilization; CAS, siderophore production; ACC, 1-aminocyclopropanecarboxylic acid degradation.

availability, organic matter content and, above all, interactions with other rhizosphere microorganisms. The interaction with the biotic factor is very important because PGPR must occupy a new niche, adhering to the plant roots, and the inoculum must compete for available nutrients released, essentially, by the root exudates, maintaining a minimum population able to exert its biological effect.

Studies of characterization of the soil microbial community activity are conducted using various techniques, such as thymidine ($[H^3]$) incorporation, radioactive DNA precursors to assess population growth and leucine ($L-[C^{14}]$) radioactive protein precursor to assess the metabolic activity of the population [37,84–86]. Stable isotope probing (SIP), based on radioactive labeling of different substrates, is considered to have enormous potential [23]. A further approach to quantifying the activity in the rhizosphere is by means of SIR [38].

1.4
Screening Strategies of PGPR

The rhizosphere of wild populations of plants is proposed as one of the optimal sources in which to isolate PGPR. This is because of the high selective pressure a plant exerts in this zone. The plant selects, among others, beneficial bacteria [4,31]. In the screening of PGPR, the different soil types, plant species, seasons and the plant's physiological moment must be considered to ensure the successful isolation of putative beneficial rhizobacteria.

The first step in obtaining a PGPR is the isolation of rhizospheric bacteria. It is generally accepted that the rhizosphere is the soil volume close to the roots (soil at 1–3 mm from the root and the soil adhering to the root). To collect this soil fraction, the root is normally shaken vigorously and soil still adhering is collected as the rhizosphere. Depending on the type of study, the root containing the endophyte bacteria is included, as some have been described as PGPR. Other researchers refer to the rhizosphere as the soil adhering to the roots after they have been washed under running water.

Rhizobacteria extraction starts with the suspension of soil in water, phosphate buffer or saline solution. Some compounds such as pyrophosphate are effective for soil disgregation, but can alter cell membranes [87]. Sample dispersion is made with chemical dispersants such as chelants that exchange monovalent ions (Na^+) for polyvalent cations (Ca^{2+}) of clay particles, reducing the electrostatic attraction between the soil and the bacterial cells. Various researchers have used ionic exchange resins derived from iminodiacetic acid, for example, Dowex A1 [88] or Chelex-100 [89,90]. Other dispersants are Tris buffer or sodium hexametaphosphate [91]. Detergents are used because the microbial cells present in the treated sample adhere by extracellular polymers to the soil particles. MacDonald [88] demonstrated that using detergents (sodium deoxycholate at 0.1%) together with Dowex A1 increased the microbial extraction from soil to 84%. This method was modified later by Herron and Wellington [89], replacing Dowex with Chelex-100 and combining with polyethylene glycol (PEG 6000) to dissolve and separate the phases. Other chemical solvents used in extraction protocols are Calgon at 0.2% for the extraction of bacteria from soil in studies of bacterial counts with acridine orange [40,85], citrate buffer used in studies of membrane phospholipids from soil microbes [92] and Winogradsky solution [54] for microbial diversity studies using molecular techniques (ARDRA, DGGE or REP-PCR) or phenotypical tests (Biolog).

Chemical extraction methods may be combined with physical methods, and these can be divided into three categories: shaking, mixing (homogenizing or grinding) and ultrasonics. Shaking is probably the least efficient method but adequate for sensitive bacteria or bacteriophages [93]. Techniques based on homogenization could damage some groups of bacteria, such as Gram-negative bacteria, and extraction would be selective. A combined method of grinding and chemical dispersants would be more effective [94]. Ultrasonic treatments are the best among methods used to break the physical forces between soil particles. In clay soils, pretreatment of the sample is necessary [95]; however, most sensitive bacteria, such as Gram-negative ones, could be damaged. This effect can be avoided using less aggressive ultrasonic treatments [96].

After rhizobacterial isolation, a screening of the putative PGPR is performed using two different strategies:

(a) Isolation, to select putative bacteria beneficial to the plant using specific culture media and specific isolation methods. For example, Founoune *et al.* [97] isolated *Pseudomonas fluorescens* from the Acacia rhizosphere as a species described as PGPR.

(b) After isolation of the maximum number of bacteria to avoid the loss of bacterial variability, different tests are performed to reduce the various types of bacteria chosen, so that only the putative beneficial ones remain. The test is performed *in vitro* to check biochemical activities that correspond with potential PGPR traits. Genetic tests may also be performed to remove genetic redundancy, that is, select different genomes that may have different putative beneficial activities [31,98,99].

Among the biochemical tests used to find putative PGPR traits, the most common are the following: (i) test for plant growth regulator production (i.e. auxins, gibberellin and cytokinins); (ii) the ACC (1-aminocyclopropanecarboxylic acid) deaminase test; this enzyme degrades the ethylene precursor ACC, causing a substantial alteration in ethylene levels in the plant, improving root system growth [100]; (iii) phosphate solubilization test, phosphate solubilization may improve phosphorous availability to the plant [101,102]; (iv) siderophore production test, which may improve plant's iron uptake [103]; (v) test for nitrogen-fixing bacteria to improve the plant's nitrogen nutrition [63]; and (vi) test for bacteria capable of producing enzymes that can degrade pathogenic fungi cell walls (i.e. chitinase or β-1,3-glucanase) preventing plant diseases [98].

The most common genetic techniques are PCR-RAPD (randomly amplified polymorphic DNA, ERIC-PCR, BOX-PCR and REP-PCR. They all compare bacterial genomes and establish a homology index among them. These techniques allow the formation of groups of bacteria with very similar genomes and thus with supposed similar PGPR abilities [31,99].

This approach of testing *in vitro* abilities has been proved to be an effective strategy to isolate PGPR; however, there are limitations. Some of the biochemical traits shown *in vitro* are inducible; that is, they are expressed in certain conditions but not in others. Therefore, a bacterial PGPR trait could be expressed in the laboratory in a culture media but not in the rhizosphere. This is true of PGPR traits related to plant nutrition, such as phosphate solubilization and siderophore production, that are not expressed in phosphorous-rich and iron-rich soils, respectively.

There are also problems with a bacterial property called phase variation, which produces strong genetic variations in bacteria by an enzyme called site-specific invertase. Hence, when these genetic variations occur, a strong phenotypical change occurs. It may be the case that a bacterial culture exhibits a PGPR trait, but after a time does not because of these phase variations [104].

After the screening process, the PGPR potential shown *in vitro* should be tested to ensure the same effect occurs in the plant. Root colonization is a necessary requirement for the bacteria to exert its effect. PGPR inoculation in distinct plant species sometimes produces erratic results [105]; however, the factors leading to failure are unclear. The competitive interactions in the rhizosphere are not well known. Independent of the factors that lead to good colonization, an inoculum screening is required to assess its impact on the rhizosphere. The introduction of a putative PGPR may alter the microbial rhizosphere communities, and this is indirectly related to plant fitness [106]. The introduced population can establish itself in the rhizosphere without changing the microbial communities or it may not establish itself but change the communities [106]. It is necessary to know how the inoculum is going to evolve to calibrate the potential risks of introducing these microorganisms, whether genetically modified or not. Several researchers have reported the alteration of these communities as a key to PGPR efficiency [34].

Biological trials may be performed in a sterile system to assess bacterial root colonization abilities. In addition, they may also be performed in either a nonsterile system or in a natural system (field trials). Competitiveness of the putative PGPR strain is a necessary requirement for colonization and to demonstrate the biological effect. On the contrary, the alteration of microbial communities present in a natural rhizosphere can also be studied in these types of systems.

A range of screening processes appeared in the recent literature. Cattelan *et al.* [98] tested several biochemical activity indicators for putative PGPR abilities in 116 bacterial strains selected from bulk soil and the rhizosphere of soybean. The indicators tested were phosphate solubilization, indole acetic acid production, siderophore production, chitinase, β-1,3-glucanase, ACC deaminase and cyanide production, putative free-living nitrogen-fixing bacteria and fungi growth inhibition. Twenty-four strains showed one or more of these activities and were assayed for traits associated with biocontrol, inhibition of rhizobial symbiosis and rhizosphere competence. These were finally tested for promotion of soybean growth. Six of the eight isolates tested positive for 1-aminocyclopropane-1-carboxylate deaminase production, four of the seven isolates were positive for siderophore production, three of the four isolates tested positive for β-1,3-glucanase production and two of the five isolates tested positive for phosphate solubilization, increasing at least one aspect of early soybean growth.

More examples of screening processes are those performed in the laboratory of Dr Gutiérrez Mañero, where PGPR have been isolated from the rhizosphere of wild plant species. For example, a screening for PGPR was performed in the mycorrhizosphere of wild populations of *Pinus pinea* and *P. pinaster* and in the mycosphere of associated *Lactarius deliciosus*, being the targeted microorganisms that are able to enhance establishment of mycorrhization. Of the 720 isolates, 50% were tested for ACC degradation, auxin and siderophore production and phosphate solubilization. One hundred and thirty-six isolates showed at least one of the evaluated activities. After PCR-RAPD analysis, 10 groups were formed with 85% similarity when all isolates were considered. One strain of each group was tested to see if it improved pine growth and eight were found to be effective. PGPR have also been isolated from the rhizosphere of *Nicotiana glauca* to improve the performance of *Lycopersicon esculentum*, a plant from the same family. The rationale was that the rhizosphere of wild populations of *N. glauca* would be a good source for putative PGPR able to induce systemic resistance and hence to be used in reducing chemical inputs of pesticides. A screening of 960 strains in the rhizosphere of *Nicotiana*, grown in three different soils (calcareous, quaternary and volcanic), was performed in both hot and cold seasons to isolate PGPR associated with this genus. A subset of 442 isolates composed of the most abundant parataxonomic groups was characterized based on their metabolic activities regarded as putative PGPR traits related to defense (siderophore and chitinase production). Fifty percent tested positive for both traits and were tested for growth promotion of *L. esculentum* seedlings and induction of resistance against *Fusarium* and *Xanthomonas*. The results were positive for 30 strains in growth, while only 6 enhanced resistance against foliar pathogen (unpublished).

Other researchers [107] have also isolated a large number of PGPR bacteria. From the rhizosphere soil of wheat plants grown at different sites, 30 isolates that showed prolific growth on agar medium were selected and evaluated for their potential to produce auxins *in vitro*. A series of laboratory experiments conducted in two cultivars of wheat under gnotobiotic (axenic) conditions exhibited increases in root elongation (up to 17.3%), root dry weight (up to 13.5%), shoot elongation (up to 37.7%) and shoot dry weight (up to 36.3%) of inoculated wheat seedlings. A positive linear correlation between *in vitro* auxin production and increase in growth parameters of inoculated seeds was found. Furthermore, auxin biosynthesis in sterilized versus nonsterilized soil inoculated with four selected PGPR was also monitored and demonstrated the superiority of the selected PGPR over indigenous microflora. Field experiments showed an increase of up to 27.5% over the control using these PGPR.

Finally, researchers such as Eleftherios *et al.* [108] have performed a direct screening to obtain endophytic bacteria able to protect against the pathogenic fungi *Verticillium dahliae*. Four hundred and thirty-eight bacteria were isolated from tomato root tips and 53 of these were found to be antagonistic against *V. dahliae* and several other soilborne pathogens in dual cultures. Significant biocontrol activity against *V. dahliae* in glasshouse trials was demonstrated in 3 of 18 evaluated antagonistic isolates. Finally, two of the most effective bacterial isolates, designated as K-165 and 5-127, were tested for rhizosphere colonization ability and chitinolytic activity, with both giving positive results.

1.5
Conclusions

Plants produce strong selective pressure in the rhizosphere and select bacteria beneficial for their growth and health. This effect results in very low bacterial diversity in the immediate area and hence is a good source from which to isolate PGPR.

Inoculation of PGPR has an impact on the rhizosphere microbial communities and this impact must be further studied because of its influence on the PGPR effect. It should also be borne in mind that communication mechanisms between bacteria (quorum sensing) should be studied, as they are involved in plant–bacteria interactions.

1.6
Prospects

Growing interest in microbial ecology reflects the importance of microorganisms in ecosystems. Soil microorganisms are essential for material and energy fluxes in the biosphere. In the rhizosphere, this is even more important because of the size of this ecosystem. Studies conducted to provide a deeper insight into this system will be of great interest to microbial ecology and will be crucial in obtaining specialized microorganisms, which can be used to solve various environmental problems. The future of PGPR ecology research depends on the development of new technologies such as DNA/RNA microarrays to provide a general view of PGPR diversity structure and function. Furthermore, quorum-sensing mechanisms of these bacteria should also be investigated to improve their performance.

References

1 Bottner, P., Pansu, M. and Sallih, Z. (1999) *Plant and Soil*, **216**, 15–25.
2 Buchenauer, H. (1998) *Journal of Plant Diseases and Protection*, **105** (4), 329–348.
3 Lynch, J.M. (1990) *The Rhizosphere* (ed. J. M. Lynch), John Wiley & Sons, Ltd, Chichester, p. 458.
4 Lucas García, J., Probanza, A., Ramos, B. and Gutiérrez Mañero, F.J. (2001) *Journal of Plant Nutrition and Soil Science*, **164**, 1–7.
5 Marilley, L. and Aragno, M. (1999) *Applied Soil Ecology*, **13**, 127–136.
6 Antoun, H. and Kloepper, J.W. (2001) Plant growth promoting rhizobacteria (PGPR) in

Encyclopedia of Genetics (eds S. Brenner and J.H. Miller), Academic Press, New York, pp. 1477–1480.

7 Zak, J.C., Willig, M.R., Moorhead, D.D.L. and Wildman, H.G. (1994) *Soil Biology & Biochemistry*, **26** (9), 1101–1108.

8 Crossley, D.A., Jr, Mueller, B.R. and Perdue, J.C. (1992) Biodiversity of microarthropods in agricultural soils: relations to processes in *Biotic Diversity in Agroecosystems* (eds M.G. Paoletti and D. Pimentel), Elsevier, Amsterdam, pp. 37–46.

9 Freckman, D.W. (1994) Life in soil. Soil biodiversity: its importance to ecosystem processes. Report of a Workshop Held at The Natural History Museum, London, UK.

10 Lavelle, P. (1996) *Biology International*, **33**, 3–16.

11 Bulla, L. (1994) *Oikos*, **70**, 167–171.

12 Kennedy, A.C. and Smith, K.L. (1995) Soil microbial diversity and the sustainability of agricultural soils, in *The Significance and Regulation of Soil Diversity* (eds H.P. Collins, G.P. Robertson and M.J. Klug), Kluwer Academic Publishers, Dordrecht, pp. 75–86.

13 Tokeshi, M. (1993) *Freshwater Biology*, **29** (3), 481–489.

14 Ekschmitt, K. and Griffiths, B.S. (1998) *Applied Soil Ecology*, **10**, 201–215.

15 Torsvik, V., Goksoyr, J.Y. and Daae, F.L. (1990) *Applied and Environmental Microbiology*, **56**, 782–787.

16 Curl, E.A. and Truelove, B. (1986) The rhizosphere. Advanced series in *Agricultural Sciences*, vol. 15 (eds D.F.R. Bommer, B.R. Sabey, Y. Vaadia, G.W. Thomas and L.D. Van Vleck), Springer-Verlag, Berlin, p. 280.

17 Whipps, J.M. (1990) Carbon economy, in *The Rhizosphere* (ed. J.M. Lynch), John Wiley & Sons, Ltd, pp. 59–97.

18 Pellet, D.M. Grunes, D.L. and Kochian, L.V. (1995) *Planta*, **196**, 788–795.

19 Ryan, P.R. Delhaize, E. and Randall, P.J. (1995) *Planta*, **196**, 103–110.

20 Lu, Y., Murase, J., Watanabe, A., Sugimoto, A. and Kimura, M. (2004) *FEMS Microbiology Ecology*, **48**, 179–186.

21 Torsvik, V. (1980) *Soil Biology & Biochemistry*, **12** (1), 15–21.

22 Ogram, A. (2000) *Soil Biology & Biochemistry*, **32**, 1499–1504.

23 Prosser, J.I. (2002) *Plant and Soil*, **244**, 9–17.

24 Fray, R.G. (2002) *Annals of Botany*, **89**, 245–253.

25 Quiñones, B., Pujol, C.J. and Lindow, S.E. (2004) *Molecular Plant–Microbe Interactions*, **17** (5), 521–531.

26 Swift, S., Throup, J.P., Williams, P., Salmond, G.P. and Stewart, G.S. (1996) *Trends in Biochemical Sciences*, **21**, 214–219.

27 Whitehead, N.A., Barnard, A.M.I., Slater, H., Simpson, N.J.L. and Salmond, G.P.C. (2001) *FEMS Microbiology Reviews*, **25**, 365–404.

28 Espinosa-Urgel, M. and Ramos, J.L. (2004) *Applied and Environmental Microbiology*, **70** (9), 5190–5198.

29 Schuhegger, R., Ihring, A., Gantner, S., Bahnweg, G., Knappe, C., Vogg, G., Hutzler, P., Schmid, M., van Breusegem, F., Eberl, L., Hartmann, A. and Langebartels, C. (2006) *Plant Cell & Environment*, **29** (5), 909–918.

30 Kloepper, J.W., Scrhoth, M.N. and Miller, T.D. (1980) *Phytopathology*, **70**, 1078–1082.

31 Barriuso, J., Pereyra, M.T., Lucas García, J. A., Megías, M., Gutiérrez Mañero, F.J. and Ramos, B. (2005) *Microbial Ecology*, **50** (1), 82–89.

32 Antoun, H. and Prévost, D. (2005) Ecology of plant growth promoting rhizobacteria in *PGPR: Biocontrol and Biofertilization* (ed. Z.A. Siddiqui), Springer, The Netherlands. pp. 1–39.

33 Lucas García, J.A., Domenech, J., Santamaría, C., Camacho, M., Daza, A. and Gutiérrez Mañero, F.J. (2004) *Environmental and Experimental Botany*, **52** (3), 239–251.

34 Ramos, B., Lucas García, J.A., Probanza, A., Barrientos, M.L. and Gutiérrez

Mañero, F.J. (2003) *Environmental and Experimental Botany*, **49**, 61–68.

35 Enebak, S.A., Wei, G. and Kloepper, J.W. (1997) *Forest Science*, **44**, 139–144.

36 Bashan, J. and Holguin, G. (1998) *Soil Biology & Biochemistry*, **30** (8/9), 1225–1228.

37 Bååth, E. and Arnebrant, K. (1994) *Soil Biology & Biochemistry*, **26** (8), 995–1001.

38 Anderson, J.P.E. and Domsch, K.H. (1978) *Soil Biology & Biochemistry*, **10**, 215–221.

39 Eiland, F. (1983) *Soil Biology & Biochemistry*, **15**, 665–670.

40 Reichardt, W., Briones, A., de Jesus, R. and Padre, B. (2001) *Applied Soil Ecology*, **17**, 151–163.

41 Hopkins, D.W., Macnaughton, S.J. and O'Donnell, A.G. (1991) *Soil Biology & Biochemistry*, **23**, 217.

42 Vance, E.D., Brookes, P.C. and Jenkinson, D.S. (1987) *Soil Biology & Biochemistry*, **19** (6), 703–707.

43 Atlas, R.M. and Bartha, R. (1993) Microbial ecology: historical development in *Microbial Ecology, Fundamentals And Applications*, The Benjamin/Cummings Publishing Company, Menlo Park, Redwood City, CA, pp. 3–20.

44 Campbell, R. and Greaves, M.P. (1990) Anatomy and community structure of the rhizosphere in *The Rizosphere* (ed. J.M. Lynch), John Wiley & Sons, Ltd, Essex, pp. 11–34.

45 Bakken, L.R. and Olsen, R.A. (1989) *Soil Biology & Biochemistry*, **21**, 789.

46 Frostegård, Å., Bååth, E. and Tunlid, A. (1993) *Soil Biology & Biochemistry*, **25** (6), 723–730.

47 Grayston, S.J., Griffith, G.S., Mawdsley, J.L., Campbell, C.D. and Bardgett, R.D. (2001) *Soil Biology & Biochemistry*, **33**, 533–551.

48 Tunlid, A. and White, D.C. (1990) Use of lipid biomarkers in environmental samples, in *Analytical Microbiology Methods* (eds A. Fox *et al.*), Plenum Press, New York, pp. 259–274.

49 Tunlid, A. and White, D.C. (1992) Biochemical analysis of biomass community structure, nutritional status, and metabolic activity of microbial communities in soil, in *Soil Biochemistry* (eds G. Stotzky and J.M. Bollag), Marcel Dekker, New York, pp. 229–262.

50 Head, I.M., Saunders, J.R. and Pickup, R.W. (1998) *Microbial Ecology*, **35**, 1–21.

51 Griffiths, B.S., Ritz, K., Ebblewhite, N. and Dobson, G. (1999) *Soil Biology & Biochemistry*, **31**, 145–153.

52 Liesack, W. and Stackerbrandt, E. (1992) *Journal of Bacteriology*, **174**, 5072–5078.

53 Clegg, C.D., Ritz, K. and Griffiths, B.S. (1998) *Antonie Van Leeuwenhoek*, **73** (1), 9–14.

54 Øvreås, L. and Torsvik, V. (1998) *Microbial Ecology*, **36**, 303–315.

55 Heuer, H. and Smalla, K. (1997) Application of denaturing gradient gel electrophoresis and temperature gradient gel electrophoresis for studying soil microbial communities, in *Modern Soil Microbiology* (eds J.D. van Elsas, J.T. Trevors and E.M.H. Wellington), Dekker, New York, pp. 353–373.

56 Kozdrój, J. and van Elsas, J.D. (2000) *Biology and Fertility of Soils*, **31**, 372–378.

57 Dunbar, J., Takala, S., Barns, S.M., Davis, J.A. and Kuske, C.R. (1999) *Applied and Environmental Microbiology*, **65** (4), 1662–1669.

58 Liu, W.T., Marsh, T.L., Cheng, H. and Forney, L.J. (1997) *Applied and Environmental Microbiology*, **63**, 4516–4522.

59 Smit, E., Leeflang, P. and Wernars, K. (1997) *FEMS Microbiology Ecology*, **23**, 249–261.

60 Guschin, D.Y., Mobarry, B.K., Proudnikov, D., Stahl, D.A., Rillmann, B. and Mirzabekov, A.D. (1997) *Applied and Environmental Microbiology*, **63**, 2397–2402.

61 Steenhoudt, O. and Vanderleyden, J. (2000) *FEMS Microbiology Reviews*, **24**, 487–506.

62 Bashan, Y., Holguin, G. and de Bashan, L.E. (2004) *Canadian Journal of Microbiology*, **50**, 521–577.

63 Vessey, J.K. (2003) *Plant and Soil*, **255**, 571–586.

64 Reinhold-Hurek, B. and Hurek, T. (2000) *International Journal of Systematic Evolutionary Microbiology*, **50**, 649–659.

65 Hurek, T. and Reinhold-Hurek, B. (2003) *Journal of Biotechnology*, **106**, 169–178.

66 Garbeva, P., van Veen, J.A. and van Elsas, J.D. (2003) *Microbial Ecology*, **45**, 302–316.

67 Timmusk, S., Nicander, B., Granhall, U. and Tillberg, E. (1999) *Soil Biology & Biochemistry*, **31**, 1847–1852.

68 Kokalis-Burelle, N., Vavrina, C.S., Roskopf, E.N. and Shelby, R.A. (2002) *Plant and Soil*, **238**, 257–266.

69 Probanza, A., Lucas García, J.A., Ruiz Palomino, M., Ramos, B. and Gutiérrez Mañero, F.J. (2002) *Applied Soil Ecology*, **20**, 75–84.

70 Patten, C.L. and Glick, B.R. (2002) *Applied and Environmental Microbiology*, **68**, 3795–3801.

71 Gutiérrez Mañero, F.J., Probanza, A., Ramos, B., Colón Flores, J.J. and Lucas García, J.A. (2003) *Journal of Plant Nutrition*, **26** (5), 1101–1115.

72 Charest, M.H., Beauchamp, C.J. and Antoun, H. (2005) *FEMS Microbiology Ecology*, **52**, 219–227.

73 Antoun, H., Beauchamp, C.J., Goussard, N., Chabot, R. and Lalande, R. (1998) *Plant and Soil*, **204**, 57–67.

74 Yanni, Y.G., Rizk, R.Y., Corich, V., Squartini, A., Ninke, K., Philip-Hollingworth, S., Orgambide, G., de Bruijn, F., Stolzfus, J., Buckley, D., Schmidt, T.M., Mateos, P.F., Ladha, J.K. and Dazzo, F.B. (1997) *Plant and Soil*, **194**, 99–114.

75 Lupwayi, N.Z., Clayton, G.W., Hanson, K.G., Rice, W.A. and Biederbeck, V.O. (2004) *Canadian Journal of Plant Science*, **84**, 37–45.

76 De Leij, F.A.A.M., Whipps, J.M. and Linch, J.M. (1993) *Microbial Ecology*, **27**, 81–97.

77 van Elsas, J.D. and van Overbeek, L.S. (1993) *Starvation in Bacteria* (ed. S. Kjelleberg), Plenum Press, New York.

78 Roszak, D.B. and Colwell, R.R. (1987) *Applied and Environmental Microbiology*, **53** (12), 2889–2893.

79 Kotlar, E., Tartakovsky, B., Argaman, Y. and Sheintuch, M. (1996) *Journal of Biotechnology*, **51**, 251–258.

80 Garland, J.L.Y. and Mills, A.L. (1991) *Applied and Environmental Microbiology*, **57**, 2351–2359.

81 Fang, C., Radosevich, M. and Fuhrmann, J.J. (2001) *Soil Biology & Biochemistry*, **33**, 679–682.

82 Handelsman, J., Rondon, M.R., Brady, S. F., Clardy, J. and Goodman, R.M. (1998) *Chemistry & Biology*, **5**, 245–249.

83 Murray, A.E., Li, G.S., Lies, D., Nealson, K.H., Zhou, J.Z. and Tiedje, J.M. (1999) Applications of DNA microarray technology to investigating gene expression in *Shewanella oneidensis* MR-I and functional diversity among related species in Abstract 38, Abstracts of the 7th Conference on Small Genomes, Arlington, USA.

84 Bååth, E. (1994) *Microbial Ecology*, **27**, 267–278.

85 Bååth, E. (1998) *Microbial Ecology*, **36**, 316–327.

86 Bååth, E., Petterson, M. and Soderberg, K.H. (2001) *Soil Biology & Biochemistry*, **33**, 1571–1574.

87 Lindahl, V. (1996) *Journal of Microbiological Methods*, **25** (3), 279–286.

88 MacDonald, R.M. (1986) *Soil Biology & Biochemistry*, **18** (4), 407–410.

89 Wellington, E.M.H. and Herron, P.R. (1990) *Applied and Environmental Microbiology*, **56** (5), 1406–1412.

90 Schloter, M., Wiehe, W., Assmus, B., Steindl, H., Becke, H., Höflich, G. and Hartmann, A. (1997) *Applied and Environmental Microbiology*, **63** (5), 2038–2046.

91 Niepold, F., Conrad, R. and Schlegel, H.G. (1979) *Antonie Van Leeuwenhoek*, **45** (3), 485–497.

92 Bååth, E. (1992) *Soil Biology & Biochemistry*, **24** (11), 1157–1165.

93 Wellington E.M.H., Marsh P., Watts J.E.M. and Burden J. (1997) Indirect approaches for studying soil microorganisms based on cell extraction and culturing, in *Modern*

Soil Microbiology (eds J.D. van Elsas, J.T. Trevors and E.M.H. Wellington), Marcel Dekker, New York, pp. 311–329.

94 Turpin, P.E., Maycroft, K.A., Rowlands, C.L. and Wellington, E.M.H. (1993) *Journal of Applied Bacteriology*, **74**, 421–427.

95 Ozawa, T. and Yamaguchi, M. (1986) *Applied and Environmental Microbiology*, **52** (4), 911–914.

96 Hopkins, D.W. and O'Donnell, A.G. (1992) *Genetic Interactions Between Microorganisms in the Environment* (eds E. M.H. Wellington and J.D. van Elsas), Manchester University Press, Manchester, UK, p. 104.

97 Founoune, H., Duponnois, R., Meyer, J.M., Thioulouse, J., Masse, D., Chotte, J.L. and Neyra, M. (2002) *FEMS Microbiology Ecology*, **1370**, 1–10.

98 Cattelan, A.J., Hartel, P.G. and Fuhrmann, J.J. (1999) *Soil Science Society of America Journal*, **63**, 1670–1680.

99 Ramos Solano, B., Pereyra de la Iglesia, M.T., Probanza, A., Lucas García, J.A., Megías, M. and Gutiérrez Mañero, F.J. (2007) *Plant and Soil*, **287**, 59–68.

100 Glick, B.R., Penrose, D.M. and Li, J. (1998) *Journal of Theoretical Biology*, **190**, 63–68.

101 De freitas, J.R., Banerjee, M.R. and Germida, J.J. (1997) *Biology and Fertility of Soils*, **24**, 358–364.

102 Richardson, A.E. (2001) *Australian Journal of Plant Physiology*, **28**, 897–906.

103 Alexander, D.B. and Zuberer, D.A. (1991) *Biology and Fertility of Soils*, **12**, 39–45.

104 van der Woude, M.W. (2006) *FEMS Microbiology Letters*, **254** (2), 190–197.

105 Germida, J.J. and Walley, F.L. (1996) *Biology and Fertility of Soils*, **23**, 113–120.

106 Gilbert, G.S., Parke, J.L., Clayton, M.K. and Handelsman, J. (1993) *Journal of Ecology*, **74** (3), 840–854.

107 Khalid, A., Arshad, M. and Zahir, Z.A. (2004) *Journal of Applied Microbiology*, **96**, 473–480.

108 Eleftherios, C.T., Dimitrios, I.T., Sotirios, E.T., Polymnia, P.A. and Panayiotis, K. (2004) *Journal of Plant Pathology*, **110**, 35–44.

2
Physicochemical Approaches to Studying Plant Growth Promoting Rhizobacteria

Alexander A. Kamnev

2.1
Introduction

Within the last few decades, the application of instrumental techniques in scientific research related to life sciences and biotechnology has been rapidly expanding. This trend largely concerns biochemistry and biophysics, thus reflecting the interpenetration and interrelationship between different fields of natural sciences in studying biological objects [1]. Albeit to a lesser degree, the use of physicochemical techniques is increasingly being put into practice in microbiology. The possibility of obtaining reliable, selective and sometimes unique information about sophisticated biological systems under study at different levels of their organization (e.g. organism, tissue, cell, cellular supramolecular structures, biomacromolecules, low-molecular-weight metabolic products, etc.) and functioning is an attractive feature of modern instrumental techniques. In addition, some techniques are nondestructive and/or provide information on intact biological matter with a minimum of sample preparation, which most closely reflects its natural state. Moreover, the bioanalytical information obtained by a combination of independent instrumental techniques may be of significant advantage, especially when comparing data on overall cellular metabolic changes (e.g. as cellular responses to some environmental factors) with analyses for microelements (e.g. trace metal uptake) and/or their chemical forms (speciation analysis).

The scientific literature of recent decades provides evidence that the field of plant growth promoting rhizobacteria (PGPR) and their interactions with host plants is highly promising for possible wide-scale applications in returning to environmentally friendly and sustainable agriculture. However, there are still a great number of problems related to the basic mechanisms of the underlying biological and chemical processes that occur both in the rhizosphere and *in vivo* (in plants and PGPR), which require systematic research at the molecular level using modern techniques.

In this chapter, some recent examples are discussed which illustrate the use of various physicochemical and spectroscopic approaches, involving a range of

Plant-Bacteria Interactions. Strategies and Techniques to Promote Plant Growth
Edited by Iqbal Ahmad, John Pichtel, and Shamsul Hayat
Copyright © 2008 WILEY-VCH Verlag GmbH & Co. KGaA, Weinheim
ISBN: 978-3-527-31901-5

instrumental techniques, aimed at obtaining structural and compositional data related to PGPR, their cellular biopolymers and secondary metabolites. Particular attention is paid to the behavior of PGPR, using the example of the widely studied ubiquitous diazotrophic plant-associated rhizobacterium *Azospirillum brasilense* [2], and the effect of various environmental factors upon it.

2.2
Application of Vibrational Spectroscopy to Studying Whole Bacterial Cells

2.2.1
Methodological Background

Cellular metabolic processes, including their alterations induced by various environmental factors, largely result in qualitative and/or quantitative compositional changes in microbial cells. Knowledge of such changes is of importance both for basic studies (e.g. on the molecular mechanisms of microbial responses to environmental stresses) and for applied research (e.g. monitoring of fermentation processes, in agricultural microbiology and biotechnology, clinical microbiology and diagnostics). In addition to 'wet' chemical analysis and biochemical methods, such changes can be controlled in microbial biomass or even in single cells using various modifications of vibrational (Fourier transform infrared [FTIR], FT-Raman) spectroscopy [3–7]. In recent years, microbiological applications of these techniques have been developed to the level of convenient and sensitive tools for monitoring both macroscopic changes in the cellular composition and fine structural rearrangements of particular cellular constituents (see, e.g. [6–10] and references therein).

An absorption spectrum in the case of conventional FTIR spectroscopy [3,4,6,10], as well as a FT-Raman scattering spectrum [3,5,7], of a sample of biomass comprising, for instance, whole bacterial cells presents a complicated summarized image. Contributions to such a spectrum are made by all the major cellular constituents (or, more exactly, their functional groups with their characteristic vibration frequencies, including also the effects of all possible molecular, atomic and/or ionic interactions). In the first instance, these are proteins and glycoproteins, polysaccharides, lipids and other biomacromolecules. As a consequence, a spectrum reflects the overall chemical composition of the cell biomass, which in certain cases can be used for identification and classification of microorganisms [3,6,9] based on differences in qualitative and/or quantitative composition of their cells.

2.2.2
Vibrational Spectroscopic Studies of *A. brasilense* Cells

2.2.2.1 **Effects of Heavy Metal Stress on *A. brasilense* Metabolism**
Bacteria of the genus *Azospirillum* have been well documented to demonstrate relatively high tolerance to moderate heavy metal stress [2,11,12]. This feature

may well be of advantage for agricultural applications of these PGPR in metal-contaminated environments [2,11] or for enhancing phytoremediation [13–17], in particular, based on using heavy metal accumulating plants. In the presence of submillimolar concentrations of conventionally toxic metals which do not significantly suppress growth of azospirillum [10,12], *A. brasilense* was found to take up and accumulate noticeable amounts of heavy metals (e.g. vanadium, cobalt, nickel, copper, zinc and lead), which were also shown to influence the uptake of essential elements (magnesium, calcium, manganese and iron) [18]. In particular, the uptake of iron (present in the medium as Fe^{2+}) was drastically reduced (by about 1 order of magnitude) in the presence of 0.2 mM Co^{2+}, Ni^{2+} or vanadium(IV) (VO^{2+}) salts, probably reflecting their competitive binding to iron chelators and transporters to the cell.

One of the conspicuous effects was an increased accumulation of the four essential cations (about two- to fivefold) in the presence of 0.2 mM copper(II) in the culture medium (while Cu was also accumulated by the bacterium up to 2 mg g^{-1} of dry cell biomass) [18]. This effect induced certain alterations in the FTIR spectra of both whole cells [19] and cell membranes (note that in *A. brasilense* membranes, in contrast to whole cells, only the Mg^{2+} content was increased approximately sixfold in the presence of Cu^{2+}) [20] as well as in electrophysical properties of the bacterial cell surface [21]. It has to be noted that the aforementioned experiments on azospirillum [18–20] were performed, besides being under moderate heavy metal stress conditions, in an NH_4^+-free phosphate–malate medium corresponding to a high C : N ratio. This kind of nutritional stress (i.e. bound nitrogen deficiency) is known to induce accumulation of a reserve storage material, poly-3-hydroxybutyrate (PHB), along with other polyhydroxyalkanoates (PHA), playing a role in stress tolerance in many bacteria, including *A. brasilense* [22,23]. Accordingly, signs of polyester accumulation were also noticeable in FTIR spectra of *A. brasilense* cells grown under nitrogen deficiency [19].

A subsequent FT-Raman spectroscopic study of whole cells of *A. brasilense* (non-endophyte strain Sp7), grown in a rich ammonium-supplemented medium in the presence of 0.2 mM Co^{2+}, Cu^{2+} or Zn^{2+} salts, suggested that some metabolic changes occur induced by the heavy metals [7]. In particular, besides some subtle changes in cellular lipid-containing constituents (to which FT-Raman spectroscopy is highly sensitive [3,5,7]), accumulation of some polymeric material could be proposed. Since the induction of PHA or PHB biosynthesis by heavy-metal stress alone, without a nutritional stress, had not been earlier described in bacteria [24] (for a recent review see [23] and references therein), an attempt was made to use FTIR spectroscopy which is more sensitive to polyester compounds [3,4,6,10].

2.2.2.2 Differences in Heavy Metal Induced Metabolic Responses in Epiphytic and Endophytic *A. brasilense* Strains

Whole cells of *A. brasilense* (non-endophyte strain Sp7) grown in a standard medium (control) and in the presence of several heavy metals (0.2 mM Co^{2+}, Cu^{2+} or Zn^{2+}) were analyzed using FTIR spectroscopy [24]. Striking differences were noticeable in the FTIR absorption profiles between the control cells (Figure 2.1a) and cells grown

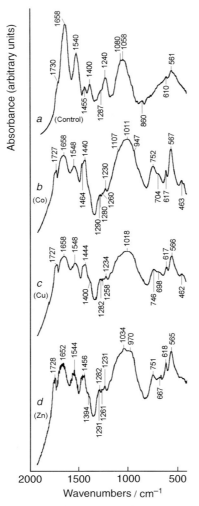

Figure 2.1 Infrared spectra of dried biomass of A. brasilense (non-endophytic strain Sp7) grown (a) in a standard phosphate–malate medium (control) and in the same medium in the presence of 0.2 mM Co^{2+} (b), Cu^{2+} (c) or Zn^{2+} (d) [10,24].

under moderate heavy metal stress (Figure 2.1b–d). The most prominent feature of the metal-stressed cells is the appearance of a relatively strong and well-resolved band at about 1727 cm^{-1} (Figure 2.1b–d) featuring polyester v(C=O) vibrations. On the contrary, in the control cells where the amide I and amide II bands of cellular proteins (at about 1650 and 1540 cm^{-1}, respectively) dominate (Figure 2.1a), there is only a weak shoulder at about 1730 cm^{-1}. Together with an increased FTIR absorption in the regions of methylene (−CH$_2$−) bending vibrations (at 1460–1440 cm^{-1}), as well as C−O−C and C−C−O vibrations (at 1150–1000 cm^{-1}) and CH$_2$ rocking vibrations (at about 750 cm^{-1}) observed in metal-stressed cells (Figure 2.1b–d), these spectroscopic changes provide unequivocal evidence for the accumulation of polyester compounds in cells of strain Sp7 as a response to metal stress.

As mentioned above, PHB has been documented to accumulate in cells of azospirilla under unfavorable conditions, playing a role in bacterial tolerance to several kinds of environmental stresses [22,23] and providing a mechanism that facilitates bacterial establishment, proliferation, survival and competition in the rhizosphere [25]. However, under normal conditions, particularly in nitrogen-supplemented media, its biosynthesis is usually suppressed [22]. Thus, the induction of biosynthesis and accumulation of PHB (and possibly other PHAs) under normal nutritional conditions by heavy metals is a novel feature for bacteria (which was for the first time documented for *A. brasilense* Sp7 [24]), which is in line with the overall strategy of bacterial responses to stresses. It has to be noted that, although cellular lipids can give similar FTIR spectroscopic signs, an accumulation of additional lipids is not physiologically appropriate for azospirilla [2,26].

Within the *A. brasilense* species, there is a unique possibility to compare the behavior of epiphytic strains (which colonize the rhizoplane only) and endophytic ones [27]. In view of that, it is of interest to compare the response of the latter to heavy metals. A comparison of FTIR spectroscopic images of another *A. brasilense* strain, Sp245 (which, in contrast to strain Sp7, is a facultative endophyte [27,28]), grown under similar conditions in the standard medium and in the presence of each of the above three cations (0.2 mM), shows no major differences between them (Figure 2.2a–d). In all four samples, there is a weak shoulder at about 1730 cm^{-1} (ester v(C=O) band), but in metal-stressed cells there occurs virtually no accumulation of PHA that was found under similar conditions in strain Sp7 (Figure 2.1). Moreover, the position of the representative v_{as}(PO$_2^-$) band of cellular phosphate moieties in strain Sp245 was constant within the relatively narrow region 1237–1240 cm^{-1}, thus confirming the relative stability of the state of these functional groups both in the control group of cells and those under metal stress (whereas in metal-stressed cells of strain Sp7, this band was found at lower frequencies, 1230–1234 cm^{-1}; Figure 2.1b–d). This finding is remarkable, especially considering the comparable uptake level of each of the cations in the bacterial cells of the two strains (0.12 and 0.13 mg Co, 0.48 and 0.44 mg Cu, 4.2 and 2.1 mg Zn per gram of dry cells for strain Sp7 and Sp245, respectively) [10].

Thus, the response of the endophytic *A. brasilense* strain Sp245 to a moderate heavy metal stress was found to be much less pronounced than that of the non-endophyte strain Sp7. These conspicuous dissimilarities in their behavior may be related to different adaptation abilities of the strains under stress conditions owing to their different ecological status and, correspondingly, different ecological niches which they can occupy in the rhizosphere. In the non-endophytic strain, PHB/PHA accumulation may be a specific flexible adaptation strategy related to the localization of the bacteria in the rhizosphere and on the rhizoplane, i.e. always in direct contact with rhizospheric soil components, in contrast to the endophyte which is somewhat more 'protected' by plant tissues [29]. This corresponds to the documented capability

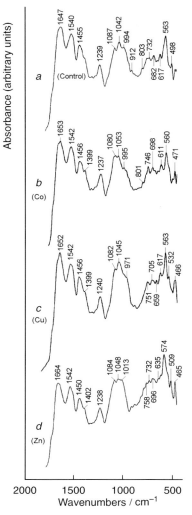

Figure 2.2 Infrared spectra of dried biomass of *A. brasilense* (facultatively endophytic strain Sp245) grown (a) in a standard phosphate–malate medium (control) and in the same medium in the presence of 0.2 mM Co^{2+} (b), Cu^{2+} (c) or Zn^{2+} (d) [10].

of strain Sp7 to outcompete other co-inoculated strains [27], as well as to a lower sensitivity of strain Sp7, as compared to strain Sp245, to copper ions, including a less pronounced copper-induced decrease in auxin production [12,30].

As noted above, the amount of cobalt accumulated by strain Sp7 (up to about 0.01% wt/wt dry biomass) induces a significant metabolic response in the bacterium, comparable to that of about fourfold higher amount of copper or about 36-fold higher amount of zinc (cf. Figure 2.1a–d). Moreover, such amounts of metal complexes per se cannot give any clearly noticeable FTIR absorption related to their intrinsic functional groups. Thus, in strain Sp7 such a moderate heavy metal stress evidently induces noticeable metabolic transformations that are revealed in their FTIR spectra as macroscopic compositional changes. In its turn, this suggests direct participation of the cations, which are taken up by the bacterial cells from the medium, in cellular processes as a result of their assimilation. However, for strain Sp245, despite the levels of metal uptake comparable with those for strain Sp7, this is not obvious, considering the lack of noticeable compositional changes revealed by FTIR spectroscopy (Figure 2.2).

In order to validate the direct involvement of metal cations in cellular metabolic processes in strain Sp245, the chemical state of accumulated trace metal species must be monitored in live cells for various periods of time. Some examples of microbiological applications of a technique, which allows such monitoring to be made specifically for cobalt ions, are discussed below.

2.3
Application of Nuclear γ-Resonance Spectroscopy to Studying Whole Bacterial Cells

2.3.1
Methodological Background

Nuclear γ-resonance (Mössbauer) spectroscopy, based on recoil-free absorption (or emission) of γ-quanta by specific nuclei (the stable ^{57}Fe isotope having been so far most widely used), is a widely applicable powerful and informative technique, providing a wealth of information on the chemical state and coordination structure of the cation influenced by its microenvironment. The ^{57}Fe absorption variant of Mössbauer spectroscopy has been extensively used in a variety of fields including biological sciences, largely for studying Fe-containing proteins or for monitoring the state of iron species in biological samples (for recent reviews see, e.g. [31,32] and references therein).

The emission variant of Mössbauer spectroscopy (EMS), with the radioactive ^{57}Co isotope as the most widely used nuclide, is several orders of magnitude more sensitive than its ^{57}Fe absorption counterpart. However, despite the incomparably higher sensitivity of the former, applications of EMS in biological fields have so far been fragmentary and sparse, primarily owing to specific difficulties related to the necessity of using radioactive ^{57}Co in samples under study [33]. Note that radioactive decay of ^{57}Co (which has a half-life of 9 months), proceeding via electron capture by its nucleus, results in the formation of ^{57}Fe in virtually the same coordination microenvironment as the parent ^{57}Co cation. The decay process is accompanied by emission of a γ-quantum, as well as by some physical and chemical aftereffects which, in particular, often lead to the partial formation of stabilized daughter ^{57}Fe cations in oxidation states other than the parent ^{57}Co ones [34,35]. This effect, although inevitably complicating the emission spectra, can provide valuable

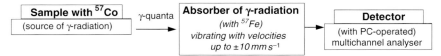

Figure 2.3 Scheme of experimental setup for measuring emission Mössbauer spectra [33].

additional information, e.g. on the electron-acceptor properties of the proximal coordination environment of the metal under study [35].

The recoil-free emission (as well as absorption) of γ-radiation (i.e. the Mössbauer effect) is observed in solids only, where the recoil energy can dissipate within the solid matrix. Therefore, solutions or liquids are usually studied when they are rapidly frozen [34]. Rapid freezing (e.g. by immersing small drops or pieces of a sample in liquid nitrogen) often allows crystallization of the liquid (solvent) to be avoided, so that the structure of the resulting glassy solid matrix represents that of the solution. Moreover, upon freezing, all the ongoing biochemical (metabolic) processes in live cells, tissues or other biological samples cease at a certain point. Thus, for live bacterial cells that have been in contact with $^{57}Co^{2+}$ traces, freezing of suspension aliquots, taken after different periods of time, allows both the initial rapid binding of the metal cation by cell-surface biopolymers and its possible further metabolic transformations to be monitored.

A scheme of the experimental setup for measuring emission Mössbauer spectra is shown in Figure 2.3 [33]. The ^{57}Co-containing sample (which in EMS is the source of γ-radiation) can be kept in a cryostat (e.g. in liquid nitrogen at $T \approx 80\,K$) with a window for the γ-ray beam, whereas the ^{57}Fe-containing standard absorber vibrates along the axis "source–absorber" at a constant acceleration value (with its sign changing periodically from $+a$ to $-a$, so that the range of velocities is usually up to $\pm 10\,mm\,s^{-1}$ relative to the sample), thus modifying the γ-quanta energy scale as a function of velocity according to the Doppler effect. EMS measurements are commonly performed using a conventional constant-acceleration Mössbauer spectrometer calibrated using a standard (e.g. α-Fe foil) and combined with a PC-operated multichannel analyzer, where each channel represents a point with a specified fixed velocity. Standard PC-based statistical analysis consists of fitting the experimental data obtained to a sum of Lorentzian-shaped lines using a least squares minimization procedure. The Mössbauer parameters calculated from the experimental data are the isomer shift (IS; relative to α-Fe), quadrupole splitting (QS), linewidth (i.e. experimentally obtained full width at half maximum, FWHM), and relative areas of subspectra (S_r) [32–34].

2.3.2
Emission Mössbauer Spectroscopic Studies of Cobalt(II) Binding and Transformations in *A. brasilense* Cells

In order to check whether in *A. brasilense* Sp245 cobalt(II) ions are merely bound by the cell surface in a purely chemical process or cobalt(II) is assimilated and somehow involved in metabolic processes, time-resolved EMS measurements could be performed using traces of $^{57}Co^{2+}$ salt. For strain Sp245, which had previously been shown to be tolerant to submillimolar concentrations of heavy metals, including cobalt(II) [10,12,18,19,30], EMS studies were, for the first time, performed on freeze-dried bacterial samples (rapidly frozen after 2–60 min of contact with $^{57}Co^{2+}$ salt and measured at $T = 80$ K) [36]. The following experiments, with the same strain were performed using suspensions of live bacteria rapidly frozen after the same periods of time (2–60 min) of contact with [^{57}Co]-cobalt(II), and EMS spectra were measured for frozen samples without drying, which more closely represented the state of cobalt in the live cells [37]. Nevertheless, comparing the data for freeze-dried bacteria [36] and for those measured in frozen aqueous suspensions [37] showed that their corresponding Mössbauer parameters were very close (both for 2 and 60 min contact with $^{57}Co^{II}$, whereas there were significant differences in the parameters between the two periods).

Typical EMS spectra of a rapidly frozen cell suspension and cell-free supernatant liquid shown in Figure 2.4 also clearly indicate differences between them. Note that two chemical forms referring to cobalt(II) were found in all samples. The [^{57}Co]-cobalt(II) forms are represented by quadrupole doublets with larger QS values (note that the third doublet with smaller IS and QS values corresponds to the aliovalent daughter [^{57}Fe]-ferric form resulting from aftereffects [33,37]). Multiple forms of cobalt(II) found in the spectra are related to the availability of different functional groups (with possibly different donor atoms) as ligands at the cell surface of *Azospirillum* [2].

In Figure 2.5, the Mössbauer parameters (IS and QS represented by points with their confidence intervals) are plotted for different cobalt(II) forms in each sample for various periods of contact (2 and 60 min) of the live bacteria with $^{57}Co^{II}$, as well as for dead bacterial cells (treated at 95 °C in the medium for 1 h in a water bath) and for the cell-free supernatant liquid [33]. Thus, each point with its confidence intervals (a rectangle) in Figure 2.5 corresponds to a separate $^{57}Co^{II}$ form with its characteristic microenvironment. Note that the parameters for both forms of [^{57}Co]-cobalt(II) show statistically significant difference for different periods of contact (2 and 60 min) of live bacteria with the metal. This shows that cobalt(II) is first rapidly absorbed by live *A. brasilense* cells in a merely chemical process, but then undergoes metabolic transformation within an hour. This finding confirms its direct involvement in bacterial metabolism, although in strain Sp245 (in contrast to strain Sp7) cobalt(II) assimilation is not accompanied by PHB accumulation (see above).

Interestingly, the parameters of the two forms of [^{57}Co]-cobalt(II) for live bacteria after 2 min, on the one hand, and for dead bacteria, on the other hand are rather close (essentially overlapping; Figure 2.5). This finding indicates that the mechanism of primary rapid Co^{2+} absorption by live cells is similar to the purely chemical binding process occurring at the surface of dead (thermally killed) cells, and is virtually unaffected by such hydrothermal treatment. Note also that the parameters for the cell-free supernatant liquid (from which the bacterial cells were removed by centrifugation) are clearly different from those for all other samples (Figure 2.5).

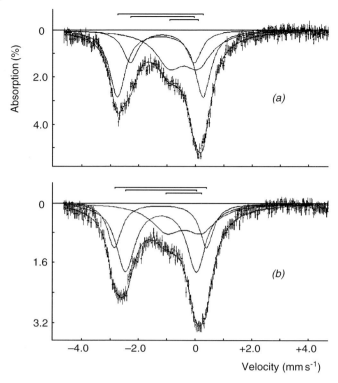

Figure 2.4 Typical emission Mössbauer spectra of (a) aqueous suspension of live cells of *A. brasilense* Sp245 in the culture medium (frozen 2 min after contact with $^{57}Co^{II}$ traces) and (b) the cell-free supernatant liquid rapidly frozen in liquid nitrogen (spectra collected at $T = 80$ K; velocity scale calibrated relative to α-Fe; intensities converted to the absorption convention) [37].

For each spectrum, the relevant subspectra (quadrupole doublets) are shown which contributed to the resulting spectrum (solid-line envelope) obtained by computer fitting to the experimental data (points with vertical error bars). The positions of the spectral components (quadrupole doublets) are indicated by horizontal square brackets above the zero lines.

It should be mentioned that primary binding of heavy metals by the cell surface in Gram-negative bacteria is mediated by capsular polysaccharide (PS, particularly carboxylated acidic PS), lipopolysaccharide (LPS, including phosphate LPS moieties), and proteinaceous materials [8,10,37]. In *A. brasilense*, these biopolymers and their covalently bound complexes characteristic of the cell surface [2] are believed to be involved in contact interactions with plant roots and in bacterial cell aggregation [2,25]. Thus, their interactions with metal ions in metal-contaminated soil can interfere with the processes of molecular plant–bacterial interactions, which must be investigated in detail. It has to be noted also that the above-described microbiological EMS studies can be applied for revealing possible biotransformations of environmentally significant ^{60}Co radionuclide traces that can result in its microbially mediated migration in soils and aquifers [10,12,38].

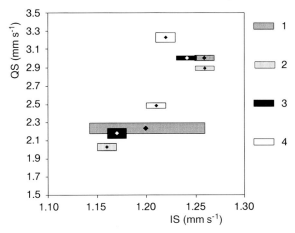

Figure 2.5 Comparison of Mössbauer parameters – isomer shift (IS, mm s^{-1}; relative to
α-Fe) and quadrupole splitting (QS, mm s^{-1}) –
calculated for the subspectra corresponding to
different forms of [^{57}Co]-cobalt(II) in aqueous
suspension of live cells of *A. brasilense*

Sp245 rapidly frozen after (1) 2 min and (2)
60 min of contact with ^{57}Co^{II}, (3) dead cells
(hydrothermally treated at 95 °C for 1 h), as well
as in cell-free supernatant liquid (4) (all spectra
measured at $T = 80$ K) [33].

2.4
Structural Studies of Glutamine Synthetase (GS) from *A. brasilense*

2.4.1
General Characterization of the Enzyme

GS (EC 6.3.1.2), which catalyzes the ATP- and metal-ion-dependent synthesis of
L-glutamine from L-glutamic acid and NH_4^+, is a key enzyme of nitrogen metabolism in many organisms from mammals to bacteria [39]. In diazotrophic PGPR,
which contribute in part to the overall soil fertility and plant–bacterial interactions by
fixing atmospheric nitrogen, basic knowledge of the structural and functional
aspects of this enzyme at the molecular level is of special importance.

Regulation of activity and biosynthesis of bacterial GSs is very complex and has so
far been investigated in detail for enteric bacteria only [39–41]. Glutamine synthetase
activity in many bacteria, including *A. brasilense*, is modulated by reversible adenylylation of its subunits in response to the cellular nitrogen status. The enzyme
is maintained in a top-active unadenylylated or slightly adenylylated form under
nitrogen-limiting conditions, while its adenylylation state (ranging from E_0 up to E_{12}
corresponding to 12 adenylylatable subunits in the GS molecule) increases under
conditions of ammonium abundance (see [40,41] and references therein). From the
structural point of view, bacterial GS molecules are dodecamers formed from two
face-to-face hexameric rings of subunits with 12 active sites formed between the
monomers [39].

Divalent cations (commonly, Mg^{2+}, Mn^{2+} or Co^{2+}) are critical for the activity of all known bacterial GSs [39–42]. According to X-ray crystallographic studies, each active center of the enzyme has two divalent cation-binding sites, n1 and n2, with the affinity of n1 for metal ions being much higher than that of n2 (both must be saturated for the GS activity to be expressed); there are also many additional metal-binding sites with relatively low affinity outside the active center of the enzyme, which are considered to be important for the conformational stability of the molecule, as well as a binding site for ammonium. A cation bound in site n1 (with a much higher affinity) is coordinated by three Glu residues (i.e. three side-chain carboxylic groups), whereas one bound in site n2 is coordinated by one His and two Glu residues (i.e. one nitrogen-donor atom of the His heterocycle and two carboxyls), and this structure is strictly conserved among different GSs [39] (note that additional nonprotein ligand(s) are one or more H_2O molecules [43]).

It has been found that the native (isolated and purified) *A. brasilense* GS shows enzymatic activity without divalent metals in the medium and therefore contains cations bound in its active centers, which is prerequisite for enzyme activity to be expressed [39,40]. However, after treating the native enzyme with 5 mM EDTA (with subsequent dialysis to remove EDTA-bound cations), a reversible loss of activity was found. Thus, while the resulting cation-free enzyme was inactive, it restored its activity after adding calculated amounts of Mg^{2+}, Mn^{2+} or Co^{2+}. The latter finding shows that the cations added are bound in the GS active centers, governed by their high affinity, so that the enzyme regains its active state. These methodological approaches [40] were useful for investigating structural changes in GS molecular conformation induced by removal or binding of activating cations as well as for probing the structural organization of the cation-binding sites in the enzyme active centers discussed below.

2.4.2
Circular Dichroism Spectroscopic Studies of the Enzyme Secondary Structure

2.4.2.1 Methodology of Circular Dichroism (CD) Spectroscopic Analysis of Protein Secondary Structure

CD spectroscopy in the UV region is one of the techniques that can be used for studying the secondary structure of proteins in solution (see [40] and references therein). The results of measurements are expressed in terms of molar ellipticity ($[\Theta]$ in $deg\,cm^2\,dmol^{-1}$), based on a mean amino acid residue weight (MRW, assuming its average weight to be equal to 115 Da), as a function of wavelength (λ, nm) determined as $[\Theta]_\lambda = \Theta \times 100(MRW)/cl$, where c is the protein concentration (in $mg\,ml^{-1}$), l is the light path length (in cm) and Θ is the measured ellipticity (in degrees) at a wavelength λ. The instrument (spectropolarimeter) is calibrated with some CD standards, e.g. (+)-10-camphorsulfonic acid, having $[\Theta]_{291} = 7820\,deg\,cm^2\,dmol^{-1}$ or nonhygroscopic ammonium (+)-10-camphorsulfonate ($[\Theta]_{290.5} = 7910\,deg\,cm^2\,dmol^{-1}$).

Calculations of the content of the protein secondary structure elements are commonly performed using a standard program. It is based on fitting the experimental spectrum to a sum of components (a negative maximum at 208 nm with molar ellipticities [Θ] around -1.6×10^4 deg $\times$ cm^2 $\times$ dmol^{-1} and a shoulder at about 222 nm are typical of predominantly α-helical proteins, whereas a similarly intensive broad negative band at about 215 nm is typical of proteins rich in β-structure).

2.4.2.2 The Effect of Divalent Cations on the Secondary Structure of GS from A. brasilense

Comparative measurements for native partly adenylylated GS ($E_{5.3}$ corresponding to 44% of adenylylated subunits) isolated from A. brasilense Sp245 [44] showed that adding 1 mM Mg^{2+}, Mn^{2+} or Co^{2+} had little effect on the shape of its CD spectrum. In contrast, the CD spectrum of the native GS changed noticeably after its treatment with 5 mM EDTA and subsequent dialysis (Figure 2.6a) reflecting changes in its molecular conformation upon removal of the bound cations.

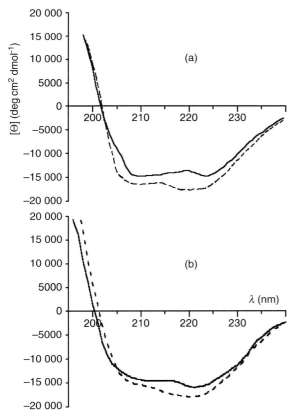

Figure 2.6 Circular dichroism spectra of native (dashed lines) and cation-free glutamine synthetase (solid lines) from A. brasilense Sp245: (a) partly adenylylated ($E_{5.3}$; 44% of adenylylated subunits) and unadenylylated (b) (E_0) [40,44].

Note that the general shape of the CD spectrum for the native *A. brasilense* GS was found to be somewhat different from those reported for a number of other bacterial GSs (see [40,44] and references therein). This may be connected with differences in the amino acid sequences of GSs obtained from different sources and the resulting differences in their secondary structures.

Calculations using the experimental CD spectroscopic data showed both the native and cation-free partly adenylylated enzyme ($E_{5.3}$) preparations to be highly structured (58 ± 2 and $49 \pm 3\%$ of the polypeptide as α-helices, 10 ± 2 and $20 \pm 2\%$ as β-structure, with only 32 ± 2 and $31 \pm 2\%$ unordered, respectively) [44]. Thus, the removal of cations from the native GS leads to lowering the proportion of α-helices and increasing that of the β-structure. These changes were found to be similar to those observed for native and cation-free unadenylylated (E_0) *A. brasilense* GS samples (Figure 2.6b) which had 59 ± 2 and $38 \pm 2\%$ α-helices, 13 ± 5 and $32 \pm 4\%$ β-structure, with 28 ± 3 and $30 \pm 3\%$ unordered, respectively [40].

It was found that in the case of unadenylylated GS, treatment of the native enzyme with EDTA at a lower concentration (1 mM instead of 5 mM) resulted in intermediate conformational changes ($43 \pm 1\%$ α-helices, $24 \pm 3\%$ β-structure and $32 \pm 2\%$ unordered) [40], evidently reflecting an incomplete removal of cations. On the contrary, adding divalent cations (Mg^{2+}, Mn^{2+} or Co^{2+}) to cation-free GS tended to change the enzyme conformation to one closer to the initial native preparation. Thus, *A. brasilense* GS appears to be most structured among all bacterial GSs known to date, with about 70% of its polypeptide chain being structured (α-helices + β-structural elements) in both unadenylylated and partly adenylylated enzyme. Upon removal of cations from the active centers, the proportions of the secondary structure elements change, but the protein remains similarly highly structured.

2.4.3
Emission Mössbauer Spectroscopic Analysis of the Structural Organization of the Cation-Binding Sites in the Enzyme Active Centers

2.4.3.1 Methodological Outlines and Prerequisites

The aforementioned reversible loss of the GS activity upon removal of native cations, with its restoration upon subsequent addition of a new cation, makes it possible, in principle, to replace the native cations by EMS-active $^{57}Co^{2+}$ under physiologically similar conditions [36]. In that case, active centers doped with $^{57}Co^{2+}$ ions can be probed using EMS. Nevertheless, for a correct analysis of the data to be obtained, several conditions should be observed [45]. First, when substituting the $^{57}Co^{2+}$ cation for the native cations, it is important to make sure that the metal is indeed bound within the active center; otherwise, the appearance of multiple binding sites and, consequently, many forms of cobalt would render the EMS data hardly interpretable. Second, the process of replacing the activating cations (e.g. by using natural Co^{2+} under identical conditions) should not result in an irreversible deactivation of the enzyme. In the latter case, the correspondence between the $^{57}Co^{2+}$ form in the enzyme sample under study and the cobalt(II) form in the physiologically active enzyme would be doubtful. Finally, the quantity of the substituted $^{57}Co^{2+}$ should conform with the overall number of the cation-binding sites in the enzyme sample. It is clear that any excessive $^{57}Co^{2+}$, binding to different functional groups of the protein macromolecule beyond the active centers, can lead to an unpredictable complication of the spectra.

Fulfillment of the above-mentioned conditions is facilitated by the fact that the affinity to the cation in the enzyme active centers is usually much higher than elsewhere on the protein globule. Moreover, when the active center contains more than one binding site with different affinities to the cation and different coordination environments (as in the case with glutamine synthetase), it may be expected that using an amount of $^{57}Co^{2+}$ under the total 'saturation limit' (but higher than that necessary to saturate half the sites) would allow one to obtain information not only on the chemical forms and coordination of the cobalt but also on its distribution between the sites. The above-discussed properties of *A. brasilense* GS were found to be suitable for using the EMS technique in studying $^{57}Co^{2+}$-doped enzyme preparations [36,44,45].

2.4.3.2 Experimental Studies of *A. brasilense* GS

Measurements were performed on *A. brasilense* GS ($E_{2.2}$ corresponding to the adenylylation state 18%) using the EMS technique (according to the scheme presented in Figure 2.3). Analysis of the emission Mössbauer spectra of $^{57}Co^{2+}$-doped GS both in rapidly frozen aqueous solution and in the dry state (Figure 2.7; both spectra measured at $T = 80$ K) indeed showed in each spectrum the presence of two forms of cobalt(II) with different affinities (in view of unequal distribution of $^{57}Co^{II}$ between the forms; cf. doublets 1 and 2 in Figure 2.7) as well as with different coordination reflected by different Mössbauer parameters (Figure 2.8). The presence of the third spectral component (doublet 3 in both spectra) is related to the aftereffects of the nuclear transformation $^{57}Co \rightarrow ^{57}Fe$ resulting in the formation of an aliovalent $^{57}Fe^{3+}$ species [34,35]. In the present case, the appearance of this component does not affect the interpretation of the data on the initial Co^{II} forms [45].

The values of isomer shifts (IS = 1.08 and 1.05–1.07 mm s^{-1} relative to α-Fe) and quadrupole splittings (QS = 3.0–3.1 and 2.3–2.4 mm s^{-1}; see Figure 2.8) obtained for doublets 1 and 2, respectively, allowed those components to be correlated with the two cation-binding sites in the GS active center (sites n2 and n1, respectively [39]). As mentioned above, these sites of bacterial GSs have different coordination environments, as well as a correspondingly lower (for site n2) and higher (for site n1) affinity to the cation. The latter difference is in line with the nonuniform distribution of $^{57}Co^{II}$ between the spectral components (as the areas of quadrupole doublets 1 and 2 in each spectrum corresponding to different $^{57}Co^{II}$ forms are significantly different; see Figure 2.7). The close (overlapping) values of the Mössbauer parameters for the corresponding $^{57}Co^{II}$ forms for GS in frozen solution and in the dry state at $T = 80$ K (see Figure 2.8) reflect the unaffected cobalt(II) microenvironment in each of the forms at the active centers in both states. This, in turn, correlates well with the conformational stability of bacterial glutamine synthetases and suggests that no significant structural changes occur upon drying the enzyme [44].

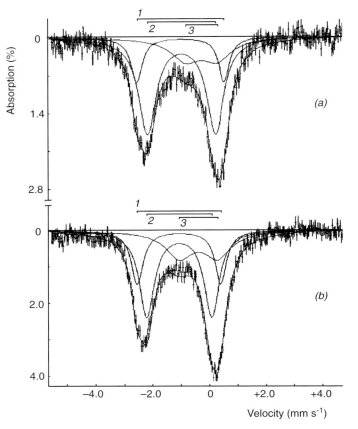

Figure 2.7 Emission Mössbauer spectra of cation-free glutamine synthetase (GS; $E_{2.2}$) from *A. brasilense* Sp245 incubated with $^{57}Co^{2+}$ for 60 min at ambient temperature (a) in rapidly frozen aqueous solution and as a dried solid (b) (measured at $T = 80$ K; intensities converted to the absorption convention). For each spectrum, the relevant subspectra (quadrupole doublets) are shown which contributed to the resulting spectrum (solid-line envelope) obtained by computer fitting to the experimental data (points with vertical error bars). The positions of the spectral components (quadrupole doublets) are indicated by horizontal square brackets above the zero lines.

The relatively low IS values for both the $^{57}Co^{II}$ forms in GS with $E_{2.2}$ (IS = 1.05 and 1.08 mm s^{-1} at $T = 80$ K; see Figure 2.8) may indicate a tetrahedral symmetry of cobalt(II) coordination. In this case, the coordination mode of all the Glu residues must be monodentate, which is often observed for cation-binding sites in metalloproteins [46] (note that there is also at least one water molecule as a ligand, according to Eads *et al.* [43]). Similarly, low IS values were found for EMS spectra of dilute frozen aqueous solutions of $^{57}Co^{II}$ complexes with amino acids (anthranilic acid and tryptophan; IS = 1.1 and 0.9 mm s^{-1}, QS = 2.7 and 2.8 mm s^{-1}, respectively), also assuming a tetrahedral symmetry [49]. Note that tetrahedral coordination (T_d) of

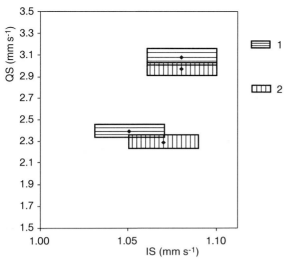

Figure 2.8 Comparison of Mössbauer parameters–isomer shift (IS, mm s^{-1}; relative to α-Fe) and quadrupole splitting (QS, mm s^{-1}) – for different forms of [^{57}Co]-cobalt(II) in ^{57}Co^{II}-doped glutamine synthetase from A. brasilense Sp245 (1) in rapidly frozen aqueous solution and in the solid (dried) state (2) (measured at $T = 80$ K).

cobalt(II) is possible [46–48], although in many proteins cobalt was found to have preference for higher coordination numbers, i.e. 5 and 6 (see [47] and references cited therein).

2.4.3.3 Conclusions and Outlook

The EMS technique, recently applied for the first time to probing cation binding in the active centers of a bacterial enzyme doped with the radioactive ^{57}Co isotope [36,44], has shown that each active center of glutamine synthetase from azospirilla has two cation-binding sites with different affinities to cobalt(II) as an activating cation and with different coordination symmetry. The results obtained are in good agreement with the current literature data on the structural organization of the active centers in bacterial adenylylatable glutamine synthetases [39,41].

For future structural investigations of the active centers in metal-containing biocomplexes and enzymes, the advantages of using the highly sensitive and selective emission variant of Mössbauer spectroscopy can hardly be overestimated. This nuclear chemistry technique has recently been shown to be sensitive also (i) to the effects of competitive binding of different activating cations ($Mn^{2+} + {}^{57}Co^{2+}$, with a redistribution of the latter between the two sites in GS) at the active centers, showing that heterobinuclear two-metal-ion catalysis by GS is principally possible, as well as (ii) to fine structural changes induced by covalent modifications of the enzyme molecule related to its activity [50]. The results obtained are highly promising for further study of the molecular mechanisms of enzymatic activity regulation and enzyme–substrate biospecific interactions using the unique possibilities of the EMS technique.

Besides cobalt-activated enzymes, EMS may be applicable to studying other metalloproteins upon substituting ^{57}Co for the native metal. For instance, substitution of Co^{2+} (as an optically active probe) for Zn^{2+} has been used extensively in optical spectroscopic methods to obtain structural information on zinc metalloproteins and is based on the fact that these two cations typically exhibit similar coordination geometries for a given ligand set [48]. Thus, the possibility of using ^{57}Co as a substituting probe, considering also the exceptionally high sensitivity of the EMS technique, can significantly expand the limits of its applicability in biochemistry and related fields in the life sciences.

2.5
General Conclusions and Future Directions of Research

The highly sophisticated field of bioscience comprising the interactions of microorganisms with their hosts (higher organisms) has been increasingly attracting attention during the past decade both in basic research and in applied fields, particularly those related to agricultural and environmental biotechnology. As for plant–microbe interactions, the subject can be reasonably classified and accordingly divided into a few major categories [51]: (i) the physiological and biochemical properties and responses of the macropartner (the host plant), (ii) the corresponding properties and behavior of the micropartner (consortia of plant-associated microorganisms, in particular, in the rhizosphere), as well as (iii) any processes or phenomena directly related to their interactions per se, including remote exchange of molecular signals and their perception, microbial quorum sensing and its inhibition (including chemical and enzymatic 'quorum quenching' or 'anti-quorum sensing', contact and intercellular interactions, the effects and role of the chemical composition and conditions of the medium, and so on (see [51] and references cited therein).

It is clear that any purely chemical (i.e. abiotic) processes, induced in the rhizosphere by the presence or formation of chemically active species (e.g. metal ions, oxidizing agents, etc.), which result in chemical depletion, inactivation or degradation of any biomolecules directly involved in plant–microbe interactions via their binding and/or redox transformation, would inevitably affect these biologically specific interactions. However, in the rapidly increasing pool of basic and applied research data related to plant–microbe interactions (see, e.g. the recent highly informative review [25]), such chemical interferences seem to have been paid significantly less attention so far than they really deserve [13,51,52] considering their possible contribution to the overall effects. This imbalance in approaching the whole problem, leading to a virtual imbalance in understanding the diversity of molecular mechanisms underlying the processes and phenomena in highly sophisticated soil–plant–microbe systems, still remains to be corrected by increasingly involving experts from chemical and physical sciences and applying a complex of relevant modern instrumental techniques.

In order to illustrate the applicability of a range of instrumental techniques in bioscience, a number of recent stimulating reviews and highly informative experimental reports may be recommended, such as: applications of vibrational spectroscopy in microbiology [6,8–10,53]; noninvasive characterization of microbial cultures and various metabolic transformations using multielement NMR spectroscopy [54]; X-ray crystallography in studying biological complexes [55]; biological, agricultural and environmental research using X-ray microscopy and microradiography [56] and X-ray absorption spectroscopy [57]; surface characterization of bacteria using X-ray photoelectron spectroscopy (XPS), time-of-flight secondary-ion mass spectrometry (ToF-SIMS) [58] and atomic force microscopy (AFM) [59]; inductively coupled plasma–mass spectrometry (ICP-MS) as a multielement and multiisotope highly sensitive analytical tool [60]; novel biochemical [33] and microbiological applications of the emission variant of Mössbauer (nuclear γ-resonance) spectroscopy (based on the use of ^{57}Co) [10] as well as its traditionally used transmission variant using the stable ^{57}Fe isotope [31,32], its combination with electron paramagnetic resonance (EPR) spectroscopy [61]; stable isotope technologies in studying plant–microbe interactions [62], and so on.

Acknowledgments

The author is grateful to many of his colleagues both at the Institute in Saratov and from other research organizations, who have contributed to the studies considered in this chapter, for their help in experimental work, long-term collaboration and many stimulating discussions. Support for the author's research in Russia and for his international collaboration, which contributed in part to the interdisciplinary fields considered in this chapter, has been provided within the recent years by grants from INTAS (EC, Brussels, Belgium; Project 96-1015), NATO (Projects LST. CLG.977664, LST.EV.980141, LST.NR.CLG.981092, CBP.NR.NREV.981748, ESP. NR.NRCLG 982857), the Russian Academy of Sciences' Commission (Grant No. 205 under the 6th Competition-Expertise of research projects) as well as under the Agreements on Scientific Cooperation between the Russian and Hungarian Academies of Sciences for 2002–2004 and 2005–2007.

References

1 Ivanov, V.T. and Gottikh, B.P. (1999) *Herald of the Russian Academy of Sciences*, **69**, 208–214.

2 Bashan, Y., Holguin, G. and de-Bashan, L.E. (2004) *Canadian Journal of Microbiology*, **50**, 521–577.

3 Naumann, D., Keller, S., Helm, D., Schultz, Ch. and Schrader, B. (1995) *Journal of Molecular Structure*, **347**, 399–405.

4 Schmitt, J. and Flemming, H.-C. (1998) *International Biodeterioration & Biodegradation*, **41**, 1–11.

5 Schrader, B., Dippel, B., Erb, I., Keller, S., Löchte, T., Schulz, H., Tatsch, E. and Wessel, S. (1999) *Journal of Molecular Structure*, **480–481**, 21–32.

6 Naumann, D. (2000) Infrared spectroscopy in microbiology, in: *Encyclopedia of Analytical Chemistry* (ed. R. A. Meyers), John Wiley & Sons, Ltd, Chichester, UK, pp. 102–131.

7 Kamnev, A.A., Tarantilis, P.A., Antonyuk, L.P., Bespalova, L.A., Polissiou, M.G., Colina, M., Gardiner, P.H.E. and Ignatov, V.V. (2001) *Journal of Molecular Structure*, **563–564**, 199–207.

8 Jiang, W., Saxena, A., Song, B., Ward, B.B., Beveridge, T.J. and Myneni, S.C.B. (2004) *Langmuir*, **20**, 11433–11442.

9 Yu, C. and Irudayaraj, J. (2005) *Biopolymers*, **77**, 368–377.

10 Kamnev, A.A., Tugarova, A.V., Antonyuk, L.P., Tarantilis, P.A., Kulikov, L.A., Perfiliev, Yu.D., Polissiou, M.G. and Gardiner, P.H.E. (2006) *Analytica Chimica Acta*, **573–574**, 445–452.

11 Belimov, A.A., Kunakova, A.M., Safronova, V.I., Stepanok, V.V., Yudkin, L.Yu., Alekseev, Yu.V. and Kozhemyakov, A.V. (2004) *Microbiology (Moscow)*, **73**, 99–106.

12 Tugarova, A.V., Kamnev, A.A., Antonyuk, L.P. and Gardiner, P.H.E. (2006) *Azospirillum brasilense* resistance to some heavy metals, in: *Metal Ions in Biology and Medicine*, vol. 9 (eds M.C. Alpoim, P.V. Morais, M.A. Santos, A.J. Cristóvão, J.A. Centeno and Ph. Collery), John Libbey Eurotext, Paris, pp. 242–245.

13 Kamnev, A.A. and van der Lelie, D. (2000) *Bioscience Reports*, **20**, 239–258.

14 Glick, B.R. (2003) *Biotechnology Advances*, **21**, 383–393.

15 Khan, A.G. (2005) *Journal of Trace Elements in Medicine and Biology*, **18**, 355–364.

16 Biró, B., Köves-Péchy, K., Tsimilli-Michael, M. and Strasser, R.J. (2006) Role of the beneficial microsymbionts in the plant performance and plant fitness, in: *Soil Biology, vol. 7: Microbial Activity in the Rhizosphere* (eds K.G. Mukerji, C. Manoharachary and J. Singh), Springer, Berlin, Heidelberg, pp. 265–296.

17 Lyubun, Ye.V., Fritzsche, A., Chernyshova, M.P., Dudel, E.G. and Fedorov, E.E. (2006) *Plant and Soil*, **286**, 219–227.

18 Kamnev, A.A., Renou-Gonnord, M.-F., Antonyuk, L.P., Colina, M., Chernyshev, A.V., Frolov, I. and Ignatov, V.V. (1997) *Biochemistry and Molecular Biology International*, **41**, 123–130.

19 Kamnev, A.A., Ristic̆, M., Antonyuk, L.P., Chernyshev, A.V. and Ignatov, V.V. (1997) *Journal of Molecular Structure*, 408/409, pp. 201–205.

20 Kamnev, A.A., Antonyuk, L.P., Matora, L.Yu., Serebrennikova, O.B., Sumaroka, M.V., Colina, M., Renou-Gonnord, M.-F. and Ignatov, V.V. (1999) *Journal of Molecular Structure*, **480–481**, 387–393.

21 Ignatov, O.V., Kamnev, A.A., Markina, L.N., Antonyuk, L.P., Colina, M. and Ignatov, V.V. (2001) *Applied Biochemistry and Microbiology (Moscow)*, **37**, 219–223.

22 Kadouri, D., Jurkevitch, E. and Okon, Y. (2003) *Applied and Environmental Microbiology*, **69**, 3244–3250.

23 Kadouri, D., Jurkevitch, E., Okon, Y. and Castro-Sowinski, S. (2005) *Critical Reviews in Microbiology*, **31**, 55–67.

24 Kamnev, A.A., Antonyuk, L.P., Tugarova, A.V., Tarantilis, P.A., Polissiou, M.G. and Gardiner, P.H.E. (2002) *Journal of Molecular Structure*, **610**, 127–131.

25 Somers, E., Vanderleyden, J. and Srinivasan, M. (2004) *Critical Reviews in Microbiology*, **30**, 205–240.

26 Olubai, O., Caudales, R., Atkinson, A. and Neyra, C.A. (1998) *Canadian Journal of Microbiology*, **44**, 386–390.

27 Kirchhof, G., Schloter, M., Aßmus, B. and Hartmann, A. (1997) *Soil Biology & Biochemistry*, **29**, 853–862.

28 Rothballer, M., Schmid, M. and Hartmann, A. (2003) *Symbiosis*, **34**, 261–279.

29 Lodewyckx, C., Vangronsveld, J., Porteous, F., Moore, E.R.B., Taghavi, S. and van der

Lelie, D. (2002) *Critical Reviews in Plant Sciences*, **21**, 583–606.

30 Kamnev, A.A., Tugarova, A.V., Antonyuk, L.P., Tarantilis, P.A., Polissiou, M.G. and Gardiner, P.H.E. (2005) *Journal of Trace Elements in Medicine and Biology*, **19**, 91–95.

31 Oshtrakh, M.I. (2004) *Spectrochimica Acta A*, **60**, 217–234.

32 Krebs, C., Price, J.C., Baldwin, J., Saleh, L., Green, M.T. and Bollinger, J.M. Jr (2005) *Inorganic Chemistry*, **44**, 742–757.

33 Kamnev, A.A. (2005) *Journal of Molecular Structure*, **744–747**, 161–167.

34 Vértes, A. and Nagy, D.L. (eds) (1990) *Mössbauer Spectroscopy of Frozen Solutions*. Akad. Kiadó, Budapest, 1990, Chapter 6 (Russian edition) (1998) (ed. Yu.D. Perfiliev), Mir, Moscow, pp. 271–293.

35 Perfiliev, Yu.D., Rusakov, V.S., Kulikov, L.A., Kamnev, A.A. and Alkhatib, K. (2006) *Hyperfine Interactions*, **167**, 881–885.

36 Kamnev, A.A., Antonyuk, L.P., Smirnova, V.E., Serebrennikova, O.B., Kulikov, L.A. and Perfiliev, Yu.D. (2002) *Analytical and Bioanalytical Chemistry*, **372**, 431–435.

37 Kamnev, A.A., Antonyuk, L.P., Kulikov, L.A. and Perfiliev, Yu.D. (2004) *BioMetals*, **17**, 457–466.

38 Lloyd, J.R. (2003) *FEMS Microbiology Reviews*, **27**, 411–425.

39 Eisenberg, D., Gill, H.S., Pfluegl, G.M.U. and Rotstein, S.H. (2000) *Biochimica et Biophysica Acta*, **1477**, 122–145.

40 Antonyuk, L.P., Smirnova, V.E., Kamnev, A.A., Serebrennikova, O.B., Vanoni, M.A., Zanetti, G., Kudelina, I.A., Sokolov, O.I. and Ignatov, V.V. (2001) *BioMetals*, **14**, 13–22.

41 Antonyuk, L.P. (2007) *Applied Biochemistry and Microbiology (Moscow)*, **43**, 244–249.

42 Bespalova, L.A., Antonyuk, L.P. and Ignatov, V.V. (1999) *BioMetals*, **12**, 115–121.

43 Eads, C.D., LoBrutto, R., Kumar, A. and Villafranca, J.J. (1988) *Biochemistry*, **27**, 165–170.

44 Kamnev, A.A., Antonyuk, L.P., Smirnova, V.E., Kulikov, L.A., Perfiliev, Yu.D., Kudelina, I.A., Kuzmann, E. and Vértes, A. (2004) *Biopolymers*, **74**, 64–68.

45 Kamnev, A.A., Antonyuk, L.P., Kulikov, L.A., Perfiliev, Yu.D., Kuzmann, E. and Vértes, A. (2005) *Bulletin of the Russian Academy of Sciences (Physics)*, **69**, 1561–1565.

46 Holm, R.H., Kennepohl, P. and Solomon, E.I. (1996) *Chemical Reviews*, **96**, 2239–2314.

47 Innocenti, A., Zimmerman, S., Ferry, J.G., Scozzafava, A. and Supuran, C.T. (2004) *Bioorganic & Medicinal Chemistry Letters*, **14**, 3327–3331.

48 Namuswe, F. and Goldberg, D.P. (2006) *Chemical Communications*, (22), 2326–2328.

49 Kamnev, A.A., Kulikov, L.A., Perfiliev, Yu.D., Antonyuk, L.P., Kuzmann, E. and Vértes, A. (2005) *Hyperfine Interactions*, **165**, 303–308.

50 Kamnev, A.A., Antonyuk, L.P., Smirnova, V.E., Kulikov, L.A., Perfiliev, Yu.D., Kuzmann, E. and Vértes, A. (2005) *FEBS Journal*, **272** (Suppl 1), 10.

51 Kamnev, A.A. (2008) Metals in soil *versus* plant-microbe interactions: biotic and chemical interferences, in: *Plant-Microbe Interaction* (eds E.A. Barka and Ch. Clément), Research Signpost, Trivandrum (Kerala, India), Chapter 13, pp. 291–318.

52 Kamnev, A.A. (2003) Phytoremediation of heavy metals: an overview, in: *Recent Advances in Marine Biotechnology, vol. 8: Bioremediation* (eds M. Fingerman and R. Nagabhushanam), Science Publishers, Inc., Enfield, NH, USA, pp. 269–317.

53 Aroca, R. (2006) *Surface-Enhanced Vibrational Spectroscopy*, John Wiley & Sons, Ltd, Chichester, 400 pp.

54 Lens, P.N.L. and Hemminga, M.A. (1998) *Biodegradation*, **9**, 393–409.

55 Sommerhalter, M., Lieberman, R.L. and Rosenzweig, A.C. (2005) *Inorganic Chemistry*, **44**, 770–778.

56 Reale, L., Lai, A., Bellucci, I., Faenov, A., Pikuz, T., Flora, F., Spanò, L., Poma, A.,

Limongi, T., Palladino, L., Ritucci, A., Tomassetti, G., Petrocelli, G. and Martellucci, S. (2006) *Microscopy Research and Technique*, **69**, 666–674.

57 Prange, A. and Modrow, H. (2002) *Reviews in Environmental Science and Biotechnology*, **1**, 259–276.

58 Pradier, C.M., Rubio, C., Poleunis, C., Bertrand, P., Marcus, P. and Compère, C. (2005) *Journal of Physical Chemistry B*, **109**, 9540–9549.

59 Teschke, O. (2005) *Microscopy Research and Technique*, **67**, 312–316.

60 Cottingham, K. (2004) *Analytical Chemistry*, **76**, 35A–38A.

61 Schünemann, V., Jung, C., Lendzian, F., Barra, A.-L., Teschner, T. and Trautwein, A.X. (2004) *Hyperfine Interactions*, **156/157**, 247–256.

62 Prosser, J.I., Rangel-Castro, J.I. and Killham, K. (2006) *Current Opinion in Biotechnology*, **17**, 98–102.

3

Physiological and Molecular Mechanisms of Plant Growth Promoting Rhizobacteria (PGPR)

Beatriz Ramos Solano, Jorge Barriuso, and Francisco J. Gutiérrez Mañero

3.1
Introduction

Plant growth promoting rhizobacteria (PGPR) include bacteria that inhabit the rhizosphere, improve plant health and may also enhance plant growth. The term PGPR was coined by Kloepper and coworkers in 1980 [35], although PGPR was first mentioned in 1978 by the same author in the *Proceedings of the Fourth International Congress of Bacterial Plant Pathogens*, conducted in France. Since then, research on PGPR has increased noticeably, with 11 reports appearing in the USDA (US Department of Agriculture) between 1980 and 1990, 34 from 1990 to 1995 and 72 from 1995 to 2000.

Currently, the number of works per year on this topic has seen a 10-fold increase, creating a new discipline that has changed the basic traditional concepts of plant physiology and microbial ecology.

Bashan and Holguin [8] proposed a revision of the original definition of the term PGPR, since there are a number of bacteria that may have a beneficial effect on the plant even though they are outside the rhizosphere environment.

Bacteria identified to be PGPR could be members of several genera such as *Azotobacter, Acetobacter, Azospirillum, Burkholderia, Pseudomonas* and *Bacillus* [3,7,13,36,37,45,46,52]. The positive effect of PGPR occurs through various mechanisms. This role involves not only the direct effect of a single bacterial strain but also that of the molecular dialogue established among soil microorganisms and between microorganisms and the plant, including *quorum-sensing* mechanisms.

3.2
PGPR Grouped According to Action Mechanisms

A thorough understanding of the PGPR action mechanisms is fundamental to manipulating the rhizosphere in order to maximize the processes within the system

Plant-Bacteria Interactions. Strategies and Techniques to Promote Plant Growth
Edited by Iqbal Ahmad, John Pichtel, and Shamsul Hayat
Copyright © 2008 WILEY-VCH Verlag GmbH & Co. KGaA, Weinheim
ISBN: 978-3-527-31901-5

that strongly influence plant productivity. PGPR action mechanisms have been grouped traditionally into direct and indirect mechanisms. Although the difference between the two is not always obvious, indirect mechanisms, as a general rule, are those that happen outside the plant, while direct mechanisms are those that occur inside the plant and directly affect the plant's metabolism. The latter, therefore, require the participation of the plant's defensive metabolic processes, which transduce the signal sent from the bacteria influencing the plant.

Accordingly, direct mechanisms include those that affect the balance of plant growth regulators, either because the microorganisms themselves release growth regulators that are integrated into the plant or because the microorganisms act as a sink of plant released hormones, and those that induce the plant's metabolism leading to an improvement in its adaptive capacity. Two important mechanisms are included in this group: induction of systemic resistance to plant pathogens and protection against high salinity conditions. According to this description, indirect mechanisms include

- those that improve nutrient availability to the plant;
- inhibition of microorganisms that have a negative effect on the plant (niche exclusion);
- free nitrogen fixation in the rhizosphere, which improves nitrogen availability.

3.2.1
PGPR Using Indirect Mechanisms

The list of indirect mechanisms used by PGPR is substantial. Some have been included here, with the most relevant being discussed in detail:

- Free nitrogen fixation.
- Production of siderophores.
- Phosphate solubilization.
- Hydrolysis of molecules released by pathogens. Toyoda and Utsumi [55] reported the ability of two strains, *Pseudomonas cepacia* and *Pseudomonas solanacearum*, to break down fusaric acid, a compound responsible for root rot caused by the fungi *Fusarium*.
- Synthesis of enzymes able to hydrolyze fungal cell walls [40].
- Synthesis of cyanhydric acid. This is a trait common to strains of the genus *Pseudomonas*; hence, it may indicate a certain antipathogenic effect [62].
- Improvement of symbiotic relationships with rhizobia and mycorrhizae [20,22,23,42].

A detailed discussion of the first three mechanisms listed above follows.

3.2.1.1 Free Nitrogen-Fixing PGPR
These types of bacteria were the first PGPR that were assayed to improve plant growth, especially in terms of crop productivity. The first report of these bacteria appeared before Word War II, when they were widely used on cereal fields in the

Soviet Union [9]. They are free-living organisms that inhabit the rhizosphere but do not establish a symbiosis with the plant. Although they do not penetrate the plant's tissues, a very close relationship is established; these bacteria live sufficiently close to the root such that the atmospheric nitrogen fixed by the bacteria that is not used for their own benefit is taken up by the plant, forming an extra supply of nitrogen. This relationship is described as an unspecific and loose symbiosis.

Free nitrogen-fixing bacteria belong to a wide array of taxa; among the most relevant bacterial genera are *Azospirillum, Azotobacter, Burkholderia, Herbaspirillum* and *Bacillus* [61]. Biological nitrogen fixation is a high-cost process in terms of energy. Bacterial strains capable of performing this process do so in order to fulfill their physiological needs and thus little nitrogen is left for the plant's use. However, growth promotion caused by nitrogen-fixing PGPR was attributed to nitrogen fixation for many years, until the use of nitrogen isotopes showed additional effects. This technique showed that the benefits of free nitrogen-fixing bacteria are due more to the production of plant growth regulators than to the nitrogen fixation [9]. Production of plant growth regulators is discussed below.

Azotobacteraceae is the most representative of bacterial genera able to perform free nitrogen fixation. Various reports describe the benefits of Azotobacteraceae on several crops [43]. According to data provided by the FAO [21], amounts of nitrogen supplied to soil are low; Bhattacharya and Chaudhuri [11] report that the amount ranges between 20 and 30 kg per hectare per year.

Azotobacter is the genus most used in agricultural trials. The first reports appeared in 1902 and it was widely used in Eastern Europe during the middle decades of the last century [29]. As previously suggested, the effect of *Azotobacter* and *Azospirillum* is attributed not only to the amounts of fixed nitrogen but also to the production of plant growth regulators (indole acetic acid, gibberellic acid, cytokinins and vitamins), which result in additional positive effects to the plant [48].

Application of inoculants in agriculture has resulted in notable increases in crop yields, especially in cereals, where *Azotobacter chroococcum* and *Azospirillum brasilense* have been very important. These two species include strains capable of releasing substances such as vitamins and plant growth regulators, which have a direct influence on plant growth [5,19,29,43,48,60]. According to González and Lluch [29], the production of these substances by *Azotobacter* strains is seriously affected by nitrogen availability, which affects auxin and gibberellin production; but when nitrate is available, auxin release is impaired while gibberellin synthesis is enhanced.

As mentioned above, the amount of nitrogen from free fixation available to the plant is low because it is used efficiently by the bacteria. Three strategies have been proposed to address this low-yield problem: (i) glutamine synthase bacterial mutants, (ii) formation of paranodules and (iii) facilitating the penetration of plant tissues by nitrogen-fixing bacterial endophytes that enhance colonization in a low competition niche. Regarding the first strategy, mutations target the glutamine synthase gene, focusing on achieving low efficiency in retaining the fixed nitrogen so that it is released for the plant. The main problem with these types of mutants is that they are not very effective in colonizing the root system [9]. To overcome this problem, a second strategy has been developed: the creation of a special environment for the nitrogen-fixing mutants called paranodules. These structures can be formed by the plant when plant growth regulators, either synthetic or bacterial, are supplied. Paranodules are small tumors that nitrogen-fixing bacteria penetrate. They colonize the intracellular spaces and fix nitrogen for the plant in a competitive environment. This strategy is defined as a formation of nitrogen-fixing nodules in nonlegumes and has already been assayed in corn and wheat [15]. The third strategy consists of enhancing the penetration of nitrogen-fixing endophytes on the plant tissue. Endophytes enter the plant upon the emergence of lateral roots when the endodermis is broken down, allowing penetration by the bacteria up to the xylem vessels. This stimulation of root branching owing to the presence of nitrogen-fixing bacteria strains results in enhanced penetration of nitrogen-fixing bacteria into the plant tissues and, hence, in an increase of fixed nitrogen available to the plant [5].

3.2.1.2 Siderophore-Producing PGPR

Iron is an essential nutrient for plants. Iron deficiency is exhibited in severe metabolic alterations because of its role as a cofactor in a number of enzymes essential to important physiological processes such as respiration, photosynthesis and nitrogen fixation. Iron is quite abundant in soils but is frequently unavailable for plants or soil microorganisms since the predominant chemical species is Fe^{3+}, the oxidized form that reacts to form insoluble oxides and hydroxides inaccessible to plants or microorganisms.

Plants have developed two strategies for efficient iron absorption. The first consists of releasing organic compounds capable of chelating iron, thus rendering it soluble. Iron diffuses toward the plant where it is reduced and absorbed by means of an enzymatic system present in the cell membrane. The second strategy consists of absorbing the complex formed by the organic compound and Fe^{3+}, where the iron is reduced inside the plant and readily absorbed. Some rhizosphere bacteria are able to release iron-chelating molecules to the rhizosphere and hence serve the same function as the plants.

Siderophores are low molecular weight compounds, usually below 1 kDa, which contain functional groups capable of binding iron in a reversible way. The most frequent functional groups are hydroximates and catechols, in which the distances among the groups involved are optimal to bind iron. Siderophore concentration in soil is approximately around 10^{-30} M. Siderophore-producing bacteria usually belong to the genus *Pseudomonas*, the most common being *Pseudomonas fluorescens*, which release pyochelin and pyoverdine. Rhizosphere bacteria release these compounds to increase their competitive potential, since these substances have an antibiotic activity and improve iron nutrition for the plant [27].

Siderophore-producing rhizobacteria improve plant health at various levels: they improve iron nutrition, inhibit growth of other microorganisms with release of their antibiotic molecule and hinder the growth of pathogens by limiting the iron available for the pathogen, generally fungi, which are unable to absorb the iron–siderophore complex.

3.2.1.3 Phosphate-Solubilizing PGPR

After nitrogen, phosphorous is the most limiting nutrient for plants. However, phosphorous reserves, although abundant, are not available in forms suitable for plants. Plants are only able to absorb the soluble forms, that is, mono- and dibasic phosphate. Besides inorganic forms of phosphorous in soil, the phosphorous present in organic matter is of considerable importance. The organic forms of phosphorous are estimated to comprise between 30 and 50% of total soil phosphorous. This reservoir can be mineralized by microorganisms, making it available to the plant as soluble phosphates. Many bacteria from different genera are capable of solubilizing phosphate and include *Pseudomonas, Bacillus, Rhizobium, Burkholderia, Achromobacter, Agrobacterium, Micrococcus, Aerobacter, Flavobacterium, Chryseobacterium* and *Erwinia*. Bacteria use two mechanisms to solubilize phosphate: (i) releasing organic acids that mobilize phosphorous by means of ionic interactions with the cations of the phosphate salt and (ii) by releasing phosphatases responsible for releasing phosphate groups bound to organic matter. Most of these bacteria are able to solubilize the Ca–P complex, and there are others that operate on the Fe–P, Mn–P and Al–P complexes. Generally, these mechanisms are more efficient in basic soils. Results with PGPR capable of solubilizing phosphate are sometimes erratic, probably due to soil composition, and to demonstrate their effect they have to be inoculated in soils with a phosphorous deficit and stored in insoluble forms. Inoculations of these types of PGPR sometimes improve plant growth, and sometimes they are completely inefficient. Without doubt, knowledge of their mechanisms and ecology in the rhizosphere will improve their use in sustainable agriculture [33].

3.2.2
PGPR Using Direct Mechanisms

Table 3.1 presents a summary of the direct mechanisms used by PGPR. In this chapter, a special emphasis has been placed on plant growth regulators, which can be considered the principal PGPR mechanism, and on the induction of systemic resistance (ISR), which has become an important mechanism in the control of pathogenic pests in recent years.

Table 3.1 Direct PGPR action mechanisms [14 Modified].

Mechanism	Effect	References
Plant growth regulator production	Biomass (aerial part and root); flowering	[31,32]
Ethylene synthesis inhibition	Root length	[25]
Induction of systemic resistance	Health	[57]
Root permeability increase	Biomass and nutrient absorption	[51]
Organic matter mineralization (nitrogen, sulfur, phosphorus)	Biomass and nutrient content	[41]
Mycorrhizal fungus association	Biomass and phosphorus content	[24,54]
Insect pest control	Health	[64]

3.2.2.1 **PGPR that Modify Plant Growth Regulator Levels**

Plant growth regulator production by bacteria was first described more than 40 years ago, determined in the 1960s using the biological assays then available. Using modern techniques, it has been demonstrated that the production of plant growth regulators such as auxins and ethylene by bacteria is a common trait. Others such as cytokinins are less common, while gibberellin production in high concentrations has only been described for two strains from the genus *Bacillus* isolated in the rhizosphere of *Alnus glutinosa* [32], the amounts being 1000 times higher than those produced by *Rhizobium* when forming the nodule.

Modification of plant physiology by producing plant growth regulators is an important mechanism, not only because it alters the principal mechanism of growth regulation and cell differentiation in the plant but also because it is based on the development of common metabolic pathways in plants and bacteria, implying interesting coevolutive aspects. Biosynthesis pathways of plant growth regulators have many steps in common with secondary metabolism pathways. This implies that the genes of both pathways have a common ancestor, which in the course of evolution has produced either a great divergence in the function, conserving the genetic homology, or the function has remained the same but there has been a great genetic divergence. This is evident in the fenolic compound biosynthesis pathway and shikimic acid pathway, which are shared by both plants and microorganisms and are essential for synthesis of amino acids such as tryptophan, the precursor in auxin synthesis. The same occurs in the biosynthetic pathway of terpenes, which are gibberellin precursors. Therefore, the existence of common biosynthetic pathways and metabolic products implies the possibility of creating a connection using receptors for these metabolites.

Production and release of plant growth regulators by bacteria causes an alteration in the endogenous levels of the plant growth regulator. This is evident depending on several factors, especially

- plant growth regulator concentration;
- proximity of the bacteria to the root surface;
- the ability of the growth regulator to diffuse in soil, across the plant cell wall and inside the plant cell;
- competitiveness of the bacteria to colonize and survive in areas of high root exudation.

The effect of bacteria on plant growth regulators depends on many factors and therefore results obtained using these types of PGPR vary. The next point to consider is the physiological activity of each growth regulator.

Auxins derive from tryptophan metabolism, and their effects depend on the concentration, the organ affected and the physiological status of the plant. Auxins synthesized by the plant and the microorganisms differ only in the biosynthetic pathway (Figure 3.1), depending on the plant and/or microorganism. More than 80% of soil bacteria in the rhizosphere are capable of producing auxins; thus, the potential of these microorganisms to affect the endogenous levels of this regulator and, therefore, its effects on plant growth are remarkable.

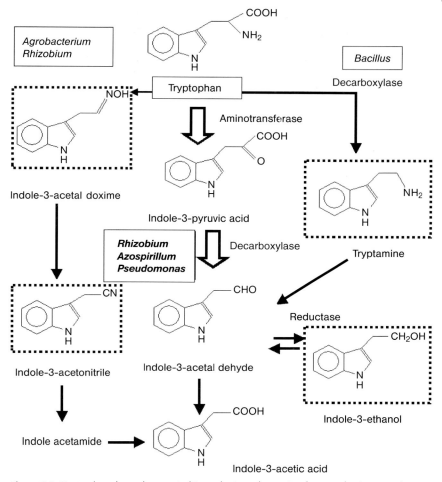

Figure 3.1 Tryptophan-dependent auxin biosynthetic pathways in plants and microorganisms.

The reason so many rhizosphere bacteria are able to produce auxins is still unknown. Some authors suggest that these bacteria have a tryptophan-related metabolism and that auxin biosynthesis is a detoxification alternative [6]. Other authors propose that auxins have some cellular function because a clear relationship has been observed between auxin and AMPc levels, which regulate many metabolic processes [34]. However, the mutualistic view of this fact could be correct, that is, auxin synthesis improves plant growth; hence, there is greater exudation and thus more nutrients for rhizobacteria. This hypothesis explains a beneficial association between rhizospheric microorganisms and the plant. The plant controls the energy flux in the system, since it is the primary producer, and contributes most of the organic matter to the rhizosphere. Auxins principally affect plant roots [50]. Those released by rhizobacteria mainly affect the root system, increasing its size and weight, branching number and the surface area in contact with soil. All these changes lead to an increase in its ability to probe the soil for nutrient exchange, therefore improving plant nutrition and growth capacity [31]. Another important result of inoculation with auxin-producing bacteria is the formation of adventitious roots, which derive from the stem. The auxins induce the stem tissues to redifferentiate as root tissue. All the above effects can vary considerably depending on the auxin concentration that reaches the root system, including an excess that could be inhibitory.

The production of hormones such as gibberellins or cytokinins is still not well documented owing to the small number of bacteria able to produce these plant growth regulators [18,50,53]. There is little information regarding microorganisms that produce gibberellins, although it is known that symbiotic bacteria existing within nodules in leguminous plants to fix nitrogen (rhizobia) are able to produce gibberellins, auxins and cytokinins in very low concentrations when the plant is forming the nodule and there is a high cell duplication rate [4]. However, the production of gibberellins by PGPR is rare, with only two strains being documented that produce gibberellins, *Bacillus pumilus* and *Bacillus licheniformis* [32]. These bacteria were isolated from the rhizosphere of *A. glutinosa* and have shown a capacity to produce large quantities of gibberellins GA_1, GA_3, GA_4 and GA_{20} *in vitro*. These types of hormones are the largest group of plant regulators, including more than 100 different molecules with various degrees of biological activity. The common structure of these diterpenic growth regulators is a skeleton of 19–20 carbon atoms, and there is a clear relationship between structure and biological effect. The reason for the pronounced effect of gibberellins is that these hormones can be translocated from the roots to the aerial parts of the plant. The effects in the aerial part are notable, and more so when the bacteria also produce auxins that stimulate the root system, enhancing the nutrient supply to the sink generated in the aerial part.

Ethylene is another growth regulator whose levels alter PGPR, consequently affecting physiological processes in the plant. Ethylene is fundamentally related to plant growth and defense systems and is also implicated in stress response. Factors such as light, temperature, salinity, pathogen attack and nutritional status cause marked variations in ethylene levels. The influence of abiotic factors on ethylene levels was deduced before those of biotic factors [1,44]. This hormone mediates in stress response and adaptive processes, thus being decisive for plant survival. It also mediates other processes not related to stress such as ripening, root growth and seed germination. Although ethylene is important as a growth regulator for normal plant development, there are examples in which ethylene does not appear to have a significant role; for example, mutant plants impaired in ethylene synthesis can survive. Application of ethylene synthesis inhibitors to several plant species makes the plant more sensitive to pathogen attack and abiotic stress [38]. The ethylene effect has been known for centuries. In China and Russia, pear and banana were stored in rooms where wood or incense was burned (producing ethylene in the combustion) to accelerate ripening [1]. In the nineteenth century, Russian researcher Neljubov deduced the ethylene effect in plants using different combustion gases [2]. It is often preferable to delay or reduce ethylene synthesis, slowing down ripening and thus extending the lifetime of the fruit [30]. As ethylene levels decrease, root systems increase their growth, with the benefits already mentioned. Using PGPR capable of reducing ethylene levels in the plant could be an interesting method to improve certain plant physiological processes. Ethylene biosynthesis starts in the methionine cycle; one aminocyclopropanecarboxylic acid (ACC) molecule results from each turn of the cycle. The enzyme responsible for ACC production is the ACC synthase, regulated by a large number of signals such as auxin, ethylene and environmental factors. This enzyme has been purified and cloned in many plants. The ACC is the substrate for the ACC oxidase, also called ethylene-forming enzyme (EFE), which produces ethylene-consuming oxygen. This enzyme has also been cloned from numerous species and belongs to a multigenic family that produces different types of ACC oxidase depending on the plant organ and developmental state.

The model proposed for ethylene regulation in the plant by a PGPR is based on the ability of some bacteria to degrade ACC, the direct precursor of ethylene [28]. The degradation of this compound creates an ACC concentration gradient between the interior and the exterior of the plant, favoring its exudation, and hence a reduction of the internal ethylene level. This, in combination with auxins that may be produced by the same microorganism, causes a considerable effect on important physiological processes such as root system development (Figure 3.2). The bacterial ACC

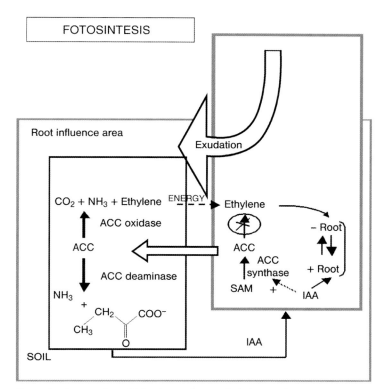

Figure 3.2 Proposed model for the regulation of ethylene in the plant by a PGPR that can degrade ACC [28].

deaminase competes with the plant's ACC oxidase. This enzyme has been isolated and identified in several bacterial and fungal genera, all having the ability to use ACC as the sole nitrogen source. Curiously, no microorganism has yet been found that is able to form ethylene from ACC. This model has been widely confirmed using mutants [26–28].

PGPR that reduce ethylene levels in plants are also able to improve processes involved in plant stress response, such as nodule formation in legumes or mycorrhiza formation. A temporary reduction of ethylene in the earlier stages of plant growth is therefore beneficial. Ethylene and auxins are two related growth regulators, and maintaining a balance between them is essential for the formation of new roots as some effects attributed to auxin-producing bacteria are actually a result of ACC degradation.

3.2.2.2 PGPR that Induce Systemic Resistance

As already mentioned, the existence of microorganisms capable of preventing diseases in plants without the plant's participation is known. This occurs by systems such as niche exclusion or pathogen-inhibiting substance production. When physical contact of the pathogen and the protecting microorganism is required, the process is known as biocontrol [12,16].

Early in the 1990s, Van Peer *et al.* and Wei *et al.* [58,63] made an important discovery regarding plant defense mechanisms and productivity. Certain nonpathogenic bacteria were able to prevent pathogen attack before the pathogen reached the plant. The difference with biocontrol is that the beneficial bacteria do not interact physically with the pathogen but trigger a response in the plant, which is effective against the subsequent attack of a pathogen. This response is systemic; that is, the bacteria interact with the plant in a restricted area but the response is extended to the whole plant. This response is mediated by metabolic changes that are sometimes not apparent. *Priming* or *biopriming* is when the plant is systemically protected by nonpathogenic bacteria against subsequent pathogen attack but the effect is not detected until pathogen challenge [17]. For the protection to be effective, an interval is necessary between the PGPR–plant contact and the pathogen attack in order for the expression of the plant genes involved in the defense. This mechanism was first known as rhizobacteria-mediated induced systemic resistance, but is now called induced systemic resistance [57]. This mechanism was discovered in the plant model *Arabidopsis thaliana*, but has now been described in many plant species, including bean, tobacco, tomato and radish. This finding is fundamental because it proposes an 'immune' response in the plant, raising the possibility of 'vaccination' for the plant.

Acquisition of resistance by the plant after a pathogen attack, causing little damage or localized necrosis and resistance to a further pathogen attack, has been known for many years. This phenomenon is called systemic acquired resistance (SAR) [49]. During a pathogen attack, reactive oxygen species (ROS) are produced in necrotic areas, causing tissue death and a blockage in pathogen expansion. The defensive responses SAR and ISR are induced by some molecules, called elicitors, present or produced in the pathogens or the PGPR, respectively. Biotic elicitors are classified into several groups: proteins, polysaccharides, lipopolysaccharides and volatile compounds [47]. The ones described most often are polysaccharides. Identification of these elicitors is essential for the practical application of defense responses for both agricultural and industrial purposes, since some of the defensive compounds are molecules with pharmacological activity.

In *A. thaliana*, the SAR and ISR responses are regulated by distinct pathways. The former response, SAR, is associated with an increase in salicylic acid levels and the translation of an ankirine-type protein called NPR1, located in the nucleus, which induce the transcription of the pathogenetic related genes. These genes codify the PR proteins and are responsible for systemic resistance in the plant [39,56]. In ISR response, salicylic acid levels are not altered but are mediated by two growth regulators, ethylene and jasmonic acid, which act as signal transductors and not as stress hormones. In ISR, the NPR1 protein is also involved, but here is induced the expression of other proteins different from PRs [17]. Responses arising from SAR and ISR lead to plant protection against different pathogen spectra, but there are spectra that overlap. The ability of a PGPR to induce systemic resistance depends on the plant–beneficial bacteria–pathogen system, a highly specific response. However, both SAR and ISR responses can coexist in the same plant at the same time [59]. Thus, the use of PGPR or PGPR mixes that are able to trigger both responses at same time would result in an important advance in the improvement of pest defense systems.

3.3
Conclusions

It may be concluded that PGPR with molecular mechanisms related to plant nutrition should be used in the appropriate soil; for example, phosphate-solubilizing bacteria will exhibit their effect in a soil with low phosphorous content and siderophore-producing bacteria will exhibit their effect in a soil deficient in available iron. If not, the bacteria will be ineffective. Furthermore, the activities expressed by the bacteria are inducible and not usually expressed in soils rich in nutrients. However, PGPR that can alter hormone balance in the plant are very efficient at improving plant fitness.

Some PGPR may interact with plant root receptors and have exhibited notable effects on the secondary metabolism of the plant. These include defensive metabolism, providing the plant with protection against pathogens and, furthermore, improving resistance to abiotic stress conditions or inducing the synthesis of molecules of pharmacological interest. Nonetheless, for PGPR to be used effectively in agriculture, it will be necessary to study each plant–PGPR–soil system individually.

3.4
Future Prospects

The future in PGPR research should be directed toward the selection of PGPR or PGPR mixes to help solve current agricultural problems such as the use of highly contaminating pesticides and fertilizers and cultivation in low-fertility soils.

In other contexts, PGPR applications may lead to the creation of functional foods, that is, foods having a beneficial effect on human health.

References

1 Abeles, F.B. (1973) *Ethylene in Plant Biology*, Academic Press, New York.

2 Abeles, F.B., Morgan, P.W. and Saltveit, M.E., Jr (1992) *Ethylene in Plant Biology*, 2nd edn, Academic Press, New York.

3 Arshad, M. and Frankenberger, W.T. (1998) *Advances in Agronomy*, **66**, 45–151.

4 Atzorn, R., Crozier, A., Wheeler, C.T. and Sandberg, G. (1988) *Planta*, **175**, 532–538.

5 Baldini, Y.J. (1997) *Soil Biology & Biochemistry*, **29** (5), 911–922.

6 Bar, T. and Okon, Y. (1992) *Symbiosis*, **13**, 191–198.

7 Bashan, Y. and Levanony, H. (1990) *Canadian Journal of Microbiology*, **36**, 591–608.

8 Bashan, J. and Holguin, G. (1998) *Soil Biology & Biochemistry*, **30** (8/9), 1225–1228.

9 Bedmar, E.J., González, J., Lluch, C. and Rodelas, B. (2006) *Fijación de nitrógeno: Fundamentos y Aplicaciones*, SEFIN, Granada.

10 Bent, E., Tzun, S., Chanway, C.P. and Eneback, S. (2001) *Canadian Journal of Microbiology*, **47**, 793–800.

11 Bhattacharya, P. and Chaudhuri, S.R. (1993) *Yohana*, **37** (9), 12–31.

12 Bloemberg, G.V. and Lugtenberg, B.J.J. (2001) *Current Opinion in Plant Biology*, **4**, 343–350.

13 Brown, M.E. (1974) *Annual Review of Phytopathology*, **12**, 181–197.

14 Chanway, C.P. (1997) *Forest Science*, **43** (1), 99–112.

15 Christiansen-Weniger, C. (1998) *Critical Reviews in Plant Sciences*, **17**, 55–76.

16 Compant, S., Duffy, B., Nowak, J., Clement, C. and Ait Barka, E. (2005) *Applied and Environmental Microbiology*, **71** (9), 4951–4959.

17 Conrath, U., Pieterse, C.M.J. and Mauch-Mani, B. (2002) *Trends in Plant Science*, **7**, 210–216.

18 de Salomone, I.E.G., Hynes, R.K. and Nelson, L.M. (2001) *Canadian Journal of Microbiology*, **47**, 404–411.

19 de Troch, P. (1993) Bacterial surface polysaccharides in relation: a genetic and chemical study of *Azospirillum brasilense*. Disertationes de la Agricultura, p. 238.

20 Duponnois, R. and Plenchette, C. (2003) *Mycorrhiza*, **13** (2), 85–91.

21 FAO((1995) Manual técnico de la fijación simbiótica del nitrógeno.

22 Founoune, H., Duponnois, R., Meyer, J.M., Thioulouse, J., Masse, D., Chotte, J.L. and Neyra, M. (2002) *FEMS Microbiology Ecology*, **1370**, 1–10.

23 Garbaye, J. (1994) *New Phytologist*, **128**, 197–210.

24 Germida, J.J. and Walley, F.L. (1996) *Biology and Fertility of Soils*, **23**, 113–120.

25 Glick, B.R., Jacobson, C.B., Schwarze, M.M.K. and Pasternak, J.J. (1994) 1-Aminocyclopropae-1-carboxylic acid deaminase play a role on plant growth by *Pseudomonas putida* GR12-2, in *Improving Plant Productivity with Rhizosphere Bacteria* (eds M.H. Ryder, P.M. Stephens and G.D. Bowen), CSIRO, Adelaide, pp. 150–152.

26 Glick, B.R., Jacobson, C.B., Schwarze, M.M.K. and Pasternak, J.J. (1994) *Canadian Journal of Microbiology*, **40**, 911–915.

27 Glick, B.R. (1995) *Canadian Journal of Microbiology*, **41** (2), 109–117.

28 Glick, B.R., Penrose, D.M. and Li, J. (1998) *Journal of Theoretical Biology*, **190**, 63–68.

29 González, J. and Lluch, C. (1992) *Biología del Nitrógeno. Interacción Planta-Microorganismo*, Rueda, Madrid, Spain.

30 Good, X., Kellog, J.A., Wagoner, W., Langhoff, D., Matsumara, W. and Bestwwick, R.K. (1994) *Plant Molecular Biology*, **26**, 781–790.

31 Gutiérrez Mañero, F.J., Acero, N., Lucas, J.A. and Probanza, A. (1996) *Plant and Soil*, **182**, 67–74.

32 Gutiérrez Mañero, F.J., Ramos, B., Probanza, A., Mehouachi, J., Tadeo, F.R. and Talón, M. (2001) *Physiologia Plantarum*, **111**, 206–211.

33 Gyaneshwar, P., Kumar, G.N., Parekh, L.J. and Poole, P.S. (2002) *Plant and Soil*, **245** (1), 83–93.

34 Katsy, E. (1997) *Russian Journal of Genetics*, **33**, 301–306.

35 Kloepper, J.W., Schroth, M.N. and Miller, T.D. (1980) *Phytopathology*, **70**, 1078–1082.

36 Kloepper, J.W., Lifshitz, R. and Schroth, M.N. (1988) *ISI Atlas of Science – Animal & Plant Sciences*, **1**, 60–64.

37 Kloepper, J.W., Lifshitz, R. and Zablotowicz, R.M. (1989) *Trends in Biotechnology*, **7**, 39–43.

38 Knoester, M., van Loon, L.C., van den Heuvel, J., Hennig, J., Bol, J.F. and Linthorst, H.J.M. (1998) *Proceedings of the National Academy of Sciences of the United States of America*, **95**, 1933–1937.

39 Lawton, K.A., Friedrich, L., Hunt, M., Weymann, K., Delaney, T., Kessmann, H., Staub, T. and Ryals, J. (1996) *Plant Journal*, **10**, 71–82.

40 Lim, H.S., Kim, Y.S. and Kim, S.D. (1991) *Applied and Environmental Microbiology*, **57**, 510–516.

41 Liu, L., Kloepper, J.W. and Tuzun, S. (1995) *Phytopathology*, **85**, 1064–1068.

42 Marek-Kozackuk, M. and Skorupska, A. (2001) *Biology and Fertility of Soils*, **33**, 146–151.

43 Mayea, S., Carone, M., Novo, R., Boado, I., Silveira, E., Soria, M., Morales, Y. and Valiño, A. (1998) *Microbiología Agropecuaria. Tomo II*, Félix Varela, La Habana, pp. 156–178.

44 Morgan, P.W. and Drew, C.D. (1997) *Physiol Plantarum*, **100**, 620–630.

45 Okon, Y. and Labandera-Gonzalez, C.A. (1994) Agronomic applications of *Azospirillum* in *Improving Plant Productivity with Rhizosphere Bacteria* (eds M.H. Ryder, P.M. Stephens and G.D. Bowen), Commonwealth Scientific and Industrial Research Organisation, Adelaide, Australia, pp. 274–278.

46 Probanza, A., Acero, N., Ramos, B. and Gutiérrez Mañero, F.J. (1996) *Plant and Soil*, **182** (1), 59–66.

47 Radman, R., Saez, T., Bucke, C. and Keshavarz, T. (2003) *Biotechnology and Applied Biochemistry*, **37**, 91–102.

48 Rodelas, M.B., González, J., Martínez, M.V., Pozo, C. and Salmeron, V. (1999) *Biology and Fertility of Soils*, **29** (2), 165–169.

49 Ryals, J.A., Neuenschwander, U.H., Willits, M.G., Molina, A., Steiner, H. and Hunt, M.D. (1996) *Plant Cell*, **8**, 1809–1819.

50 Salisbury, F.B. (1994) The role of plant hormones, in *Plant–Environment Interactions* (ed. R.E. Wilkinson), Marcel Dekker, New York, USA, pp. 39–81.

51 Sumner, M.E. (1990) Crop responses to *Azospirillum inoculation. Advances in Soil Sciences*, **12**, 53–168.

52 Tang, W.H. (1994) Yield-increasing bacteria (YIB) and biocontrol of sheath blight of rice, in *Improving Plant Productivity with Rhizosphere Bacteria* (eds M.J. Ryder, B.H.P.J. Thomma,K. Erggemont, F.M.J. Tierens and W.F. Broekaert), *Plant Physiology*, **121**, 1093–1101.

53 Timmusk, S., Nicander, B., Granhall, U. and Tillberg, E. (1999) *Soil Biology & Biochemistry*, **31**, 1847–1852.

54 Toro, M., Azcón, R. and Barca, J.M. (1998) *New Phytologist*, **138**, 265–273.

55 Toyoda, H. and Utsumi, R. (1991) Method for the prevention of *Fusarium* diseases and microorganisms used for the same. US Patent 4,988,586.

56 Uknes, S., Winter, A.M., Delaney, T.P., Vy, B., Morse, A., Friedrich, L., Nye, G., Potter, S., Ward, E. and Ryals, J. (1993) Biological induction of systemic acquired resistance in *Arabidopsis. Molecular Plant–Microbe Interactions*, **6**, 692–698.

57 Van Loon, L.C., Bakker, P.A.H.M. and Pieterse, C.M.J. (1998) Systemic resistance induced by rhizosphere bacteria. *Annual Review of Phytopathology*, **36**, 453–483.

58 Van Peer, R., Niemann, G.J. and Schippers, B. (1991) *Phytopathology*, **91**, 728–734.

59 van Wees, S.C.M., de Swart, E.A.M., van
Pelt, J.A., van Loon, L.C. and Pieterse,
C.M.J. (2000) *Proceedings of the National
Academy of Sciences of the United States of
America*, **97**, 8711–8716.

60 Velazco, A. and Castro, R. (1999) Estudio
de la inoculación de *Azospirillum brasilense*
en el cultivo del arroz (*Oryza sativa*) var.
A82 en condiciones de macetas.*Cultivos
Tropicales*, **20** (1), 5–9.

61 Vessey, J.K. (2003) *Plant and Soil*, **255**,
571–586.

62 Voisard, C., Keel, C., Haas, D. and
Defago, G. (1989) *EMBO Journal*, **8**,
351–358.

63 Wei, G., Kloepper, J.W. and Tuzun, S.
(1991) *Phytopathology*, **81**, 1508–1512.

64 Zehnder, G., Kloepper, J., Yao, C. and Wei,
G. (1997) *Journal of Economic Entomology*,
90 (2), 391–396.

4

A Review on the Taxonomy and Possible Screening Traits of Plant Growth Promoting Rhizobacteria

Marina Rodríguez-Díaz, Belén Rodelas-Gonzalés, Clementina Pozo-Clemente, Maria Victoria Martínez-Toledo, and Jesús González-López

4.1
Introduction

The term plant growth promoting rhizobacteria (PGPR) was coined by Kloepper and Schroth [1] to encompass those bacteria that are able to colonize plant root systems and promote plant growth. At the time, the ones known were mainly pseudomonads that acted as control agents of soilborne plant pathogens. Yet another group of plant growth promoters exists that includes species that directly affect plant metabolism and their consequent growth. The term PGPR includes neither those bacteria that act as biocontrol agents in the phyllosphere [2] nor the intracellular nematode pathogen *Pasteuria penetrans* [3], which sporulates inside the nematode, preventing it from reproducing and hence protecting the plant against damage. The lack of nomenclature for this latter group and the characterization of new plant growth promoting bacteria (PGPB) that did not belong to any previously defined group led to confusion in classification and terminology. Bashan and Holguin [2] compiled all terms published up to 1998 that referred to bacteria exhibiting positive effects on plants, and coined the term PGPB, making a distinction between biocontrol-PGPB and PGPB. The biocontrol-PGPB group encompassed bacteria that suppress plant pathogens by either producing plant pathogen inhibitory substances or by increasing the natural resistance of the plant, and the PGPB group encompassed those bacteria that affect plants by means other than suppression of other microorganisms.

Nonetheless, the formerly proposed nomenclature was not widely accepted. For instance, other classifications maintain the terminology used by Kloepper and Schroth [1] or just subclassify PGPR by their mechanisms of action as (i) direct PGPR (bacteria whose metabolites are used as growth regulators or their precursors) and (ii) indirect PGPR (bacteria whose metabolites are involved in biological control, antagonistic determinants or those that hinder/inhibit microorganisms causing harm to plants by means of antibiotics, siderophores, lytic enzymes or induction of plant-systemic resistance [4]. Gray and Smith [5] subdivided the PGPR

Plant-Bacteria Interactions. Strategies and Techniques to Promote Plant Growth
Edited by Iqbal Ahmad, John Pichtel, and Shamsul Hayat
Copyright © 2008 WILEY-VCH Verlag GmbH & Co. KGaA, Weinheim
ISBN: 978-3-527-31901-5

in two categories: extracellular PGPR (e-PGPR) and intracellular PGPR (i-PGPR). The first category, e-PGPR, includes bacteria existing in the rhizosphere, on the rhizoplane or in the spaces between cells of the root cortex, stimulating plant growth by producing phytohormones, improving plant disease resistance or improving mobilization of soil nutrients. Picard and Bosco [6] state that noninvasive rhizobacteria (e-PGPR) represent the most common plant–microbe interactions in healthy plants.

i-PGPR are those bacteria that exist inside root cells, generally in specialized nodular structures where they fix nitrogen. The latter, i-PGPR, are therefore those bacteria that are also known as rhizobia, a name derived from the genus *Rhizobium* [7] and refer to bacteria that induce nodules in legumes and fix atmospheric nitrogen in symbiosis with them. However, bacteria showing this ability also belong to genera other than *Rhizobium*, thus the term legume-nodulating bacteria (LNB) has been proposed [7]. Notwithstanding, LNB are not the only bacteria able to fix atmospheric nitrogen in symbiosis as can be seen in Table 4.1.

4.2
Taxonomy of PGPR

The term taxonomy is defined as the science dedicated to the study of relationships among organisms and has to do with their classification, nomenclature and identification. The accurate comparison of organisms at different times by different scientists depends on a reliable taxonomic system that allows a precise classification of the organisms under study.

Since its beginnings in the late nineteenth century, bacterial taxonomy relied on phenotypic traits such as cell and colonial morphologies and biochemical, physiological and immunological tests. Taxonomy was revolutionized thanks to the discovery of the polymerase chain reaction (PCR) technique in 1983. Since then, the search for a trait that would be in congruence with the evolutionary divergence of organisms, that is their phylogeny, was mandatory. This trait was found in the ribosomal RNA [25], a molecule used by all living cells. The gene sequences of the 16S subunit of the ribosomal RNA have been used since to compare evolutionary similitudes among strains. At present, and by correlation with experimental data obtained in the comparison of total genomic DNA (DNA–DNA hybridization), it is proposed that a similarity below 98.7–99% on an UPGMA analysis of the 16S rDNA sequences of two bacterial strains is sufficient to consider them as belonging to different species [26]. Notwithstanding, it is possible that two strains showing 16S rDNA sequence similarities above the 98.7% threshold may represent two different species [27]. In these cases, total genome DNA–DNA hybridizations must be performed and those strains for which similarities are below 70% are considered to belong to different species. In this context, and given the fact that no taxonomic technique is absolutely accurate, the use of a polyphasic approach to taxonomy [28] was implemented in bacterial taxonomy.

Table 4.1 Genera that are named plant growth promoters in the literature, their classification according to their mechanisms of action by Bashan and Holguin [2] and Gray and Smith [5] and their plant growth promoter capabilities.

	Definition by																
	Bashan and Holguin [2]														Gray and Smith [5]		
	PGPB									Biocontrol-PGPB							References
										PPIS				INRP			
Genus	PS	NFF	AD	AP	VO	QUO	STR	PP	NFS	AB	FU	SID	NE	SRI	i-PGPR	e-PGPR	
Nostoc									+						+		[8]
Anabaena		+													+		[9]
Frankia									+						+		[8]
Curtobacterium	+															+	[5]
Arthrobacter				+								+				+	[8,10]
Micrococcus	+															+	[8,11]
Streptomyces			+a										+			+	[12,13]
Flavobacterium	+															+	[8,11]
Bacillus	+			+	+	+					+		+	+		+	[5,8,10,11,13–16]
Paenibacillus	+	+									+					+	[8]
Staphylococcus	+		+													+	[5]
Clostridium													+			+	[13]
Caulobacter																+	[8]
Blastobacter									+						+		[3,7,17]
Bradyrhizobium									+						+		[8]
Ochrobactrum									+						+		[3,17]
Devosia									+						+		[3,18]
Hyphomicrobium																+	[8]

(continued)

Table 4.1 (Continued)

Genus	PGPB — PS	NFF	AD	AP	VO	QUO	STR	PP	NFS	Biocontrol-PGPB / PPIS — AB	FU	SID	NE	INRP / SRI	Gray and Smith [5] i-PGPR	e-PGPR	References
Methylobacterium									+						+		[3]
Mesorhizobium									+						+		[8]
Phyllobacterium																+	[8]
Agrobacterium										+						+	[8,11,13,14]
Rhizobium	+								+b				+		+		[8,11,19]
Azorhizobium									+						+		[8]
Sinorhizobium/Ensifer									+						+		[8]
Acetobacter		+	+			+										+	[8,20]
Gluconacetobacter		+	+	+												+	[21]
Swaminathania		+	+	+												+	[22]
Azospirillum	+	+	+					+								+	[11,20]
Achromobacter	+						++									+	[11]
Alcaligenes									+				+			+	[3,8,13]
Burkholderia	+	+	+						+							+	[5,8,11,16]
Herbaspirillum	+	+												+		+	[8]
Ralstonia	+	+						+								+	[21]
Chromobacterium																+	[8]
Azoarcus	+															+	[20]

										Ref.
Desulfovibrio							+			[13]
Aeromonas					+				+	[8]
Citrobacter	+								+	[11]
Enterobacter	+								+	[8,11,23]
Erwinia	+								+	[8,11,24]
Klebsiella	+	+[c]	+[c]						+	[11,23]
Kluyvera									+	[23]
Pantoea				+					+	[14]
Serratia	+				+				+	[8,11,13]
Pseudomonas	+	+	+	+	+	+	+		+	[8,11,13,14,16]
Azotobacter		+							+	[20]
Acinetobacter	+								+	[8,11]

AB: antibacterial; AD: aminocyclopropanecarboxylic acid (ACC) degradation; AP: auxin production; biocontrol-PGPB: suppression of plant pathogens; e-PGPR: extracellular plant growth promoting rhizobacteria; FU: fungicide; i-PGPR: intracellular plant growth promoting rhizobacteria; INRP: increases the natural resistance of the plant; NE: nematicidal; NFF: nitrogen fixation as free cells; NFS: nitrogen fixation in symbiosis; PGPB: plant growth promoting bacteria; PP: production of phytohormones; PPIS: production of plant-pathogen inhibitory substances; PS: phosphate solubilizing; QUO: degradation of pathogen quorum-sensing molecules; SID: siderophore production; SRI: systemic resistance inducer; STR: resistance to stress conferral; VO: production of volatiles.

[a] Production of a nitrogenase that is insensitive to oxygen.
[b] Symbiotic extrachromosomal plasmid.
[c] Harbors the ACC deaminase gene.

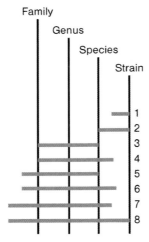

Figure 4.1 Taxonomic resolution of some of the currently used techniques in taxonomy (from Vandamme *et al.* [29]). Phage and bacteriocin typing, serological (monoclonal, polyclonal) techniques and genomic studies such as restriction fragment length polymorphism (RFLP) and low-frequency restriction fragment analysis (PFGE); **2**: Zymograms (multilocus enzyme polymorphism), total cellular protein electrophoretic patterns and genomic techni- ques such as ribotyping, DNA amplification (AFLP, AP-PCR, rep-PCR, RAPD, etc.); **3**: Genomic techniques such as DNA–DNA hybridization, mol% G + C, DNA amplification (ARDRA) and tDNA-PCR; **4**: Chemotaxonomic markers (polyamines, quinones, etc.); **5**: Cellular fatty acid fingerprinting (FAME); **6**: Cell wall structure; **7**: Phenotype (classical, API, Biolog, etc.); **7**: rRNA sequencing; **8**, DNA probes and DNA sequencing.

An effective polyphasic taxonomy will encompass the study of different molecule types within the bacterial cell, for instance membrane fatty acids, proteins, ribosomes/ribosomal gene sequences and other biochemical and phenotypic characteristics that will produce a robust characterization of the strains and achieve a sound classification system in order to obtain the most reliable strain identifications. A good review on the modern polyphasic bacterial taxonomy is offered by Zakhia and de Lajudie [7] (Figure 4.1).

The phenotypic techniques used nowadays embrace morphological, physiological and biochemical characteristics of the strains studied using classical methodologies or commercialized systems, plus analyses of cellular components such as cellular fatty acids by the 'fatty acid methyl ester' (FAME) technique, evaluation of total cellular proteins by the 'sodium dodecyl sulfate-polyacrylamide gel electrophoresis' (SDS-PAGE) technique and 'multilocus enzyme electrophoresis' (MLEE).

This work intends to give the reader an insight into the present taxonomical status of the main groups of PGPR, as reviewed in Table 4.2. In later sections, symbiotic and asymbiotic plant growth promoting bacteria are described according to their taxonomic positions.

Table 4.2 Taxonomic affiliation of validated genera [30] containing PGPR strains as described in the literature.

Genus	Phylum	Class	Order	Suborder	Family
Nostoc	*Cyanobacteria*		*Nostocales*		*Nostocaceae*
Anabaena					
Frankia	*Actinobacteria*	*Actinobacteria*	*Actinomycetales*	*Frankineae*	*Frankiaceae*
Curtobacterium				*Micrococcineae*	*Microbacteriaceae*
Arthrobacter					*Micrococcaceae*
Micrococcus					
Streptomyces				*Streptomycineae*	*Streptomycetaceae*
Flavobacterium	*Bacteroidetes*	*Flavobacteria*	*Flavobacteriales*		*Flavobacteriaceae*
Bacillus	*Firmicutes*	*Bacilli*	*Bacillales*		*Bacillaceae*
Paenibacillus					*Paenibacillaceae*
Staphylococcus					*Staphylococcaceae*
Clostridium		*Clostridia*	*Clostridiales*		*Clostridiaceae*
Caulobacter	*Proteobacteria*	*Alphaproteobacteria*	*Caulobacterales*		*Caulobacteraceae*
Blastobacter			*Rhizobiales*		*Bradyrhizobiaceae*
Bradyrhizobium					*Bradyrhizobiaceae*
Ochrobactrum					*Brucellaceae*
Devosia					*Hyphomicrobiaceae*
Hyphomicrobium					*Hyphomicrobiaceae*
Methylobacterium					*Methylobacteraceae*
Mesorhizobium					*Phyllobacteraceae*
Phyllobacterium					*Phyllobacteraceae*
Agrobacterium					*Rhizobiaceae*
Rhizobium					*Rhizobiaceae*
Azorhizobium					*Rhizobiaceae*
Sinorhizobium/Ensifer					*Rhizobiaceae*
Acetobacter			*Rhodospirillales*		*Acetobacteraceae*

(continued)

Table 4.2 (*Continued*)

Genus	Class	Order	Family
Gluconacetobacter			*Acetobacteraceae*
Swaminathania			*Acetobacteraceae*
Azospirillum			*Rhodospirillaceae*
Achromobacter	*Betaproteobacteria*	*Burkholderiales*	*Alcaligenaceae*
Alcaligenes			*Alcaligenaceae*
Burkholderia			*Burkholderiaceae*
Herbaspirillum			*Oxalobacteraceae*
Ralstonia			*Ralstoniaceae*
Chromobacterium		*Neisseriales*	*Neisseriaceae*
Azoarcus		*"Rhodocyclales"*	*"Rhodocyclaceae"*
Desulfovibrio	*Deltaproteobacteria*	*Desulfovibrionales*	*Desulfovibrionaceae*
Aeromonas	*Gammaproteobacteria*	*Aeromonadales*	*Aeromonadaceae*
Citrobacter		*Enterobacteriales*	*Enterobacteriaceae*
Enterobacter			
Erwinia			
Klebsiella			
Kluyvera			
Pantoea			
Serratia			
Pseudomonas		*Pseudomonadales*	*Pseudomonadaceae*
Azotobacter			
Acinetobacter			*Moraxellaceae*

4.3
Symbiotic Plant Growth Promoting Bacteria

Microorganisms included in this section could be subdivided into PGPR capable of legume nodulation and PGPR capable of nodulating plants other than legumes.

4.3.1
LNB

Microorganisms capable of legume nodulation are mainly found in the class Alphaproteobacteria, with all the families included in the order Rhizobiales. However, other families belonging to the order Burkholderiales in the class *Betaproteobacteria* have recently been reported to contain species with the ability to fix nitrogen in symbiosis with legumes.

4.3.1.1 Alphaproteobacteria

This class contains those PGPR known as rhizobia. Rhizobia were initially classified according to phenotypic traits, starting with their ability to fix dinitrogen and nodulate plants. At the early stage of bacterial taxonomy, all legume symbionts were classified into the single genus *Rhizobium*, based on their ability to fix dinitrogen by forming symbiotic associations with legumes [7]. In addition, the intrageneric classification of the genus *Rhizobium* at the species level was based primarily on the types of host plants (host specificity) infected by legume symbionts [9]. However, it has been established that classification cannot be based on the specificity to the symbiotic host plant, as characteristics such as nodulation, host specificity, and in some cases pathogenicity, are due to particular strains carrying plasmids. Such plasmids may be lost or acquired and with them those specific characteristics of the bacterium [16]. This is clearly not a stable platform for any taxonomic nomenclature. Even more, the plasmids may represent almost 50% of the total DNA in rhizobia [7], for which DNA–DNA hybridization (being the crucial technique for species delineation) is not totally reliable when applied to these LNB. For that reason, the strategy of studying several loci, such as *atpD* and *recA* [31,32] to estimate phylogenetic relationships among the genomes of Alphaproteobacteria with emphasis on the rhizobacterial genera has been suggested [33]. The application of the polyphasic approach, however, has enabled a reassessment of the taxonomic relationships between the genera comprised in *Rhizobiaceae* [9,34]. Indeed, the genus *Rhizobium* contained strains that were later reclassified as the new genera *Bradyrhizobium*, *Sinorhizobium* and *Mesorhizobium* [9].

The *Rhizobium–Agrobacterium* Group Historically, *Rhizobium* and *Agrobacterium* were regarded as distinct genera on the basis of their respective symbiotic and pathogenic relationships with host plants, but molecular data have undermined this concept and allowed the rationalization of these two taxa into a single group [16]. Indeed, early analyses on 16S rDNA gene sequences showed that bacteria within the family Rhizobiaceae did not form clear cut taxonomic units but the two belonged to what became to be referred to as the *Agrobacterium–Rhizobium* complex [35]. A more recent tree by Sawada *et al.* [9], representing the phylogenetic relationships among legume symbionts and their relatives on the basis of the 16S rDNA sequence divergence, delineates several groups of genera that will probably have their taxonomy revised (Figure 4.2).

The groups delineated by Sawada *et al.* [9] are the *Rhizobium–Agrobacterium* group (containing also the genera *Allorhizobium* and *Blastoblaster*), the *Sinorhizobium–Ensifer* group, the *Mesorhizobium* group (containing also the non-LNB genera *Aminobacter* and '*Pseudoaminobacter*') and the *Bradyrhizobium* group (containing also strains of the

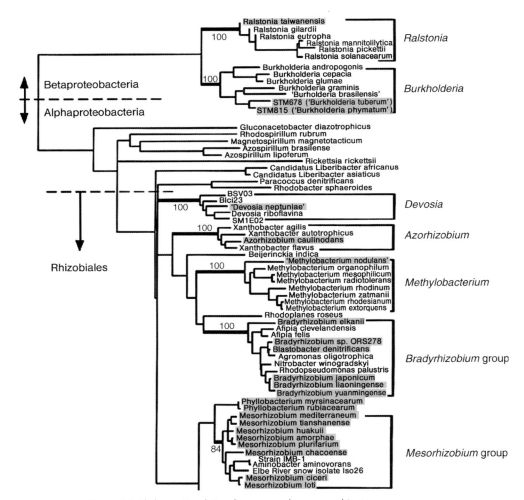

Figure 4.2 Phylogenetic relationships among legume symbionts and their relatives inferred on the basis of the 16S rDNA sequence divergence. From Sawada *et al.* [9].

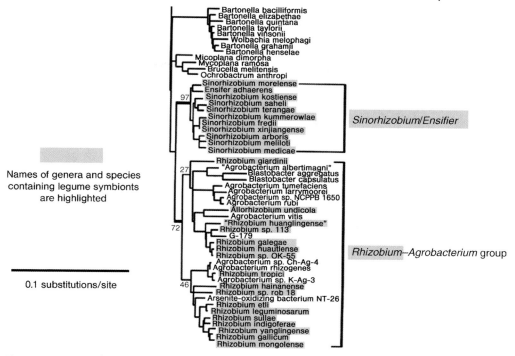

Names of genera and species
containing legume symbionts
are highlighted

0.1 substitutions/site

Figure 4.2 *(Continued)*

genera *Agromonas, Nitrobacter, Rhodopseudomonas* and *Afipia* and the species *Blastobla-ster denitrificans*).

A phylogenetic study based on combined *atpD* and *recA* sequences in addition to the *rrs* gene was obtained for strains in the *Rhizobium–Agrobacterium* group by Gaunt *et al.* [31], and their analyses corroborate that the group proposed by Sawada *et al.* [9] is a stable taxon. These preliminary works are also supported by data obtained with similar analyses by Vinuesa *et al.* [32]. In opposition, other phylogenetic studies of this group support its being divided into several groups [36]. Notwithstanding, only a thorough polyphasic taxonomy will validate this group as a single genus – *Rhizobium* [9].

Sinorhizobium/Ensifer *Sinorhizobium* was first identified as fast-growing soybean isolates originating from China. The *Shinorhizobium* and *Rhizobium* groups share phenotypic characters, but the question as to whether or not a generic boundary should be established between these two groups has been a controversial issue since the genus *Sinorhizobium* was first proposed. It was observed that *Sinorhizobium* and *Rhizobium* did not differ sufficiently to warrant their separation into two genera when analysis of their partial 23S rRNA sequences were performed [37]. It has been clarified since that these two groups are separate monophyletic groups, based on phylogenetic analyses using various markers and that they also differ in details of fatty acid composition; hence, they are considered to represent separate genera. Nonetheless, the work of Sawada *et al.* [9] showed the genus *Sinorhizobium* clustering with the genus *Ensifer* (a non-LNB), and further data on sequence comparison of 16S–23S rDNA internal transcribed spacer regions has led to the proposal to join the genera *Ensifer* and *Sinorhizobium* into one genus. Still, a consensus over whether the genus should be named *Sinorhizobium* [38] or *Ensifer* [39] is yet to be achieved.

The *Mesorhizobium* Group   *Mesorhizobium* species cluster on the basis of 16S rDNA sequence alignment with high bootstrap values [40], and on the basis of such an analysis are distinct from other genera of the family *Rhizobiaceae*. This genus is also distinguishable in terms of DNA homology and phenotypic differentiation from other genera by a distinct fatty acid profile and a slower growth rate [41]. At present, these characteristics are sufficient to justify *Mesorhizobium* as a distinct genus [34], but it is no longer a member of the family *Rhizobiaceae*. It has been placed into the family *Phyllobacteriaceae* along with the leaf-nodulating *Phyllobacterium*, and recent works highlight its status as a group of three genera [9]. The *Mesorhizobium* group contains also the non-LNB genera *Aminobacter* and '*Pseudoaminobacter*'.

The *Bradyrhizobium* Group   The genus *Bradyrhizobium*, another of the LNB genera outside the Rhizobiaceae, belongs to the family *Bradyrhizobiaceae* of the *Rhizobiales* (http://www.cme.msu.edu/Bergeys). Strains located in this genus can be distinguished from other legume symbionts on the basis of slow growth and the production of an alkaline reaction by no serum zone in litmus milk. The difficulties in performing DNA–DNA hybridizations in LNB are noted along with the difficulty of obtaining DNA from *Bradyrhizobium* in quantity and quality to perform extensive DNA–DNA hybridizations [42]. Doignon-Bourcier *et al.* [42] chose to characterize *Bradyrhizobium* strains by intergenic spacer (IGS) PCR–RFLP, amplified fragment length polymorphism (AFLP) and 16S ARDRA. For especially diverse strains, the study of their tRNA^{ala} gene sequence has also been performed [38].

The review of Sawada *et al.* [9] describes the *Bradyrhizobium* group as also containing strains of the genera *Agromonas*, *Nitrobacter*, *Rhodopseudomonas* and *Afipia* and the species *Blastoblaster denitrificans*. A proposal to merge all the involved species of the different genera would be justified; however, phenotypic differences and the convenience of maintaining the current genera in their present form led to this proposal being rejected [43]. At present, only the transfer of the species *Blastobacter denitrificans* to the genus *Bradyrhizobium* as *Bradyrhizobium denitrificans* has been proposed [44].

Azorhizobium   *Azorhizobium* nodulates the stem of *Sesbania rostrata* and molecular data indicate it is distinct from other members of *Rhizobiaceae*. It does share many molecular systematic characteristics with *Xanthobacter* and *Aquabacter*. Indeed, it has been suggested that all three genera could be combined into a single genus – *Xanthobacter* [45]. The genus is now found in the family *Hyphomicrobiaceae* of the Rhizobiales.

Methylobacterium This genus is formed by facultatively methylotrophic bacteria and is located in the family *Methylobacteraceae*. Only one species, *Methylobacterium nodulans* [46,47], is able to fix dinitrogen in symbiosis with legumes (*Crotalaria* spp.). No *Methylobacterium* species, other than *M. nodulans*, have been confirmed to have the ability of symbiotic dinitrogen fixation and the structures of NodA of *M. nodulans* and *Bradyrhizobium* species are similar. These two facts have led to the inference that *M. nodulans* gained this ability by obtaining symbiotic genes that were horizontally transferred from *Bradyrhizobium* species [9].

Devosia The genus *Devosia* is included in the family *Hyphomicrobiaceae* of Rhizobiales. This genus contains a species *D. neptuniae* [48,49] that is capable of nitrogen fixation in symbiosis with *Neptunia natans*.

4.3.1.2 Betaproteobacteria

Burkholderia Members of the genus *Burkholderia* of the family Burkholderiaceae form a discrete and compact monophyletic group with a high bootstrap value, which is composed only of *Burkholderia* species and its chemotaxonomic and phenotypic characters are specific of this genus [9]. The genus *Burkholderia* contains diverse species with different physiological and ecological properties, which were isolated from soils, plants, animals and humans. However, only two strains (STM678 and STM815) have been confirmed to have the ability of symbiotic nitrogen fixation [50]. Based on the fact that high *nodAB* similarity is observed between these strains and legume symbionts of the class *Alphaproteobacteria*, it seems that the symbiotic genes have been horizontally transferred among strains, crossing the boundary between these classes [50].

Four LNB species of *Burkholderia* have been published, namely *Burkholderia tuberum* sp. nov. [51,52], *B. phymatum* sp. nov. [51,52], *B. mimosarium* sp. nov. [53] and *B. nodosa* sp. nov. [54].

Ralstonia This genus is also allocated in the family *Burkholderiaceae* and, as the genus *Burkholderia*, it is ubiquitous. The definitions and circumscriptions of the genus *Ralstonia* are widely accepted because the monophyletic group formed by its species is homogeneous and presents a high bootstrap value [9]. Furthermore, the chemotaxonomic and phenotypic characters of its species are specific to the group [55].

Only one species of the genus, *Ralstonia taiwanensis* [56] is capable of nitrogen fixation in symbiosis, but no insight on the acquisition of this trait has yet been obtained.

4.3.2
Bacteria Capable of Fixing Dinitrogen in Symbiosis with Plants Other Than Legumes

Special attention has been given to LNB for their agricultural importance. However, other organisms and symbioses are increasingly seen as major contributors to overall nitrogen fixation and into sustaining diverse agricultural, forest and ecosystem settings. These include actinorhizal symbioses (e.g. between *Casuarina* and *Frankia*) and associative relationships including sugarcane (*Saccharum officinarum*) and coffee plants with *Gluconacetobacter* spp. [57,58].

4.3.2.1 Actinobacteria

Frankia Symbiotic nitrogen fixation is also a characteristic of the genus *Frankia*: these bacteria are associated with actinorhizal plants that pioneer the colonization of poor soil or disturbed ground, although those plants are not as commercially important as grain and pasture legumes with rhizobia in their root nodules. *Frankia* is a rather poorly defined genus within the Actinobacteria lineage of descent. However, host specificity was useful in species designation in rhizobia, this phenomenon has not been as effective in *Frankia*, and even with the abundance of molecular tools at the disposal of the scientific community, differentiation of species within the genus is still a difficult task [59].

Other Actinobacteria are also plant growth promoters but not involved in symbiosis. They belong to the genera *Arthrobacter* and *Micrococcus* [5], *Curtobacterium* [10] and *Streptomyces* [3].

4.3.2.2 Cyanobacteria

Many plants have developed symbiotic associations with N_2-fixing cyanobacteria, particularly of the genera *Nostoc* and *Anabaena*, achieving nitrogen autotrophy. The plant partners involved are phylogenetically diverse, ranging from unicellular algae to angiosperms. There is also a great variety in these symbioses in terms of location of the cyanobiont (extracellular or intracellular) and the host organs and tissues involved (cells, bladders, cavities in gametophyte thallus or sporophyte fronds, root nodules or stem glands) [60].

Cyanobacteria symbiosis with plants represent a range and variety far larger than those encountered in rhizobial (LNB) or actinorhizal (Actinobacteria) symbioses, which are restricted to the roots or stems of a few angiosperm families. Cyanobacteria–plant symbioses occur throughout the world and might even be the dominant bacteria–plant association in some regions. The combination of carbon and nitrogen autotrophy enables these systems to colonize a wide range of nutrient-poor habitats. In terms of nitrogen fixation they might not be considered to be of global importance, but symbioses such as diatoms, lichens and cycads are of considerable significance in the nitrogen economy of those areas in which they form the dominant vegetation [60]. Furthermore, the *Azolla* symbiosis has been shown to be of major agronomic importance, particularly in rice cultivation [61]. For an extensive review on N_2-fixation by cyanobacteria symbionts and their use as artificial providers of fixed nitrogen to cereals, see Rai *et al.* [60].

Cyanobacterial systematics have traditionally been based on morphological characteristics, which have led to confusion of strains across genera, and some authors complain about the continued use of traditional nomenclature and the botanical code [62]. Examples of the inaccurate taxonomy applied so far to cyanobacteria are as follows: strain PCC 7120, reported to belong to *Anabaena*, happened to be a member of the genus *Nostoc* on the basis of DNA–DNA hybridization [62]), and there is also evidence to suggest that *Anabaena azollae* might be a *Nostoc* [60]. Based on morphological characteristics, cyanobionts have mostly been classified as *Nostoc*, although others have been identified (e.g. *Calothrix* spp. and *Anabaena* spp.) [59]. Little work has been done to elucidate the host specificity and diversity of the cycad cyanobionts, and only few studies used molecular techniques for discriminating between different cyanobionts [63]. In addition, the molecular techniques used by different authors are very diverse, and it is difficult to compare such methods, for example RFLP and PCR amplification techniques of different segments of the genome. Additionally, several studies analyzed cultured symbionts which, in the case of *Azolla*, has been shown to be problematic, since the cyanobiont obtained in culture is not the same organism as the main strain in symbiosis [63].

The molecular techniques most recently used to classify cyanobacteria have been DNA–DNA hybridization and hybridization of highly repetitive (STRR) DNA sequences [62] and amplification of the tRNA^{Leu} (UAA) intron from the cyanobacterial symbionts of cycads.

4.3.2.3 *Gluconacetobacter*

This genus, included in the family Acetobacteraceae of the class Alphaproteobacteria is composed of obligate endophytic bacteria. *Gluconacetobacter diazotrophicus* colonizes sugarcane roots, stem and leaves, where it is present in the intercellular space of parenchyma, and is considered an obligate endophyte [57]. *G. diazotrophicus*, originally described as *Acetobacter diazotrophicus* and later transferred to the genus *Gluconoacetobacter* [64] which was subsequently corrected to *Gluconacetobacter* [65], was the first nitrogen-fixing *Acetobacteraceae* species described [66].

Two other nitrogen-fixing species have been described in this genus, *G. johannae* and *G. azotocaptans* [58]. The distribution of these species ranges from sugar-rich plants such as sugarcane, sweet sorghum, sweet potato and pineapple to sugar-poor plants such as coffee and ragi [66] and more recently from Kombucha tea [67].

4.4
Asymbiotic Plant Growth Promoting Bacteria

4.4.1
Alphaproteobacteria: Genera *Acetobacter*, *Swaminathania* and *Azospirillum*

4.4.1.1 *Acetobacter* and *Swaminathania*

The six nitrogen-fixing bacterial species so far described in the family *Acetobacteraceae* belong to the genera *Acetobacter*, *Gluconacetobacter* and *Swaminathania*. The genus *Gluconacetobacter* has been briefly described above. Of the two species of *Acetobacter*, one is *A. nitrogenifigens* [68], a new species isolated from Kombucha tea. The other nitrogen-fixing species in this genus is *A. peroxydans* [69] and this has lately been reported as a diazotroph species after a study on strains associated with wetland rice [66].

Recently, a novel genus has been described from strains isolated from the rhizosphere, roots and stems of salt-tolerant, mangrove-associated wild rice (*Porteresia coarctata Tateoka*) [22]. The isolates were able to fix nitrogen and solubilize phosphate in the presence of NaCl and belonged to a well-defined taxon, for which the species name proposed was *Swaminathania salitolerans* gen. nov., sp. nov. [22].

4.4.1.2 *Azospirillum*

Only the genus *Azospirillum* contains species reported as plant growth promoters of all genera described within the family *Rhodospirillaceae*. Strains belonging to this genus occur as free cells in the soil or associated with the roots, stems, leaves and seeds mainly of cereals and forage grasses, although they have also been isolated from coconut plants, vegetables, fruits, legumes and tuber plants. Notwithstanding, root nodules are not induced by strains in this genus. 16S rDNA sequencing studies show a high degree of relatedness among *Azospirillum* species and that they form a cohesive phylogenetic cluster within the Alphaproteobacteria [70]. Notwithstanding, strains of *Azospirillum* may still be misnamed in other genera, as shown by the recent reclassification of *Rosseomonas fauriae* as *Azospirillum brasilense* [71].

4.4.2
Gammaproteobacteria

4.4.2.1 **Enterobacteria**

The family *Enterobacteriaceae*, in the class Gammaproteobacteria, encompass a wide range of microorganisms including 42 genera in the last edition of *Bergey's Manual of Bacteriology* [30], but half of the isolates of new or unusual *Enterobacteriaceae* seem to be misidentified [72]. To avoid misidentification of species in future, Paradis *et al.* [73] propose the classification of species within *Enterobacteriaceae* by studying the genes encoding the elongation factor Tu and F-ATPase-β-subunit additionally to the gene encoding the 16S small ribosomal subunit.

The genera within the family *Enterobacteriaceae* that feature members described as plant growth promoting bacteria are *Citrobacter*, *Enterobacter*, *Erwinia*, *Klebsiella*, *Kluyvera*, *Pantoea* and *Serratia* although some of these genera also contain species reported to be plant pathogens, for example *Erwinia carotovora*. The seven genera mentioned above have undergone changes in their taxonomy in the time elapsed between the two most recent releases of *Bergey's Manual of Systematic Bacteriology*. These seven genera are reviewed below.

4.4.2.2 *Citrobacter*

Eight new *Citrobacter* species have been described out of already known species within this same genus, and two other species have been found to be synonyms of *C. koseri* [74].

4.4.2.3 *Enterobacter*

This genus contains the *Enterobacter agglomerans* group, which was extremely heterogeneous. Strains previously included in the *Enterobacter agglomerans* group have been proposed to be relocated into the genera *Erwinia*, *Leclercia* and *Pantoea*. Another species, *Enterobacter intermedius*, was first described as a senior subjective synonym for the species *Kluyvera cochleae* as shown by DNA–DNA hybridization [74] but was later transferred to the genus *Kluyvera* as the species *K. intermedia* comb. nov., and *K. cochleae* was demonstrated to be a later synonym of *K. intermedia* [75].

4.4.2.4 *Erwinia*
Members of this genus are mainly plant isolates, and human or animal isolates are rarely reported, although it might be a result of improper isolation and enrichment procedures or failure in their identification. *Erwinia* is quite a heterogeneous genus as shown by DNA-relatedness studies. Former members of the genus have been proposed as the new genus *Brenneria* and others have been relocated within the genera *Pantoea*, or proposed to be relocated in the genus *Pectobacterium* [74].

4.4.2.5 The *Klebsiella* Complex
The genus *Klebsiella* was found to be polyphyletic by Drancourt *et al.* [76]. These authors found *Klebsiella* species to form three DNA-relatedness clusters: Cluster I contained *Klebsiella pneumoniae* subsp. *pneumoniae*, *K. pneumoniae* subsp. *ozaenae*, *K. pneumoniae* subsp. *rhinoscleromatis* and *K. granulomatis*. Cluster II contained *K. planticola*, *K. ornithinolytica* and *K. terrigena*, and was proposed to constitute a new genus, *Raoultella* [76]. Cluster III contained *K. oxytoca*. The position of *Klebsiella mobilis* (formerly *Enterobacter aerogenes*) was very close to Cluster II, although the proposal by Drancourt *et al.* [76] for the new genus did not include this species [77]. A somewhat different structure was found by Brisse and Verhoef [78], who uncovered two groups. The first group contained *K. pneumoniae* with its three subspecies, and the second contained *K. oxytoca*, *K. planticola*, *K. ornithinolytica*, *K. terrigena* and *K. mobilis*. Furthermore, three clusters were evidenced in *K. pneumoniae*, which did not correlate with the named subspecies. These clusters may have different habitats and different physiological properties (e.g. D-adonitol fermentation). *Klebsiella oxytoca* was composed of two clusters, the significance of which is as yet unknown [77].

As the works of Dracount *et al.* [76] and Brisse and Verhoef [78] evidence, the taxonomy of *Klebsiella* is still in need of much clarification. This taxonomic unsoundness has led to phenotypic properties ruling over DNA-relatedness when species allocation of strains is carried out in the practice, hence, worsening the prospects for tidying up this group.

4.4.2.6 *Kluyvera*
Kluyvera is reportedly a genus with a turbulent history [72]. It was first described in 1956, abolished in 1962 and redefined in 1981 by Farmer *et al.* [79]. The authors retained the name as it was still used in the literature and the genus was well represented in culture collections by established strains of long standing [72].

Farmer [72] describes *Kluyvera* as a well-defined genus of the *Enterobacteriaceae* on the basis of DNA–DNA hybridization, and suggests the possibility of three new species in the genus that could be delineated out of a group of 21 strains held at the Enteric Reference Laboratory's collection. *Bergey's Manual of Bacteriology* also points out that *Kluyvera cochleae* is a junior subjective synonym for *Enterobacter intermedius* and it hence discusses this species in the genus *Enterobacter* [74]. Notwithstanding, Pavan *et al.* [75] have shown *Enterobacter intermedius* to be a member of the genus *Kluyvera*, now named as *Kluyvera intermedia*; *K. cochleae* is hence a later subjective synonym for *K. intermedia* [75].

4.4.2.7 *Pantoea*

The genus *Pantoea* contains strains previously allocated in the *Enterobacter agglomerans* complex and species formerly named as *Erwinia* (*E. herbicola*, *E. lathyri*, *E. ananas*, *E. uredovora*, *E. milletiae* and *E. stewartii*), and future prospects point to the possibility of more species within the genera *Enterobacter* and *Erwinia* being relocated to this genus.

4.4.2.8 *Serratia*

Most species currently known in the genus *Serratia* have been shown to be homogeneous and discrete genomospecies: *S. entomophila*, *S. ficaria*, *S. marcescens*, *S. odorifera*, *S. plymuthica* and *S. rubidaea*. *S. liquefaciens sensu lato* was shown to be heterogeneous and later found to be composed of three genomospecies: *S. liquefaciens sensu stricto*, *S. proteamaculans* and *S. grimessi*. Another genomospecies composed of a group of strains referred to as 'Citrobacter-like' was described as *S. fonticola*; however, this species does not possess the key characteristics of the genus *Serratia*, *Serratia* phages that are active on strains of any *Serratia* species are inactive on all *S. fonticola* strains tested, and so are bacteriocins of *Serratia* [80]. Despite these data, the 16S rRNA gene sequence of *S. fonticola* branches within the psychrotolerant *Serratia* cluster (*S. liquefaciens*, *S. proteamaculans*, *S. grimesii* and *S. plymuthica*), and this justifies the inclusion of *S. fonticola* in the genus *Serratia* [80].

4.4.2.9 *Pseudomonas*

Strains belonging to the genus *Pseudomonas* might be plant pathogenic or PGPR. The latter do not form nodules but proliferate in the surroundings of the roots, using the root exudates as sources of carbon and energy. In exchange, these strains protect the plant from other plant pathogenic microorganisms due to their production of siderophores and other molecules [5,18]. The taxonomy of the strains allocated to the genus *Pseudomonas* was reviewed by De Vos *et al.* [35], Anzai *et al.* [81] and Picard and Bosco [6]. All these studies uncovered groups of strains belonging to the Alphaproteobacteria, Betaproteobacteria and Gammaproteobacteria. The difficulty of obtaining sound groupings using the 16S rDNA gene as a phylogenetic marker seems to be due to the presence of several different copies of this gene in the same strain [82,83]. Notwithstanding, a division of the genus into five RNA similarity groups has been proposed and confirmed by workers in many different laboratories [83].

4.4.2.10 *Azotobacter* (*Azomonas*, *Beijerinckia* and *Derxia*)

The genus *Azotobacter*, in the family *Pseudomonadaceae* of the Gammaproteobacteria, is composed of bacteria that promote plant growth mainly due to their ability to fix dinitrogen from the atmosphere and do not nodulate plants [20,84]. The genus has undergone revision between the last two editions of the *Bergey's Manual of Systematic Bacteriology* and has been moved from its previous family, Azotobacteraceae [85] to the family *Pseudomonadaceae* after studies on the phylogeny of its members. The genus *Azomonas*, included in the former family *Azotobacteraceae*, has also been moved into the family Pseudomonadaceae with a comment on the phylogenetic heterogeneity of its members, which will probably result in their separation into more than one genus [86]. It was once proposed that the genera

Beijerinckia and *Derxia* were to be included in Azotobacteraceae, but *Beijerinckia* remained in the family *Beijerinckiaceae* of Alphaproteobacteria and *Derxia* has been placed in the family *Alcaligenaceae* of Betaproteobacteria [87].

4.4.3
Firmicutes. Genera *Bacillus* and *Paenibacillus*

A review on the applications and systematics of *Bacillus* and related genera presented at a meeting held in Brugges (Belgium) in August 2000 ('*Bacillus* 2000' Meeting) was published in 2002 [88].

4.4.3.1 *Bacillus*
Numerous *Bacillus* strains have been reported to be PGPR [3,5,11,15,18,24,89] and these employ the widest range of plant growth promoting mechanisms found for any genera as revealed in Table 4.1 and reviewed by Chanway [90].

The history of the genus *Bacillus* originated early in the history of bacteriology, when it was proposed by Cohn in 1872 and subsequently experienced great fluctuations in the number of valid *Bacillus* species recognized in *Bergey's Manual of Bacteriology*, ranging from a peak of 146 species in the fifth edition [91] to the lowest number (22) in the eighth edition [92]. The establishment of the phylogenetic relationships among the different type strains of *Bacillus* species and the use of polyphasic taxonomy applied to the genus, made possible the splitting of *Bacillus sensu lato* into 11 genera [93]. However, far from solving its taxonomy, the application of phylogenetic studies to the genus made evident that groupings used in the traditional phenotypically based schemes for *Bacillus* [94,95] did not always correlate with current, phylogenetically led classifications [96] and taxonomic progress has not yet revealed readily determinable features characteristic of each genus. Many species described recently represent genomic groups disclosed by DNA–DNA pairing experiments, and routine phenotypic characteristics to distinguish some of these species are very few and of unproven value [97]. In relation to the taxonomy of *Bacillus* strains described as PGPR, it is important to emphasize the possibility of strains identified as *B. circulans* or *B. firmus* as having been incorrectly classified given the difficult taxonomic position of members in these two species, whose taxonomy is under revision at present (Dr Rodríguez-Díaz, personal communication).

4.4.3.2 *Paenibacillus*
The genus *Paenibacillus* was described by Ash *et al.* [98] following the study of 16S rRNA gene sequences of the type strains of many *Bacillus* species. Since then, the genus was reassessed by Heyndrickx *et al.* [99] to accommodate former *Bacillus* species and the transfer of *Bacillus* species to *Paenibacillus* seems to have no end. For example, numerous strains of *Bacillus circulans sensu lato* have been found to belong to nine *Paenibacillus* species [98–101]. Examples of PGPR described as *Bacillus* species indeed being members of the genus *Paenibacillus* are found in the early studies of plant growth promotion by bacteria, mainly with regard to nitrogen fixation. Nonsymbiotic nitrogen fixation by a strain of the genus *Bacillus* was first reported by Bredemann in 1908, but this claim was later discredited due to uncertainties in the determination of the fixed nitrogen

[102]. In 1958, nitrogen fixation was conclusively proven for a *Bacillus* strain, later identified as *B. polymyxa* (now *Paenibacillus polymyxa*) [102]. Classically, all *Bacillus* strains capable of fixing molecular nitrogen were found to belong to *B. polymyxa* or *B. macerans* (now *Paenibacillus macerans*), as well as some strains identified as *B. circulans*, until Seldin *et al.* [103] described *B. azotofixans* (now *Paenibacillus durus* [104]). Achouak *et al.* [105] made a comparative phylogenetic study of 16S rDNA and nifH genes in the family *Bacillaceae*, and concluded that nitrogen fixation among aerobic endospore-forming bacteria is restricted to the genus *Paenibacillus*. To date, nitrogen fixation has been demonstrated for 11 species of *Paenibacillus*: *P. polymyxa*, *P. macerans*, *P. durus*, *P. peoriae*, *P. borealis*, *P. brasilensis*, *P. graminis*, *P. odorifer* [106], *P. wynnii* [106], *P. massiliensis* [107] and *P. sabinae* [108]. Strains belonging to these species act as plant growth promoting rhizobacteria because of their N_2-fixation ability, production of phytohormones, provision of nutrients and/or by the suppression of deleterious microorganisms through antagonistic function [109].

4.5
Screening Methods of PGPR

4.5.1
Culture-Dependent Screening Methods

Traditionally, a search for PGPR involves screening a large number of isolates and identifying a desired phenotypic trait. Once isolates are purified, the main goal is to maintain the maximum genetic diversity in the minimum number of isolates for identifying the desired phenotypic trait or for performing further biological assays in order to achieve results that are representative of the diversity occurring in Nature. This goal may be achieved through the use of intergenic transcribed sequence (ITS)-PCR, AFLP and arbitrarily primed (AP)-PCR/PCR–RAPDs techniques that define differences at the strain level [10,110].

Once the set of strains to test is defined, the screening methods used will differ accordingly to the trait of relevance in the study. Screening methods have been described for the detection of a range of single molecules (e.g. auxins production by colorimetry [6]) or an array of characteristic traits of the PGPR (e.g. screening for PGPR by testing for aminocyclopropanecarboxylic acid (ACC), auxin and siderophore production and phosphate solubilization [10]). Three rapid plate assays have been developed that allow for screening of those PGPR capable of inducing plant-systemic resistance based on their ability to attack certain plant-pathogenic fungi [89] by targeting the fungal pathogenesis-related proteins chitinase and β-1,3-glucanase, and their biphasic hydrogen production.

Other screening techniques for isolated strains rely on molecular methods to search for diverse marker genes, such as the (ACC) deaminase gene [23], or genes associated with plant–LNB interactions, such as the nodA [110] or gusA and celB genes. The marker genes gusA and celB were used to study plant–*Rhizobium* interactions, and the competition within inoculated strains and between inoculated strains and indigenous rhizobia. A GUS Gene Marking Kit has also been developed that enables microbiologists and agronomists in developing countries to carry out competition studies without sophisticated equipment [111].

4.5.2
Culture-Independent Screening Methods

Culture-independent molecular techniques are based on direct artificial chromosome or expression cloning systems, thus providing new insight into the diversity of rhizosphere microbial communities, the heterogeneity of the root environment, and the importance of environmental and biological factors in determining community structure [112]. These are usually DGGE or temperature gradient gel electrophoresis (TGGE) of PCR-amplified DNA fragments, either at a gross taxonomic level or at more refined levels, for example genus [113]. Studies carried out by de Oliveira *et al.* [113] used primers specific to *Rhizobium leguminosarum sensu lato* and *R. tropici* in a nested PCR amplification of 16S–23S ribosomal RNA gene IGS, concluding that this approach would be useful for monitoring the effect of agricultural practices on these and related rhizobial subpopulations in soils.

Possibly the most recently published screening method for PGPR is a laboratory-made microarray initially designed for analyzing the genetic diversity of nitrogen-fixing symbionts, *Sinorhizobium meliloti* and *S. medicae* [114]. The authors refer to this microarray as a low-cost alternative for 'medium-scale' projects of population genetics, accessible to any laboratory equipped for molecular biology, and state their intention to enlarge the number of polymorphic loci represented on the arrays in order to increase both the cost-effectiveness and time-effectiveness of the procedure.

4.6
Conclusions and Remarks

Systematic identification, enumeration and characterization of PGPR microbiota in environmental samples by traditional procedures are difficult. The application of molecular techniques to the study of PGPR bacteria now enables us to solve these problems and to obtain information on their phylogenetic relationships, as well as their ecological roles.

Molecular techniques such as DGGE of PCR-amplified DNA fragments, DNA–DNA hybridization and/or analysis of 16S rDNA sequences have been used to evaluate the ecological significance of PGPR, to detect and identify new PGPR microorganisms, and to monitor the success of isolation of these new species. In this context, different oligonucleotide probes have been designed and applied to identify different PGPR and to determine their spatial distribution in environmental samples such as soils and bacterial biofilms. Perhaps, in the near future, the amplification of mRNA genes by RT-PCR, or *in situ* hybridization of mRNA, would be feasible for detection of gene expression as an indicator for metabolic activities of individual bacterial species.

More genetic and ecological studies are necessary to advance our understanding of the relevance of the PGPR in soil and the rhizosphere, and also to explain the interaction of these microbial groups with plants. In this context, new molecular techniques, such as microarrays carrying diverse copies of known genes present in PGPR, will likely facilitate the screening of natural plant growth promoting communities.

The number of PGPR described in the literature and the difficulty of their taxonomy as reviewed in this chapter is remarkable. Some other problems have been encountered when searching for information regarding the nomenclature used for PGPR in those papers not involved in taxonomy, that is misnamed genera that could not be found in the official reference points for taxonomy such as *Bergey's Manual of Systematic Bacteriology*, and the DSMZ Web page for taxonomy (http://www.dsmz. de/microorganisms/main.php?contentleft_id=14), or the NCBI Web page for taxonomy (http://www.ncbi.nlm.nih.gov/entrez/query.fcgi?CMD=search&DB=taxonomy). As examples, we would like to highlight the names 'Actinobacter' and 'Aereobacter' as invalid names supposedly referring to the genera *Acinetobacter* and *Aerobacter*, respectively. The source for the misnaming of the genus *Aerobacter* (a genus within the family *Aurantimonadaceae*, in the class Rhizobiales of Alphaproteobacteria could not be tracked. For the genus *Acinetobacter* (family *Moraxellaceae*, class Pseudomonadales of Gammaproteobacteria), the root of the misnaming could be a typographical error for a sequence of a plasmid of strain BW3 representative of the genus (http://www.ebi.ac.uk/ebisearch/search.ebi?db=nucleotideSequences&query=actinobacter), as this 'Actinobacter' name is only found there, and a link exists naming strain BW3 to the correct genus name. The data matched as many reference articles refer to 'Actinobacter' as a member of Gammaproteobacteria, correlating well with the taxonomic classification of the genus *Acinetobacter*.

References

1 Kloepper, J.W. and Schroth, M.N. (1978) Proceedings of the 4th International Conference on Plant Pathogenic Bacteria, Station de Pathologie Vegetal et Phytobacteriologie, Angers. 2, pp. 879–882.

2 Bashan, Y. and Holguin, G. (1998) *Soil Biology & Biochemistry*, **30**, 1225–1228.

3 Siddiqui, Z.A. and Mahmood, I. (1999) *Bioresource Technology*, **69**, 167–179.

4 Jiménez-Delgadillo, R., Virgen-Calleros, G., Tabares-Franco, S. and Olalde-Portugal, V. (2001) *Avance y Perspectiva*, **20**, 395–400.

5 Gray, E.J. and Smith, D.L. (2005) *Soil Biology & Biochemistry*, **37**, 395–412.

6 Picard, C. and Bosco, M. (2005) *FEMS Microbiology Ecology*, **53**, 349–357.

7 Zakhia, F. and de Lajudie, P. (2006) *Canadian Journal of Microbiology*, **52**, 169–181.

8 Mavingui, P., Flores, M., Romero, D., Martinez-Romero, E. and Palacios, R. (1997) *Nature Biotechnology*, **15**, 564–569.

9 Sawada, H., Kuykendall, L.D. and Young, J.M. (2003) *Journal of General and Applied Microbiology*, **49**, 155–179.

10 Barriuso, J., Pereyra, M.T., García, J.A.L., Megías, M., Mañero, F.J.G. and Ramos, B. (2005) *Microbial Ecology*, **50**, 82–89.

11 Molina, L., Constantinescu, F., Michel, L., Reimmann, C., Duffy, B. and Défago, G. (2003) *FEMS Microbiology Ecology*, **45**, 71–81.

12 Rivas, R., Velázquez, E., Zurdo-Piñeiro, J. L., Mateos, P.F. and Martínez-Molina, E. (2004) *Journal of Microbiological Methods*, **56**, 413–426.

13 Bloemberg, G.V. and Lugtenberg, B.J.J. (2001) *Current Opinion in Plant Biology*, **4**, 343–350.

14 Thaning, C., Welch, C.J., Borowicz, J.J., Hedman, R. and Gerhardson, B. (2001) *Soil Biology and Biochemistry*, **33**, 1817–1826.

15 Compant, S., Duffy, B., Nowak, J., Clément, C. and Barka, E.A. (2005) *Applied and Environmental Microbiology*, **71**, 4951–4959.

16 Cummings, S.P., Humphry, D.R. and Andrews, M. (2001) in *Aspects of Applied Biology*, vol. 63, The Association of Applied Biologists, Warwick.

17 van Berkum, P. and Eardly, B.D. (2002) *Applied and Environmental Microbiology*, **68**, 1132–1136.

18 Spadaro, D. and Gullino, M.L. (2005) *Crop Protection*, **24**, 601–613.

19 Ribbe, M., Gadakari, D. and Meyer, O. (1997) *Journal of Biochemistry*, **272**, 26627–26633.

20 Sturz, A.V. and Christie, B.R. (2003) *Soil & Tillage Research*, **72**, 103–107.

21 Mayak, S., Tirosh, T. and Glick, B.R. (2004) *Plant Physiology and Biochemistry*, **42**, 565–572.

22 Loganathan, P. and Nair, S. (2004) *International Journal of Systematic and Evolutionary Microbiology*, **54**, 1185–1190.

23 Babalola, O.O., Osir, E.O., Sanni, A.I., Odhiambo, G.D. and Bulimo, W.D. (2003) *African Journal of Biotechnology*, **2**, 157–160.

24 Rodríguez, H. and Fraga, R. (1999) *Biotechnology Advances*, **17**, 319–339.

25 Woese, C.R. (1987) *Microbiology Reviews*, **51**, 221–271.

26 Stackebrandt, E. and Ebers, J. (2006) *Microbiology Today*, **33**, 152–155.

27 Jaspers, E. and Overmann, J. (2004) *Applied and Environmental Microbiology*, **70**, 4831–4839.

28 Vandamme, P., Pot, B., Gillis, M., De Vos, P., Kersters, K. and Swings, K. (1996) *Microbiological Reviews*, **60**, 407–438.

29 Gillis, M., Vandamme, P., De Vos, P., Swings, J. and Kersters, K. (2005) Polyphasic taxonomy in *Bergey's Manual of Systematic Bacteriology*, 2nd edn, vol. 2 (ed. G.M. Garrity), Springer-Verlag, Berlin, Heidelberg, New York, pp. 43–48.

30 Garrity, G.M. (2005) *Bergey's Manual of Systematic Bacteriology*, 2nd edn, vol. 2, Springer-Verlag, Berlin, Heidelberg, New York, pp. 1–1085.

31 Gaunt, M.W., Turner, S.L., Rigottier-Gois, L., Lloyd-Macgilp, S.A. and Young, J.P.W. (2001) *International Journal of Systematic Bacteriology*, **51**, 2037–2048.

32 Vinuesa, P., Silva, C., Lorite, M.J., Izaguirre-Mayoral, M.L., Bedmar, E.J. and Martínez-Romero, E. (2005) *Systematic and Applied Microbiology*, **28**, 702–716.

33 van Berkum, P., Terefework, Z., Paulin, L., Suomalainen, S., Lindström, K. and Eardly, B.D. (2003) *Journal of Bacteriology*, **185**, 2988–2998.

34 Young, J.M., Kuykendall, L.D., Martínez-Romero, E., Kerr, A. and Sawada, H. (2001) *International Journal of Systematic and Evolutionary Microbiology*, **51**, 89–103.

35 De Vos, P., Van Landschoot, A., Segers, P., Tytgat, T., Gillis, M., Bauwens, M., Rossau, R., Goor, M., Pot, B., Kersters, K., Lizzaraga, P. and De Ley, J. (1989) *International Journal of Systematic Bacteriology*, **39**, 35–49.

36 Kwon, S.W., Park, J.Y., Kim, J.S., Kang, J. W., Cho, Y.H., Lim, C.K., Parker, M.A. and Lee, G.B. (2005) *International Journal of Systematic and Evolutionary Microbiology*, **55**, 263–270.

37 van Berkum, P., Ruihua, F., Campbell, T.A. and Eardly, B.D. (1999) *Highlights of nitrogen fixation research*, Martínez & Hernández, Plenum Publishing Co., New York.

38 Willems, A., Munive, A., de Lajudie, P. and Gillis, M. (2003) *Systematic and Applied Microbiology*, **26**, 203–210.

39 Young, J.M. (2003) *International Journal of Systematic and Evolutionary Microbiology*, **53**, 2107–2110.

40 Nour, S.M., Cleyet-Marel, J.C., Normand, P. and Fernández, M.P. (1995) *International Journal of Systematic Bacteriology*, **45**, 640–648.

41 Jarvis, B.D.W., Sivakumaran, S., Tighe, S. W. and Gillis, M. (1996) *Plant and Soil*, **184**, 143–158.

42 Doignon-Bourcier, F., Willems, A., Coopman, R., Laguerre, G., Gillis, M. and de Lajudie, P. (2000) *Applied and Environmental Microbiology*, **66**, 3987–3997.

43 Willems, A., Coopman, R. and Gillis, M. (2001) *International Journal of Systematic and Evolutionary Microbiology*, **51**, 111–117.

44 van Berkum, P., Leibold, J.M. and Eardly, B.D. (2006) *Systematic and Applied Microbiology*, **29**, 207–215.

45 van Berkum, P. and Eardly, B.D. (1998) *The Rhizobiaceae: Molecular Biology of Model Plant-Associated Bacteria*, Spaink, Kondorosi & Hooykaas, Dordrecht, pp. 1–24.

46 Sy, A., Giraud, E., Jourand, P., Garcia, N., Willems, A., de Lajudie, P., Prin, Y., Neyra, M., Gillis, M., Boivin-Masson, C. and Dreyfus, B. (2001) *Journal of Bacteriology*, **183**, 214–220.

47 Jourand, P., Giraud, E., Béna, G., Sy, A., Willems, A., Gillis, M., Dreyfus, B. and de Lajudie, P. (2004) *International Journal of Systematic and Evolutionary Microbiology*, **54**, 2269–2273.

48 Rivas, R., Willems, A., Subba-Rao, N.S., Mateos, P.F., Dazzo, F.B., Kroppenstedt, R.M., Martínez-Molina, E., Gillis, M. and Velázquez, E. (2003) *Systematic and Applied Microbiology*, **26**, 47–53.

49 Anonymous (2003) *International Journal of Systematic and Evolutionary Microbiology*, **53**, 935–937.

50 Moulin, L., Munive, A., Dreifus, B. and Boivin-Masson, C. (2001) *Nature*, **411**, 948–950.

51 Anonymous (2003) *International Journal of Systematic and Evolutionary Microbiology*, **53**, 627–628.

52 Vandamme, P., Goris, J., Chen, W.-M., De Vos, P. and Willems, A. (2002) *Systematic and Applied Microbiology*, **25**, 507–512.

53 Chen, W.M., James, E.K., Coenye, T., Chou, J.H., Barrios, E., de Faria, S.M., Elliott, G.N., Sheu, S.Y., Sprent, J.I. and Vandamme, P. (2006) *International Journal of Systematic and Evolutionary Microbiology*, **56**, 1847–1851.

54 Chen, W.-M., de Faria, S.M., James, E.K., Elliott, G.N., Lin, K.Y., Chou, J.H., Sheu, S.Y., Cnockaert, M., Sprent, J.I. and Vandamme, P. (2007) *International Journal of Systematic and Evolutionary Microbiology*, **57**, 1055–1059.

55 Yabuuchi, E., Kosako, Y., Yano, I., Hotta, H. and Nishiuchi, Y. (1995) *Microbiology and Immunology*, **39**, 897–904.

56 Chen, W.M., Laevens, S., Lee, J.M., Coenye, T., De Vos, P., Mergeay, M. and Vandamme, P. (2001) *International Journal of Systematic and Evolutionary Microbiology*, **51**, 1729–1735.

57 Tejera, N.A., Ortega, E., González-López, J. and Lluch, C. (2003) *Journal of Applied Microbiology*, **95**, 528–535.

58 Fuentes-Ramírez, L.E., Bustillos-Cristales, R., Tapia-Hernández, A., Jiménez-Salgado, T., Wang, E.T., Martínez-Romero, E. and Caballero-Melado, J. (2001) *International Journal of Systematic and Evolutionary Microbiology*, **51**, 1305–1314.

59 Venssey, K.J., Pawlowski, K. and Bergman, B. (2005) *Plant and Soil*, **274**, 51–78.

60 Rai, A.N., Söderbäck, E. and Bergman, B. (2000) *The New Phytologist*, **147**, 449–481.

61 Nilsson, M., Bhattacharya, J., Rai, A.N. and Bergman, B. (2002) *The New Phytologist*, **156**, 517–525.

62 Lu, W., Evans, E.H., McColl, S.M. and Saunders, V.A. (1997) *FEMS Microbiology Ecology*, **153**, 141–149.

63 Costa, J.L., Paulsrud, P. and Lindblad, P. (1999) *FEMS Microbiology Ecology*, **28**, 85–91.

64 Yamada, Y., Hoshino, K. and Ishikawa, T. (1998) *International Journal of Systematic Bacteriology*, **48**, 327–328.

65 Yamada, Y., Hoshino, K. and Ishikawa, T. (1997) *Bioscience Biotechnology, and Biochemistry*, **6**, 1244–1251.

66 Muthukumarasamy, R., Cleenwerck, I., Revathi, G., Vadivelu, M., Janssens, D., Hoste, B., Gum, K.U., Park, K.D., Son, C.Y., Sa, T. and Caballero-Mellado, J. (2005) *Systematic and Applied Microbiology*, **28**, 277–286.

67 Dutta, D. and Gachhui, R. (2007) *International Journal of Systematic and Evolutionary Microbiology*, **57**, 353–357.

68 Dutta, D. and Gachhui, R. (2006) *International Journal of Systematic and Evolutionary Microbiology*, **56**, 1899–1903

69 Visser't Hooft, F. (1925) Biochemische onderzoekingen over het geslacht Acetobacter, Technische Universiteit Delft, Ph.D. Thesis.

70 Baldani, J.I., Krieg, N.R., Divan-Baldani, V.L., Hartmann, A. and Döbereiner, J. (2005) in *Bergey's Manual of Systematic Bacteriology*, 2nd edn, vol. 2, Springer-Verlag, Garrity, New York, Berlin, Heidelberg, pp. 7–26.

71 Helsel, L.O., Hollis, D.G., Steigerwalt, A. G. and Levett, P.N. (2006) *International Journal of Systematic and Evolutionary Microbiology*, **56**, 2753–2755.

72 Farmer, J.Jr (2005) in *Bergey's Manual of Systematic Bacteriology*, 2nd edn, vol. 2B (ed. G.M. Garrity), Springer-Verlag, New York, Berlin, Heidelberg,694–698.

73 Paradis, S., Boissinot, M., Paquette, N., Bélanger, S.D., Martel, E.A., Boudreau, D.K., Picard, F.J., Ouellette, M., Roy, P.H. and Bergeron, M.G. (2005) *International Journal of Systematic and Evolutionary Microbiology*, **55**, 2013–2025.

74 Brenner, D.J. and Farmer, J.Jr (2005) in *Bergey's Manual of Systematic Bacteriology*, 2nd edn, vol. 2 (ed. G.M. Garrity)Springer-Verlag, New York, Berlin, Heidelberg, pp. 587–850.

75 Pavan, M.E., Franco, R.J., Rodriguez, J. M., Gadaleta, P., Abbott, S.L., Janda, J.M. and Zorzópulos, J. (2005) *International Journal of Systematic and Evolutionary Microbiology*, **55**, 437–442.

76 Drancourt, M., Bollet, C., Carta, A. and Rousselier, P. (2001) *International Journal of Systematic and Evolutionary Microbiology*, **51**, 925–932.

77 Grimont, P.A.D. and Grimont, F. (2005) in *Bergey's Manual of Systematic Bacteriology*, 2nd edn, vol. 2 (ed. G.M. Garrity), Springer-Verlag, New York, Berlin, Heidelberg, pp. 684–685.

78 Brisse, S. and Verhoef, J. (2001) *International Journal of Systematic and Evolutionary Microbiology*, **51**, 915–924.

79 Farmer, J.Jr, Fanning, G.R., Huntley-Carter, G.P., Holmes, B., Hickman, F.W., Richard, C. and Brenner, D.J. (1981) *Journal of Clinical Microbiology*, **13**, 919–933.

80 Grimont, F. and Grimont, P.A.D. (2005) in *Bergey's Manual of Systematic Bacteriology*, 2nd edn, vol. 2 (ed. G.M. Garrity), Springer-Verlag, New York, Berlin, Heidelberg, pp. 799–811.

81 Anzai, Y., Kim, H., Park, J.Y., Wakabayashi, H. and Oyaizu, H. (2000) *International Journal of Systematic and Evolutionary Microbiology*, **50**, 1563–1589.

82 Milyutina, I.A., Bobrova, V.K., Matveeva, E.V., Schaad, N.W. and Troitsky, A.V. (2004) *FEMS Microbiology Letters*, **239**, 17–23.

83 Palleroni, N.J. (2005) in *Bergey's Manual of Systematic Bacteriology*, 2nd edn, vol. 2 (ed. G.M. Garrity),Springer-Verlag, New York, Berlin, Heidelberg, pp. 322–379.

84 Rodelas, B., González-López, J., Pozo, C., Salmerón, V. and Martínez-Toledo, M.V. (1999) *Applied Soil Ecology*, **12**, 51–59.

85 Tchan, Y.T. (1984) in *Bergey's Manual of Determinative Bacteriology*, 9th edn, vol. 1, Krieg & Holt, Baltimore, pp. 219–220.

86 Kennedy, C. and Rudnick, P. (2005) in *Bergey's Manual of Systematic Bacteriology*, 2nd edn, vol. 2 (ed. G.M. Garrity), Springer-Verlag, New York, Berlin, Heidelberg, pp. 379–384.

87 Kennedy, C., Rudnick, P., MacDonald, M.L. and Melton, T. (2005) in *Bergey's Manual of Systematic Bacteriology*, 2nd edn, vol. 2 (ed. G.M. Garrity), Springer-Verlag, New York, Berlin, Heidelberg, pp. 384–402.

88 Berkeley, R.C.W., Heyndrickx, M., Logan, N.A. and De Vos, P. (2002) *Applications and Systematics of Bacillus and Relatives*, Blackwell Science, Oxford.

89 Bargabus, R.L., Zidack, N.K., Sherwood, J.E. and Jacobsen, B.J. (2004) *Biological Control*, **30**, 342–350.

90 Chanway, C.P. (2002) (eds R.C.W. Berkeley, M. Heyndrickx,N.A. Logan

and P. De Vos), *Applications and Systematics of Bacillus and Relatives*, Blackwell Publishing, Oxford.

91 Bergey, D.H., Breed, R.S. and Murray, E. G.D. (1939) *Bergey's Manual of Determinative Bacteriology*, 5th edn, The Williams and Willkins Co., Baltimore.

92 Gibson, T. and Gordon, R.E. (1974) *Bergey's Manual of Systematic Bacteriology*, Buchanan & Gibbons, Baltimore, pp. 529–550.

93 Rodríguez-Díaz, M. (2005) Application of miniaturized identification systems to the polyphasic taxonomy of Bacillus and relatives, Ph.D. Thesis, Glasgow Caledonian University, Glasgow.

94 Logan, N.A. and Berkeley, R.C.W. (1984) *Journal of General Microbiology*, **130**, 1871–1882.

95 Smith, N.R.S., Gordon, R.E. and Clark, F. E. (1952) *Aerobic Spore-Forming Bacteria. Monograph No. 16*, United States Department of Agriculture, Washington, DC.

96 Logan, N.A. (2002) *Applications and Systematics of Bacillus and Relatives*, Berkeley, Heyndrickx, Logan & De Vos, Oxford.

97 Logan, N.A. and Rodríguez-Díaz, M. (2006) *Principles and Practice of Clinical Bacteriology*, 2nd edn,Emmerson, Hawkey & Gillespie, Chichester.

98 Ash, C., Priest, F.G. and Collins, M.D. (1993) *Antononie Leeuwnhoek International Journal of General and Molecular Microbiology*, **64**, 253–260.

99 Heyndrickx, M., Vandemeulebroecke, K., Scheldeman, P., Kersters, K., De Vos, P., Logan, N.A., Aziz, A.M., Ali, N. and Berkeley, R.C.W. (1996) *International Journal of Systematic Bacteriology*, **46**, 988–1003.

100 Shida, O., Takagi, H., Kadowaki, K., Nakamura, L.K. and Komagata, K. (1997) *International Journal of Systematic Bacteriology*, **47**, 289–298.

101 Shida, O., Takagi, H., Kadowai, K., Nakamura, L.K. and Komagata, K. (1997) *International Journal of Systematic Bacteriology*, **47**, 299–306.

102 Seldin, L., van Elsas, J.D. and Penido, E. G.C. (1983) *Plant and Soil*, **70**, 243–255.

103 Seldin, L., van Elsas, J.D. and Penido, E. G.C. (1984) *International Journal of Systematic Bacteriology*, **34**, 451–456.

104 Anonymous((2003) *International Journal of Systematic Bacteriology*, **53**, 931.

105 Achouak, W., Normand, P. and Heulin, T. (1999) *International Journal of Systematic Bacteriology*, **49**, 961–967.

106 Rodríguez-Díaz, M., Lebbe, L., Rodelas, B., Heyrman, J., De Vos, P. and Logan, N. A. (2005) *International Journal of Systematic and Evolutionary Microbiology*, **55**, 2093–2099.

107 Zhao, H., Xie, B. and Chen, S. (2006) *Science in China*, **49**, 115–122.

108 Ma, Y., Xia, Z., Liu, X. and Chen, S. (2007) *International Journal of Systematic and Evolutionary Microbiology*, **57**, 6–11.

109 Rodrigues Coelho, M.R., von der Wied, I., Zahner, V. and Seldin, L. (2003) *FEMS Microbiology Letters*, **222**, 243–250.

110 Rasolomampianina, R., Bailly, X., Fetiarison, R., Rabevohitra, R., Bena, G., Ramaroson, L., Raherimandimby, M., Moulin, L., De Lajudie, P., Dreyfus, B. and Avarre, J.C. (2005) *Molecular Ecology*, **14**, 4135–4146

111 Sonnleitner, R. (2000) *Geoderma*, **96**, 360–362.

112 Nelson, L.M. Plant growth-promoting rhizobacteria (PGPR): prospects for new inoculants. Crop Management Volume, doi: 10:10.1094/CM-2004-0301-05-RV.

113 de Oliveira, V.M., Manfio, G.P., da Costa Coutinho, H.L., Keijzer-Wolters, A.C. and van Elsas, J.D. (2006) *Journal of Microbiological Methods*, **64**, 366–379.

114 Bailly, X., Béna, G., Lenief, V., de Lajudie, P. and Avarre, J.C. (2006) *Journal of Microbiological Methods*, **67**, 114–124.

5
Diversity and Potential of Nonsymbiotic Diazotrophic Bacteria in Promoting Plant Growth

Farah Ahmad, Iqbal Ahmad, Farrukh Aqil, Mohammad Saghir Khan, and Samsul Hayat

5.1
Introduction

Plant roots support the growth and activities of an array of microorganisms that may impart profound effects on growth and health of plants. Diversity of such microorganisms is studied for certain culturable microorganisms, including bacteria, fungi, actinomycetes and other eukaryotic microorganisms. Among rhizobacteria, a high diversity has been identified and categorized as deleterious, beneficial or neutral with respect to plants [1]. Microbial ecologists have, in particular, studied microbial community composition since it exerts significant control over soil processes and overall soil health. Diversity and community structure in the rhizosphere are under investigation and have been found to be influenced by both plant and soil types. The use of molecular techniques to study soil microbial diversity, such as terminal restriction fragment length polymorphism (TRFLP), single-strand confirmation polymorphism (SSCP) and denaturation gradient gel electrophoresis (DGGE)/temperature gradient gel electrophoresis (TGGE), 16S or 18S rDNA analysis and DNA microarray, has resulted in the identification of more novel strains from rhizospheric populations and appreciation of their genetic diversity [2].

Numerous species of soil bacteria flourish in the rhizosphere of plants, and this results in promotion of plant growth by a plethora of growth promotion mechanisms. These bacteria are generally termed plant growth promoting rhizobacteria (PGPR), coined by Kloepper in 1978. However, utilization of soil microorganisms to stimulate plant growth in agriculture has been known and studied since ancient times. Research on PGPR, especially on fluorescent *Pseudomonas*, *Bacillus* and many other diazotrophic bacteria, has been able to explain various mechanisms that can be grouped as follows: direct PGP mechanisms such as direct growth-promoting activities involving asymbiotic fixation of atmospheric nitrogen, solubilization of minerals such as phosphates, and production of plant growth regulators, for example auxins, gibberellins, cytokinin and ethylene [3–5], and indirect mechanisms such as the production of hydrogen cyanide, antibiotics,

Plant-Bacteria Interactions. Strategies and Techniques to Promote Plant Growth
Edited by Iqbal Ahmad, John Pichtel, and Shamsul Hayat
Copyright © 2008 WILEY-VCH Verlag GmbH & Co. KGaA, Weinheim
ISBN: 978-3-527-31901-5

siderophores, synthesis of cell wall lysing enzymes and competition with detrimental microorganisms for sites on plant roots [6–8]. In addition to the mechanisms described above, some rhizobacteria may promote plant growth indirectly by affecting symbiotic nitrogen fixation and nodulation [9]. 1-Aminocylopropane l-carboxylic deaminase (ACC deaminase) activity and quorum sensing (QS) in cell–cell communication of the expression of several rhizobacterial traits as well as bacteria–host interactions can play a significant role in the overall outcome of plant–bacteria interactions as discussed in Chapter 7 by Ahmad *et al.*, by inducing systemic resistance and improving plant tolerance to stress (drought, high salinity, metal toxicity and pesticide load) [10,11].

Among commonly known and studied groups of PGPR, diazotrophic bacteria occupy a unique position owing to their ability to fix nitrogen both symbiotically and asymbiotically. Symbiotic nitrogen fixation (legume–*Rhizobium* symbiosis) is the most widely studied area, and its contribution to global nitrogen fixation and crop productivity is well known. In asymbiotic nitrogen fixers, free-living associations as well as associative and endophytic relationships exist. The set of common asymbiotic diazotrophic PGPR for which evidence exists includes *Azotobacter*, *Azospirillum*, *Azoarcus*, *Burkholderia*, *Gluconoacetobacter diazotrophicus*, *Herbaspirillum* sp. and *Paenibacillus* (*Bacillus polymyxa* and *Bacillus* sp.) [8,9,12,13].

It is interesting that several free-living or associative PGPR have the ability to fix nitrogen, yet rarely does their mode of action for plant growth promotion derive credit from biological nitrogen fixation (BNF). This has led to investigation of other mechanisms in asymbiotic diazotrophic bacteria and their contribution to promoting plant growth in a number of agricultural crops.

The objectives of this chapter are initially to describe the diversity of rhizospheric diazotrophs and to assess the PGP potential of nonsymbiotic nitrogen fixers, primarily free-living diazotrophic rhizobacteria; the interaction with other microorganisms in relation to plant growth promotion and major constraints and future directions of PGPR research are briefly discussed.

5.2
Rhizosphere and Bacterial Diversity

Microbial diversity is an essential component of biological diversity and ecosystem conservation. Such diversity can be considered an invisible national resource of any country. Recent developments using modern technology in microbial diversity research indicate that the majority of naturally occurring microorganisms worldwide are as yet undiscovered and their ecological role is unknown. Soil is considered a storehouse of microbial activity, although the space occupied by living microorganisms is estimated to be less than 5% of the total soil volume. Therefore, major microbial activity is confined to 'hot spots', that is, aggregates with accumulated organic matter and within the rhizosphere [14].

Soil microbial communities are often difficult to characterize, mainly because of their immense phenotypic and genotypic diversity, heterogeneity and crypticity.

With respect to the latter, bacterial populations in upper layers of the soil can contain as many as 10^9 cells per gram of soil [15]. Most of these cells are unculturable. The fraction of the cells making up soil microbial biomass that have been cultured and studied in detail is negligible and often comprises less than 5% of the total population [16,17].

Stimulation of microbial growth around plant roots by the release of different organic compounds is known as the rhizospheric effect. The ability to secrete a vast array of compounds into the rhizosphere is one of the most remarkable metabolic features of plant roots, with nearly 5–21% of all photosynthetically fixed carbon being transferred to rhizosphere through root exudates [18]. The nature of root exudates is chemically diverse and can be grouped as low and high molecular weight compounds (Table 5.1).

The microbial population in and around the roots includes bacteria, fungi, yeasts and protozoa. Some are free living while others form symbiotic associations with various plants. Rhizosphere microbial populations could be regarded as a stable community around a particular plant species in a specific soil, or alternatively, as a succession of populations. The interaction between these microorganisms and the roots of the plant may be beneficial, harmful or neutral for the plant, and sometimes the effect of microorganisms may vary as a consequence of soil conditions [21].

Table 5.1 Compounds and enzymes identified in plant root exudates.[a]

Class of compounds	Type of compounds
Amino acids	Alanine, α-aminoadipic acid, γ-aminobutyric acid, arginine, asparagine, aspartic acid, cysteine, cystine, glutamic acid, glutamine, glycine, histidine, homoserine, isoleucine, leucine, lysine, methionine, ornithine, phenylalanine, proline, serine, therionine, tryptophan, tyrosine, valine
Organic acids	Acetic acid, aconitic acid, aldonic acid, butyric acid, citric acid, erythronic acid, formic acid, fumaric acid, glutaric acid, glycolic acid, lactic acid, malic acid, malonic acid, oxalic acid, piscidic acid, propionic acid, pyruvic acid, succinic acid, tartaric acid, tartronic acid, valeric acid
Sugars	Arabinose, deoxyribose, fructose, galactose, glucose, maltose, oligosaccharides, raffinose, rhamnose, ribose, sucrose, xylose
Vitamins	p-Aminobenzoic acid, biotin, choline, n-methylnicotinic acid, niacin, panthothenate, pyridoxine, riboflavin, thiamine
Purines/pyrimidines	Adenine, guanine, uridine, cytidine
Enzymes	Amylase, invertase, phosphatase, polygalactouronase, proteases
Inorganic ions/gaseous molecules	HCO_3^-, OH^-, H^+, CO_2, H_2
Miscellaneous	Auxins, flavonones, glycosides, saponin, scopoletin

[a]Partially adopted from Refs [19,20].

There are various approaches to studying microbial diversity, which can be broadly divided into (i) cultivation-based methods and (ii) cultivation-independent methods. Both approaches have their own unique limitations and advantages. Traditional methods to study microbial diversity have been based on cultivation and isolation [22]. For this purpose, a wide variety of culture media have been designed to maximize the variety and populations of microorganisms. A Biolog-based method applied for directly analyzing the potential activity of soil microbial communities, called community-level physiological profiling (CLPP), was used to study microbial diversity [23].

Molecular technology has helped to better understand microbial diversity. These molecular techniques include polymerase chain reaction (PCR) or real-time polymerase chain reaction (RT-PCR), which is used to target the specific DNA or RNA in soil. The 16S or 18S ribosomal RNA (rRNA) or their genes (rDNA) represent useful markers for prokaryotes and eukaryotes, respectively. PCR products generated using primers based on conserved regions of the 16S or 18S rDNA from soil DNA or RNA yield a mixture of DNA fragments representing all PCR-accessible species present in the soil. The mixed PCR products can be used for (a) preparing clone libraries [16,24] and (b) a range of microbial community fingerprinting techniques. Such clone libraries are useful to identify and characterize the dominant bacterial or fungal types in soil and thereby provide a picture of diversity [2]. Moreover, a range of other techniques have been developed to fingerprint soil microbial communities, for example, DGGE/TGGE [25,26], amplified rDNA restriction analysis (ARDRA) [27], T-RFLP [28], SSCP [29] and ribosomal intergenic spacer length polymorphism (RISA) [30].

5.2.1
Diazotrophic Bacteria

Free-living prokaryotes with the ability to fix atmospheric dinitrogen (diazotrophs) are ubiquitous in soil. The capacity for nitrogen fixation is widespread among Bacteria and Archaea. The great diversity of diazotrophs also extends to their physiological characteristics, as nitrogen fixation is performed by chemotrophs and phototrophs and by autotrophs as well as heterotrophs [31]. In natural ecosystems, biological nitrogen fixation (by free-living, associated and symbiotic diazotrophs) is the most important source of nitrogen [32]. The estimated contribution of free-living nitrogen-fixing prokaryotes to the nitrogen input of soils ranges from 0 to 60 kg $ha^{-1} year^{-1}$ [32]. The ability of free-living diazotrophs to take advantage of their capacity to perform nitrogen fixation depends on a number of conditions that vary for each organism, such as the availabilities of carbon and nitrogen and oxygen partial pressures [31]. Because of the direct link of diazotroph populations to the carbon/nitrogen balance of a soil and their high diversity associated with different physiological properties, they are of interest as potential bioindicators for the nitrogen status of soils. Reliable tools for the description of diazotroph communities would contribute greatly to our understanding of the role diazotrophs play in the soil nitrogen cycle. Environmental variables that can influence diazotrophy, including

Table 5.2 Diversity of diazotrophs.

Group of bacteria		Example	Nature
Cyanobacteria		*Anabaena*	Free living
		Nostoc	Free living
Actinobacteria		*Frankia*	Symbiotic
Gram-positive bacteria		*Bacillus*	Facultative microaerophilic
		Paenibacillus	Facultative microaerophilic
		Clostridium	Anaerobic
Proteobacteria	α	*Acetobacter*	Associative nitrogen fixer (endophytic)
		Azospirillum	Microaerophilic, asymbiotic
		Beijerinkia	Asymbiotic
		Bradyrhizobium	Symbiotic
		Rhizobium	Symbiotic
	β	*Azocarus*	Aerobic/microaerophilic
		Burkholderia	Associative nitrogen fixer (endophytic)
		Herbaspirillum	Associative nitrogen fixer (endophytic)
	γ	*Azotobacter*	Asymbiotic, free living

host primary production and root exudation and edaphic physicochemical parameters [33], have also been intensively studied, as have the diazotrophic organisms themselves. Various diazotrophic bacteria are listed in Table 5.2.

Due to the physiological diversity of diazotrophs and the documented unculturability of many prokaryotes [34,35], cultivation-based strategies have severe limitations for the description of the diversity of free-living soil diazotrophs. Therefore, molecular approaches have been developed as discussed above. These molecular approaches to study the diversity of diazotrophic organisms are primarily based on PCR amplification of a marker gene (*nifH*) for nitrogen fixation.

5.2.1.1 Symbiotic Diazotrophic Bacteria

Two groups of nitrogen-fixing bacteria, that is rhizobia and *Frankia*, have been studied extensively. *Frankia* forms root nodules on species of *Alnus* and *Casuarina*. Symbiotic nitrogen-fixing rhizobia are now classified into 36 species distributed among seven genera (*Allorhizobium, Azorhizobium, Bradyrhizobium, Mesorhizobium, Methylobacterium, Rhizobium* and *Sinorhizobium*) [36]. Legume–rhizobia symbiosis and nitrogen fixation are not the focal points of this chapter, however. Rhizobia have been widely studied and their contribution to sustainable crop production is well acknowledged. Other dimensions of rhizobial research include their application in rice plants, as discussed in Chapter 11. However, recent trends also indicate that *Rhizobium* as free-living rhizospheric bacteria can promote plant growth even in nonlegume (maize, lettuce and pine) crops by their PGP activities. It can also produce indole acetic acid (IAA) and siderophores and can solubilize phosphate [37,38].

5.2.1.2 Asymbiotic Diazotrophic Bacteria

Nonsymbiotic nitrogen fixation is known to be of great agronomic significance. The main limitation to nonsymbiotic nitrogen fixation is the availability of carbon and other energy sources for the energy-intensive nitrogen fixation process. This limitation can be compensated by moving closer to or inside the plants, namely in diazotrophs present in the rhizosphere or rhizoplane, or those growing endophytically. Some important nonsymbiotic nitrogen-fixing bacteria include *Achromobacter, Acetobacter, Alcaligenes, Arthrobacter, Azospirillum, Azotobacter, Azomonas, Bacillus, Beijerinckia, Clostridium, Corynebacterium, Derxia, Enterobacter, Herbaspirillum, Klebsiella, Pseudomonas, Rhodospirillum, Rhodopseudomonas* and *Xanthobacter* [39].

Various diazotrophic bacteria including species of *Azospirillum, Azotobacter, Bacillus, Beijerinckia* and *Clostridium* have been commonly associated with higher plants. The widespread distribution of dinitrogen-fixing associative symbiosis has led to interest in determining their relative importance in agricultural systems [40].

In natural ecosystems, biological nitrogen fixation (by free-living associative and symbiotic diazotrophs) is the most important source of nitrogen. The estimated contribution of free-living nitrogen-fixing prokaryotes to the nitrogen input of soil ranges from 0 to 60 kg ha^{-1} year^{-1}. The contribution of asymbiotic and symbiotic nitrogen fixation varies greatly but in some terrestrial ecosystems asymbiotic nitrogen fixation may be the dominant source [32,41].

Several new nitrogen-fixing bacteria associated with grasses and cereals, including sugarcane, have been described by many workers, namely *Pseudomonas* sp. [42], *Enterobacter, Klebsiella, Pseudomonas* sp., *Azospirillum* [43], *Campylobacter* sp., *Bacillus azotofixans* [44] and *Herbaspirillum seropedicae* [45].

Azotobacter The family Azotobacteraceae comprises two genera, namely *Azomonas* (noncyst forming) with three species (*A. agilis, A. insignis* and *A. macrocytogenes*) and *Azotobacter* (cyst forming) comprising six species [46], namely *A. chroococcum, A. vinelandii, A. beijerinckii, A. nigricans, A. armeniacus* and *A. paspali*. *Azotobacter* is generally regarded as a free-living aerobic nitrogen fixer. The genus *Azotobacter* comprises large Gram-negative bacteria, obligatory aerobic rods to oval shape, capable of fixing nitrogen nonsymbiotically. Phylogenetically, it is identified as β-proteobacteria. *Azotobacter* can form resting structures called cysts that are resistant to desiccation, mechanical disintegration and ultraviolet and ionizing radiation [47].

Application of *Azotobacter* and *Azospirillum* has been reported to improve yields of both annual and perennial grasses. Saikia and Bezbaruah [48] reported increased seed germination of *Cicer arietinum, Phaseolus mungo, Vigna catjung* and *Zea mays*; however, yield improvement is attributed more to the ability of *Azotobacter* to produce plant growth promoting substances such as phytohormone IAA and siderophore azotobactin, rather than to diazotrophic activity.

Azospirillum Members of the genus *Azospirillum* fix nitrogen under microaerophilic conditions and are frequently associated with the roots and rhizospheres of a large number of agriculturally important crops and cereals. These bacteria are helically curved rods and are motile by means of polar flagella, usually tufts at both poles. These are phylogenetically listed as α-proteobacteria. They occur as micro-aerophilic rods associated with plants. *Azospirillum* is able to enhance plant growth and yields in a wide range of economically important crops in different soils and climatic regions. Plant beneficial effects of *Azospirillum* have mainly been attributed to the production of phytohormones, nitrate reduction and nitrogen fixation, which have been the subject of extensive research [49,50]. Due to their frequent occurrence in the rhizosphere, these are known as associative diazotrophs. This organism came into focus with the work of Dobereiner and associates from Brazil [51], followed closely by reports from India [52,53].

Despite their nitrogen-fixing capability ($1–10 \, kg \, N \, ha^{-1}$), the increase in yield is mainly attributed to improved root development due to the production of growth-promoting substances and consequently increased rates of water and mineral uptake [54]. Azospirilla proliferate in the rhizosphere of numerous plant species and the genus *Azospirillum* now contains seven species: *A. brasilense*, *A. lipoferum*, *A. amazonense*, *A. halopraeferens*, *A. irakense*, *A. dobereinerae* and *A. largimobile*. An understanding of the mechanism of osmoadaptation in *Azospirillum* sp. can contribute toward the long-term goal of improving plant–microbe interactions for salinity-affected fields and crop productivity. The synthesis and activity of nitrogenases in *A. brasilense* are inhibited by salinity stress. Tripathi *et al.* [55,56] reported accumulation of compatible solutes such as glutamate, proline, glycine betaine and trehalose in response to salinity/osmolarity in *Azospirillum* sp. Usually, proline plays a major role in osmoadaptation through increase in osmotic stress that shifts the dominant osmolyte from glutamate to proline in *A. brasilense*. Saleena *et al.* [57] have studied the diversity of indigenous *Azospirillum* sp. associated with rice cultivated along the coastline of Tamil Nadu. On the basis of mutational studies of *Azospirillum*, some workers suggested a role of PHB synthesis and accumulation in enduring various stresses, namely UV irradiation, heat, osmotic pressure, osmotic shock and desiccation [58].

Acetobacter *Acetobacter diazotrophicus* (family Acetobacteraceae), isolated from roots and stems of sugarcane, was first reported as a nitrogen-fixing bacterium from Brazil [59] and subsequently from Australia [60], India [61–63], Mexico [64] and Uruguay [65]. The bacterium is an endophytic diazotroph, previously known as *Acetobacter diazotrophicus* and now known as *Gluconoacetobacter diazotrophicus*. This species is able to fix nitrogen and transfer fixed nitrogen to the host plant with much greater efficiency than diazotrophs occurring in the plant rhizosphere. The acetic acid bacteria comprise a group of Gram-negative, aerobic, motile rods that carry out incomplete oxidation of alcohols and sugars leading to the accumulation of organic acids as end products. This bacterium has very interesting phenotypes such as the ability to (i) fix nitrogen inside sugarcane, (ii) survive in very acidic conditions, (iii) grow on 30% sucrose, (iv) produce significant amounts of plant growth hormones in culture, (v) solubilize phosphate and (vi) enhance growth of sugarcane in the presence of nitrogen fertilizer. With ethanol as a substrate, acetic acid is produced, so the bacterium is named acetic acid bacteria. It is phylogenetically described as α-proteobacteria. *Gluconoacetobacter diazotrophicus* is isolated from sugarcane (*Saccharum officinarum*) roots and stems and endophytically fixes nitrogen.

The family Acetobacteraceae includes the genera *Acetobacter, Gluconobacter, Gluconoacetobacter* and *Acidomonas*. Based on 16S rRNA sequence analysis, the name *Acetobacter diazotrophicus* has been changed to *Gluconoacetobacter diazotrophicus* [66]. In addition to *G. diazotrophicus*, two more diazotrophs, *G. johannae* and *G. azotocaptans* have been included in the list [64]. The genetic diversity of *G. diazotrophicus* isolated from various sources does not exhibit much variation [67]. However, Suman *et al.* [68] found that the diversity of the isolates of *G. diazotrophicus* by RAPD analysis was more conspicuous than that reported on the basis of morphological and biochemical characters. On the basis of DNA fingerprinting studies, the existence of genetically distinct *G. diazotrophicus* strains in sugarcane cultivars has been reported from Louisiana.

Azoarcus *Azoarcus* gen. nov., an aerobic/microaerophilic nitrogen-fixing bacterium, was isolated from surface-sterilized tissues of Kallar grass (*Leptochloa fusca* (L.) Kunth) [69] and can infect roots of rice plants as well. Kallar grass is a salt-tolerant grass used as a pioneer plant in Pakistan on salt-affected low-fertility soils. Repeated isolation of one group of diazotrophic rods [70] from Kallar grass roots and the results of polyphasic taxonomy led to the identification of genus *Azoarcus*, with two species, *A. indigens* and *A. communis,* and three additional unnamed groups, which were distinct at species level. Nitrogen fixation by *Azoarcus* is extremely efficient (specific nitrogenase activity, one order of magnitude higher than that found for bacteroids). Such hyperinduced cells contain tubular arrays of internal membrane stacks that can cover a large proportion of the intercellular volume. These structures are considered vital for high-efficiency nitrogen fixation [69].

Burkholderia *Burkholderia* is identified as β-proteobacteria. Some strains have the ability to fix atmospheric nitrogen. Presently, the genus *Burkholderia* includes 30 species with valid published names [71], with *Burkholderia cepacia* as the typical species. For a long time, nitrogen-fixing ability in bacteria of the genus *Burkholderia* was recognized only in the species *Burkholderia vietnamiensis* [72]. Recently, two nodulating strains recovered from legume plants were assigned to the genus *Burkholderia* as two novel species, *Burkholderia tuberum* and *Burkholderia phymatum* [73].

Bacillus *Bacillus* exists as Gram-positive rods. They are endospore formers that can allow them to survive for extended periods under stressed environmental conditions. *Bacillus* is considered a well-established PGPR [74]. Some *Bacillus* species can fix nitrogen [75]. Many bacilli can produce antibiotics including bacitracin, polymyxin, tyrocidin, gramicidin and circulin. In most cases, antibiotics are released during sporulation, when the culture enters the stationary phase of growth and after it is committed to sporulation [76].

Paenibacillus These are Gram-positive aerobic or facultative short rods, which produce endospores. *Paenibacillus* can produce phytohormones, suppress phytopathogens through antagonistic functions and solubilize organic phosphate [77–80]. Some *Paenibacillus* species have been identified as nitrogen fixers; however, little information is available as to whether they may be termed PGPR [81,82].

5.3
Asymbiotic Nitrogen Fixation and Its Significance to Plant Growth

The first associative diazotroph was reported by Beijerinck in 1925 under the name *Spirillum lipoferum*. However, it was only about a half-century later, after the discovery of the highly specific *Azotobacter paspali–Paspalum notatum* association and the rediscovery of *Spirillum lipoferum* (now called *Azospirillum*) by the group of Döbereiner [83], that scientists became increasingly interested in diazotrophic bacteria associated with graminaceous plants. Because the benefit of nitrogen fixation from nodulated legumes to agriculture was already established at that time, it was expected that the associative diazotrophs would favor nonleguminous plants in the same way. Several genera of bacteria have now been reported to contain diazotrophs, which may be loosely or more intimately (i.e. endophytes) associated with plants, including *Acetobacter*, *Azoarcus*, *Azospirillum*, *Azotobacter*, *Beijerinckia*, *Burkholderia*, *Enterobacter*, *Herbaspirillum*, *Klebsiella*, *Paenibacillus* and *Pseudomonas*. An extensive phylogenetic classification of nitrogen-fixing organisms was made by Young [84]. While the capability of these organisms to fix nitrogen *in vitro* can readily be demonstrated, efforts to quantify nitrogen fixation in natural associations with plants have produced widely varying results. In many crop inoculation studies, coupled to acetylene reduction measurements, nitrogen balance and ^{15}N isotope dilution experiments have been conducted with root-associated bacteria to determine whether the bacteria supply significant amounts of nitrogen to cultivated plants [85,86].

The most useful methods for examining nitrogen fixation in the field and in large greenhouse experiments are still the ^{15}N isotope dilution and ^{15}N natural abundance techniques [86]. Using these methods, it was reported that certain Brazilian sugarcane varieties can derive 50–80% of plant nitrogen from BNF, equivalent to 150–170 kg N ha^{-1} year^{-1} [87]. However, the amount of nitrogen fixed is highly variable and dependent on plant genotype and environmental conditions [88]. Studies conducted at the International Rice Research Institute in the Philippines suggest that on the whole 20–25% of the total nitrogen needs of rice can be derived from associative fixation [89,90]. Using the ^{15}N isotope dilution technique, it was estimated that Kallar grass may fix up to 26% of its nitrogen content.

One study of nitrogen fixation with maize suggested that some cultivars fix up to 60% of their nitrogen after inoculation with appropriate strains of *Azospirillum* [91] while other cultivars showed decreased grain yield and plant nitrogen accumulation [92]. On the whole, greenhouse studies with maize, sorghum and Setaria did not show substantial nitrogen fixation in *Azospirillum*-inoculated plants [93]. In the case of wetland rice, interpretation is even more difficult because a proportion of this nitrogen may be derived from free-living nitrogen-fixing cyanobacteria in floodwater or heterotrophic nitrogen fixers in the soil [94]. To provide direct evidence that the plant benefits from the nitrogen fixed by the assumed diazotroph, plant inoculation experiments with nonnitrogen-fixing (Nif⁻) mutants as negative controls are required, coupled with careful ^{15}N-based balance studies. With the use of such mutants in inoculation experiments, it becomes clear that, in most cases, BNF is not involved in plant growth promotion. Nif⁻ mutants of *Azospirillum*, *Azoarcus* sp. strain BH72 or *Pseudomonas putida* GR12-2 have been shown to still be capable of stimulating plant growth [95,96]. No ^{15}N isotope dilution or nitrogen balance experiments have been carried out with these *Nif⁻* mutants. The fact that BNF is apparently not involved in plant growth promotion by these strains cannot be simply attributed to the absence of nitrogenase expression. Using a translational *nifH–gusA* fusion, it was observed that *Azospirillum nif* genes are expressed during the association with wheat roots [97]. However, some host specificity of BNF has been reported; for example, when the *nifK* mutant of *Azoarcus* sp. strain BH72, which has a Nif⁻ phenotype, was used to inoculate rice seedlings in a gnotobiotic system, the same increase in plant biomass and total protein content was found as after inoculation with the wild-type strain, strongly suggesting that nitrogen fixation was not involved in the observed plant growth promotion [96]. Nevertheless, immunogold labeling as well as reporter gene studies revealed high nitrogenase gene expression levels of the endophyte *Azoarcus* sp. BH72 inside roots of rice seedlings, suggesting that environmental conditions inside rice roots will allow endophytic nitrogen fixation in bacterial microcolonies in the aerenchyma [98]. However, when this Nif⁻ mutant was inoculated onto Kallar grass plantlets, these plants showed significantly lower dry weight and accumulated less nitrogen than those inoculated with the wild-type strain *Azoarcus* sp. BH72 [99,100]. The associative symbiosis (using *Azospirillum*) was observed in paranodules of maize and nonnodulated maize plants. An increase in nitrogenase activity (acetylene reduction assay) and leghemoglobin content was observed in plants treated with *Azospirillum* [101]. The five bacterial isolates (two of *Stenotrophomonas maltophilia*, two of *Bacillus fusiformis* and one of *Pseudomonas fluorescens*) were showing nitrogenase activity above $150 \, \mathrm{nmol \, h^{-1} \, mg^{-1}}$ protein [102]. Three bacterial species of *Bacillus* could fix nitrogen and significantly increased the growth of barley seedlings [103].

5.4
Plant Growth Promoting Mechanisms of Diazotrophic PGPR

The plant growth promoting rhizobacteria may promote plant growth either directly or indirectly. Direct mechanisms include (i) the ability to produce plant growth regulators (indole acetic acid, gibberellins (GAs), cytokinins (CTKs) and ethylene) [91,104,105] and (ii) solubilization of mineral nutrients such as phosphates [106–109]. Indirect mechanisms involve (i) antagonism against phytopathogens [110,111], (ii) production of siderophores [110,112], (iii) production of extracellular cell wall degrading enzymes for phytopathogens, for example β-1,3-glucanase [113]

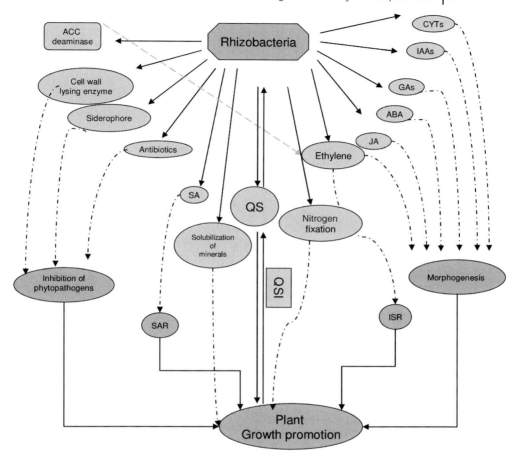

Figure 5.1 Various mechanisms involved in plant growth promotion by rhizobacteria. Some bacterial species produce phytohormones (blue ovals) such as CTKs, IAA and GAs. Indirect mechanisms concerned with inhibition of growth of phytopathogens include the production of antibiotics, siderophores and cell wall lysing enzymes. Salicylic acid, jasmonic acid (JA or analogues) and ethylene increase plant immunity by activating defense programs. ACC deaminase reduces growth-retarding ethylene production. Abbreviations: ABA, abscisic acid; ISR, induced systemic resistance; SAR, systemic acquired resistance, QSI, quorum-sensing interference.

and chitinase [114], (iv) antibiotic production [115] and (v) cyanide production [116]. The overall influence and interaction of multiple PGP traits have possible positive effects on plant growth promotion as indicated in Figure 5.1.

Diazotrophic bacteria, by their ability to convert nitrogen into ammonia that can be used by the plant, also belong to the PGPR. Because of their competitive advantages in a carbon-rich, nitrogen-poor environment, diazotrophs may become selectively enriched in the rhizosphere.

It is now clear that associative diazotrophs exert their positive effects on plant growth through different mechanisms. Apart from fixing nitrogen, diazotrophs can affect plant growth directly by the synthesis of phytohormones [13,49] and vitamins [47], inhibition of plant ethylene synthesis [3], improved nutrient uptake (microbial cooperation in rhizosphere), enhanced stress resistance [50], solubilization of inorganic phosphate and mineralization of organic phosphate [8]. However, diazotrophs are able to decrease or prevent the deleterious effects of plant pathogens mostly through the synthesis of antibiotic and fungicidal compounds [117,77], through competition for nutrients (for instance, by siderophore production) or by the induction of systematic resistance to pathogens [80]. Some of the well-known mechanisms of PGP by diazotrophic bacteria are listed in Table 5.3.

Table 5.3 Mechanisms of plant growth promotion by diazotrophic bacteria.

Mechanisms	Organisms	Effect on plant growth	References
Production of plant growth promoting substances			
Auxins	Azotobacter Azospirillum Acetobacter diazotrophicus Herbaspirillum Bacillus Paenibacillus Burkholderia Bradyrhizobium Rhizobium	Increased root length, number of lateral roots and number of roots. Significantly increased seedling (root and shoot weight)	[78,103,118–126]
Gibberellins	Azotobacter Azospirillum Acetobacter diazotrophicus Herbaspirillum Bacillus Rhizobium	Increased shoot growth of dwarf plants of maize and rice. Increased shoot growth of alder seedlings	[120,126–133]
Cytokinins	Azotobacter Azospirillum Paenibacillus Rhizobium	Affect morphology of radish and maize	[127–129,134–138]
Phosphate solubilization	Bacillus Paenibacillus	Enhanced growth and yield but not phosphorus solubilization of canola Increased percent germination and growth emergence of wheat	[37,103,109,121,123,139]
	Azotobacter Rhizobium	Increased dry matter and yield but not phosphorus uptake of lettuce and barley	

Table 5.3 (*Continued*)

Mechanisms	Organisms	Effect on plant growth	References
Augmented nutrient uptake	*Azospirillum* spp.	Pectinolytic activity may increase mineral uptake by the hydrolysis of middle lamellae of roots, also caused enhanced uptake of IAA by roots of wheat and maize Bacterial nitrate reductase increased reduced nitrogen in roots and total and dry weight in wheat	[45,140–143]
Biocontrol			
Siderophore	*Azospirillum* *Azotobacter* *Rhizobium*	Reduces iron availability, thus making it unavailable to phytopathogens. They may act	[112,144–147]
Cell wall lysing enzyme	*Paenibacillus*	as an important source of iron for higher plants in alkaline and calcareous soil Chitinase and antifungal activity Cellulase and mannanase	[77,148]
Antibiotics	*Azotobacter*	Antifungal compound – inhibits the production of conidia of fungus (*Botrytis cinerea*)	[117,149–151]
	Azospirillum *Paenibacillus* *Bacillus*	Bacteriocins Polymyxin – active against bacteria and fungus Coproduction of antifungal (surfactrin and iturin like) compounds	

5.5
Interaction of Diazotrophic PGPR with Other Microorganisms

The colonization of roots by inoculated bacteria is an important step in the interaction between beneficial bacteria and the host plant. However, it is a complex phenomenon influenced by many biotic and abiotic parameters, some of which are now apparent. In order to successfully utilize PGPR in agriculture, it is important to understand the mechanisms that enable them to colonize the rhizosphere and the factors that lead to the stimulation of their beneficial effects. It is reasonable to assume that PGPR must colonize the rhizosphere of the host plant to be most beneficial [152,153]. Root colonization is a complex phenomenon under the influence of many parameters (Figure 5.2). Various techniques that may be classified as classical, immunological and molecular are applied for monitoring the inoculant strains in the rhizosphere in relation to survival and colonization in the rhizosphere.

5.5.1
Interaction of Diazotrophic PGPR with Rhizobia

Symbiotic nitrogen fixation in legumes is accomplished by rhizobia occurring within root nodules. This process is dependent on the efficiency of the *Rhizobium* strain

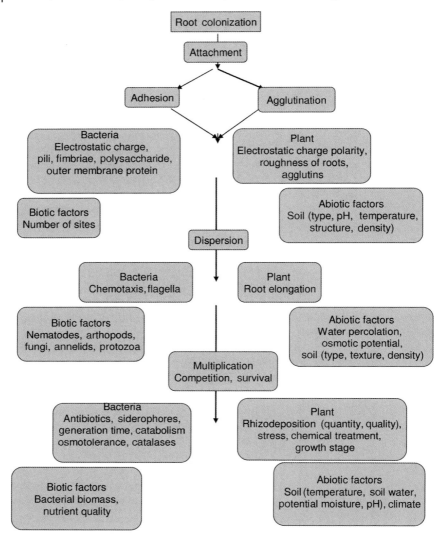

Figure 5.2 Factors of root colonization. Main steps of root colonization. Factors involved in every step of root colonization by
bacteria. Plant, biotic factors and abiotic factors.

involved and its competitiveness for nodulation against indigenous rhizobia. Symbiotic nitrogen fixation is also influenced by environmental factors. Increasing
symbiotic nitrogen fixation is rational, since leguminous crops are an important
source of protein and environmentally safe, avoiding the use of nitrogen fertilizers.
Rhizobial strain selection and legume breeding are conventional approaches to
improve this process and, more recently, molecular approaches have demonstrated

their potential. The exploitation of PGPR in combination with *Rhizobium* also constitutes an interesting alternative to improve nitrogen fixation. In addition to exploiting their individual plant growth promoting capacity, the potential of selected diazotrophs can be improved further through dual inoculation with other micro-organisms for additive and/or synergistic effects. Bacterial diazotrophs that are able to colonize the root zones of leguminous plants, for instance, could stimulate the performance of a leguminous species by affecting symbiotic nitrogen fixation. Combined inoculation of *Rhizobium* with *Azospirillum* or with *Azotobacter* has been demonstrated to increase dry matter production, grain yield and nitrogen content of several legumes when compared with inoculation with *Rhizobium* alone [154–157]. These positive results of dually inoculated legumes have been attributed to early nodulation, increased number of nodules, higher nitrogen fixation rates and a general improvement of root development [158,159]. The greater number of active nodules can be expected to contribute fixed nitrogen for higher yields under field conditions. However, concomitant application of *Azospirillum* and *Rhizobium* did not always result in promotion of nodulation and under some circumstances even inhibited the ability of the *Rhizobium* to nodulate its host. Stimulation or inhibition was found to be dependent on bacterial concentration and timing of inoculation [160,161]. Mixed inoculations of *Vicia faba* L. with four different *Rhizobium/Azospirillum* and *Rhizobium/Azotobacter* combinations led to changes in total concentration and/or distribution of mineral macro- and micronutrients when compared with plants that had been inoculated with *Rhizobium* alone [162].

As most of the diazotrophs have been shown to produce phytohormones, the stimulation of nodulation may occur as a result of a direct response of the plant root to these compounds. Similar to what was observed in several grasses and cereals [163], inoculation with *A. brasilense* was found to promote root hair formation of bean and alfalfa [164,165]. As *Rhizobium* infection takes place by the formation of infection threads in root hairs, the stimulation of a greater number of epidermal cells to differentiate into root hair cells capable of being infected may increase the probability of infection by *Rhizobium*, thereby increasing root potential for nodule initiation [161]. Apart from their direct effect on root morphology, phytohormones may also influence the nodulation process itself [161,166]. Experiments carried out in a hydroponic system showed that inoculation with *A. brasilense* increased the secretion of flavonoids by seedling roots of common bean [165].

Coinoculation of PGPR and *Bradyrhizobium* in sterile soil increased the nodule occupancy in green gram [167]. In contrast, when *Rhizobium* and *Azotobacter chroococcum* were coinoculated, there was no significant increase in plant biomass, nodulation and yield of chickpea [168]. Soybean plants showed increased weight when they were coinoculated with *Bacillus* isolates and *Bradyrhizobium japonicum* in nitrogen-free conditions as compared with the plants inoculated with *Bradyrhizobium* alone [169]. *Bacillus* isolates were further coinoculated with *Bradyrhizobium japonicum* to assess the improvement in nodulation [170]. *Azospirillum brasilense* and *Azotobacter chroococcum* with *Rhizobium* inoculation resulted in increased nitrogen fixation of fresh nodule density/plant, nodule dry weight and shoot nitrogen content of pigeon pea [171].

5.5.2
Interaction of Diazotrophic PGPR with Arbuscular Mycorrhizae

The major groups of microbial plant mutualistic symbionts are the fungi that establish a mycorrhizal symbiosis with the roots of most plant species. The soil-borne mycorrhizal fungi colonize the root cortex biotrophically, and then develop a mycelium that is a bridge connecting the root to the surrounding soil microhabitats. Arbuscular mycorrhizae are also known as biofertilizers, bioprotectants and biodegraders [109]. Mycorrhizal symbiosis can be found in all ecosystems and improves plant fitness and soil quality through key ecological processes such as phytoremediation. Their potential role in phytoremediation of heavy metal contaminated soils is becoming evident. There is a need, however, to enhance phytoremediation as a viable strategy; for example, fast-growing plants with high metal uptake ability and rapid biomass gain are needed. Most of the major plant families form associations with the most common mycorrhizal type [172]. The AM fungi are obligate microbial symbionts, unable to complete their life cycles without colonizing a host plant.

Interactions of AM fungi with other soil organisms have been described with reference to their effect on mycorrhizal development and functioning. Rhizobacteria showing a beneficial effect on mycorrhizae are often referred as 'mycorrhizae helper microorganisms'. PGPR sometimes enhance plant growth indirectly by stimulating the relationships between the host plant and AM fungi. By themselves, AM fungi are well known to enhance the uptake of various soil nutrients, especially phosphorus [18,173]. Studies suggest that coinoculation with some PGPR can enhance the relationship between plant and fungal symbionts. Tripartite relationships among PGPR, fungal symbionts (both AM and ectomycorrhizal fungi) and forest tree species are coming to the fore [174,175]. There are several reports on the interactions between AM fungi and *Rhizobium* species [176], which suggest that the interaction is synergistic; that is, AM fungi improved nodulation by means of enhanced phosphorus uptake by the plant. In addition to this principal effect of AM fungi on phosphorus-mediated nodulation, other secondary effects include supplying trace elements and plant hormones that play an important role in nodulation and nitrogen fixation [109].

The process of AM inoculation improves plant growth through increased uptake of phosphorus and other mineral nutrients, especially in soils of low fertility. Indeed, plants inoculated with *A. brasilense* and AM fungi and grown without fertilizer had nitrogen and phosphorus contents comparable to those of noninoculated plants supplemented with nitrogen and phosphorus fertilizers [177]. The diazotrophs may enhance mycorrhizal development by supplying vitamins to the rhizosphere, because mycorrhizal fungi have been shown to be dependent on or stimulated by certain vitamins [178]. Thus, inoculation with mycorrhizal fungi and vitamin-producing diazotrophs could result in improved plant growth. Petersen *et al.* [179] showed that *P. polymyxa* caused an increase in both early and final rhizobial root populations when coinoculated with *Rhizobium etli* on *Phaseolus vulgaris*, when compared with a single inoculation with *R. etli*. In contrast to the *in planta* results, population enhancements were not observed when *R. etli* and *P. polymyxa* were cocultured *in vitro* using minimal media in the absence of the seedling. The addition of seed exudates to the growth media also failed to stimulate the population increases. Mutants of *A. brasilense* and *R. leguminosarum* altered in the production of extracellular polysaccharides. Binaciotto *et al.* [180] showed the involvement of these polysaccharides in the attachment of these bacteria to the structures of AM fungi. In soil, an extensive network of AM fungi develops and PGPR are usually associated with fungal surfaces [181]. *Azotobacter chroococcum* and *Pseudomonas fluorescens* were attracted toward tomato roots colonized by *Glomus fasciculatum* compared to nonvesicular-arbuscular mycorrhizal tomato roots [182]. The presence of *G. clarum* decreased or did not significantly affect plant growth under the different culture conditions. The presence of AM fungi stimulated the nitrogen-fixing bacterial population of upland rice. Bacterial species had different effects, under both culture conditions, and some genera of nitrogen-fixing bacteria increased root and shoot growth at different plant growth stages. The level of mycorrhiza colonization had no influence on plant growth [183].

5.6
Other Dimensions of Plant Growth Promoting Activities

5.6.1
ACC Deaminase Activity

Ethylene is the only gaseous hormone produced by plants. It is also known as the 'wounding hormone' because its production in the plant can be induced by physical or chemical perturbation of plant tissues. Among its myriad effects on plant growth and development, ethylene production can inhibit root growth. In some cases, the growth promotion effects of ACC deaminase producing PGPR appear to be best expressed in stressful situations such as in heavy metal contaminated soils [184]. The enzyme ACC deaminase plays a key role in degrading ACC. The products of this hydrolysis, ammonia and α-ketobutyrate, can be used by the bacterium as a source of nitrogen and carbon for growth [185]. In this way, the bacterium acts as a sink for ACC and as such is lowering the ethylene level in plants, preventing some of the potentially deleterious consequences of high ethylene concentrations. In nature, ACC deaminase has been commonly found in soil bacteria that colonize plant roots [186]. Many of these microorganisms were identified by their ability to grow on minimal media containing ACC as its sole nitrogen source. In this way, *Azospirillum* spp., *Herbaspirillum* spp., *Azoarcus*, *Azorhizobium caulinodans*, *Gluconoacetobacter diazotrophicus*, *Burkholderia vietnamiensis*, *Azotobacter* spp., *Azorhizophilus* and *Pseudomonas* spp. were all found to be able to use ACC as the sole nitrogen source for growth. An example of such an ACC deaminase containing bacterium is the PGPR *Pseudomonas putida* GR12-2 [187] that stimulates root growth of a number of different plants (canola, lettuce and tomato) under gnotobiotic conditions [188–191]. Experiments with other (nondiazotrophic) bacteria show that PGPR expressing ACC deaminase activity can also decrease the deleterious effects of different environmental stresses such as heavy metals and flooding on plants, probably by reducing the concentration of plant stress ethylene [189,184,192]. Flooded tomato plants treated with *A. brasilense* containing the ACC deaminase structural gene (*acdS*) from *Enterobacter cloacae* UW4 showed lower levels of epinasty than plants treated with the untransformed wild-type strain [193]. Based on the proposed model for plant growth promotion by means of lowering plant ethylene levels, it is predicted that any rhizosphere bacterium that actively expresses ACC deaminase can promote the elongation of seedling roots; that is, it can act as a PGPR [194]. In this model, IAA synthesized by the PGPR is taken up by the plant and can stimulate cell proliferation and/or elongation or the activity of the enzyme ACC synthase. Plants inoculated with ACC deaminase bacteria or transgenic plants that express bacterial ACC deaminase genes can regulate their ethylene levels and consequently contribute to a more extensive root system [194]. The role of ACC deaminase production in plant growth promotion by free-living diazotrophs is less explored.

5.6.2
Induced Systemic Resistance (ISR)

Plant growth promoting rhizobacteria can suppress diseases through antagonism between the bacteria and soilborne pathogens, as well as by inducing a systemic resistance in the plant against both root and foliar pathogens. The generally non-specific character of induced resistance constitutes an increase in the level of basal resistance to several pathogens simultaneously, which is of benefit under natural conditions where multiple pathogens remain present. Specific PGPR such as *Pseudomonas* strains induce systemic resistance in carnation, radish and *Arabidopsis* (the O antigenic side chain of the bacterial outer membrane lipopolysaccharides acts as an inducing determinant), tobacco and *Arabidopsis* (pseudobactin siderophores), radish (pseudomanine siderophore) are correlated with salicylic acid (SA) production as reviewed by Van Loon and Bakker [11]. However, such a mechanism in diazotrophs is less frequently described.

5.6.3
Improved Stress Tolerance

Diazotrophic PGPR can improve a plant's tolerance to stresses such as drought, high salinity, metal toxicity and pesticide load [50]. Sarig *et al.* [195] reported that sorghum plants inoculated with *Azospirillum* were less drought stressed, having more water in their foliage, higher leaf water potential and lower canopy temperature than non-inoculated plants. Total extraction of soil moisture by *Azospirillum*-inoculated plants was greater and water was extracted from deeper layers in the soil profile. Therefore, sorghum yield increase in inoculated plants was attributed primarily to improved utilization of soil moisture. Foliar application of a diazotrophic *Klebsiella* sp. could ameliorate drought stress effects on wetland rice, as grain yield increased, together with increased nutrient uptake and proline content [196]. Proline is an important osmoregulator, accumulated as a consequence of drought stress. Creus *et al.* [197] studied the effects of *A. brasilense* Sp245 inoculation on water relations in two wheat cultivars. They found that *Azospirillum* stimulated growth of wheat seedlings grown in darkness under osmotic stress, together with a significant decrease in osmotic potential and relative water content at zero turgor, in volumetric cell wall modulus of elasticity and in absolute symplastic water volume and a significant rise in apoplastic water fraction parameters. These are known physiological mechanisms of adaptation that give plants the ability to tolerate a restricted water supply [198]. As in this hydroponic test system no nutrients were present, therefore, the improved water status of the wheat seedlings cannot be attributed to enhanced mineral uptake and consequently growth promotion. Similarly, in a hydroponic system without nutrients, *A. brasilense* Sp245 was found to partially reverse the negative effects that drought stress had on wheat seedlings, as it was observed in the growth rate of coleoptiles [199]. Apart from alleviating osmotic stress in plants, inoculation with diazotrophs can also enhance oxidative stress tolerance. Oxidative stress is defined as the oxidative damage caused by reactive oxygen species (ROS) such as the superoxide anion radical, hydrogen peroxide, the hydroxyl radical and singlet oxygen [200,205]. These highly reactive oxygen species can be generated by the oxidative metabolism of normal cells and by different stress situations. Although ROS contribute to plant defense against pathogens, they are potentially harmful to plant viability [201]. With the production of antioxidant enzymes such as superoxide dismutase (SOD), peroxidase and catalase, the cell can neutralize and thus control free radical formation. Also, pigments such as carotenoids could be involved in scavenging singlet oxygen and thus decrease oxidative stress [202]. Inoculation with *Azotobacter chroococcum* was reported to improve oxidative stress defense ability in sugar beet leaves since inoculated plants showed increased activities of superoxide dismutase, peroxidase and catalase and increased chlorophyll and carotenoid content [203]. High activities of antioxidant enzymes (especially SOD) are linked with oxidative stress tolerance [204]. However, the observed effects have not been linked yet to certain traits of diazotrophic bacteria. Therefore, it is not clear whether this increase in oxidative stress tolerance is a direct result of inoculation or rather an indirect consequence of an overall increase in plant health because of inoculation with *Azotobacter*.

5.6.4
Quorum Sensing

It has been recognized that bacteria not only can behave as individual cells but under appropriate conditions, when their number reaches a critical level, can also modify their behavior to act as multicellular entities. This phenomenon is based on the dynamics of a natural ecosystem, since bacteria do not exist as solitary cells but are typically colonial organisms that live as consortia to exploit the elaborate system of intracellular communication that facilitates adaptations to changing environmental conditions. When the bacterial population reaches a threshold, the microbial cells begin to release small signaling molecule mediated sensing response pathways. This effect has been defined as quorum sensing [206]. These microbially derived signal molecules are placed into two main categories: (i) amino acids and short peptide pheromones commonly utilized by Gram-positive bacteria [207,208] and (ii) fatty acid derivatives such as acyl homoserine lactone (AHL) utilized by Gram-negative bacteria. Cellular processes regulated by QS in bacteria are diverse and includes genetic competence development. Quorum-sensing signals and identical two-component regulatory systems are used by plant-interacting bacteria (mutualistic or pathogenic associations) to coordinate, in a cell density dependent manner or in response to changing environmental conditions, the expression of important factors for host colonization and infection. The success of invasion and survival within the host also requires that rhizobia and pathogens suppress and/or overcome plant defense responses triggered after microbial recognition, a process in which surface polysaccharides, antioxidant systems, ethylene biosynthesis inhibitors and virulence genes are involved [209]. The role of AHL and AHL analogues was also reported in *Rhizobium*–legume symbiosis and *Pseudomonas fluorescens* [210,211]. Similarly, QS systems are widespread mechanisms of gene regulation in both pathogenic and plant associative bacteria, thus requiring in-depth investigation in modulating various PGP traits and plant–bacteria interactions.

5.7
Critical Gaps in PGPR Research and Future Directions

The inoculated PGPR may release various secondary metabolites as plant growth promoting substances. The bioproduction of these substances in contact with roots is most likely subject to direct uptake by plant roots before being catabolized by soil microbes or being immobilized in soil. It has been demonstrated that these microbially derived plant growth promoting substances can promote plant growth and development. Therefore, there is a need to provide evidence and their role. This can be explored by monitoring the synthesis of PGP substances in the rhizosphere by developing analytical techniques for the separation and detection of PGP substances such as plant growth regulator in the soil and screening of microbes for the production of PGP substances in the absence or presence of a precursor.

In vitro activities exhibited by various PGPR may not give the expected results under field conditions. The failure of PGPR to produce the desired effects after seed/seedling inoculation is frequently associated with their inability to colonize plant roots. The process of root colonization is complex. Several traits associated with the survivability, tolerance, competence with indigenous rhizospheric microorganisms, expression of root colonizing traits and so on are important [49]. In many agroclimatic situations such as harsh climates, population pressure, land constraints and decline of traditional soil management practices, reduced soil fertility often exists. Therefore, considering the varied agroclimatic conditions, continuous research is needed to develop region-specific bioinoculants with *rhizospheric competence* and

PGP traits. One possible approach may involve isolation of PGPR from the indigenous soil–plant system and use them in the same agroclimatic conditions. The selection of PGPR may be based on number and types of PGP traits present. Therefore, potential PGPR adapted to particular soil and plant soil conditions and harboring multiple PGP traits should be selected and evaluated in field conditions. In order to determine the successful establishment of a PGPR in field conditions, its identity and activities must be continuously monitored (Table 5.4). Considering the cost of molecular techniques, more simple and reliable methods must be developed for rapid detection and monitoring of inoculant strains in the rhizosphere.

The extensive research data generated on plant growth promoting rhizobacteria clearly indicate that the plant–bacteria interaction leads to rhizosphere colonization and its influence on plant health is a complex process. Various mechanisms of plant growth promotion, both direct and indirect, by diazotrophic bacteria and their interactive effects have been investigated. However, evidence for the contribution of individual mechanisms of plant growth promotion is less prevalent. What is urgently needed in this direction is listed in the following:

(i) Develop more productive analytical and bioassay-based techniques for the identification and uptake of microbial products/nutrients by plant roots.

Table 5.4 Techniques used for the detection and quantification of inoculated PGPR.

Techniques	References
Classical and immunological	
• Selective media containing the appropriate toxic substances (antibiotics, heavy metals and herbicides)	[214,222]
• Immuno fluorescence colony staining approach (enumeration of colonies marked with antibodies conjugated with fluorescein isothiocyanate)	
• Immunomagnetic attraction (enumeration of bacteria captured with a supermagnet)	[212]
Molecular	
• Specific rRNA probes, coupled with PCR [using probes labeled with a fluorochrome (hybridization *in situ* coupled with confocal laser microscopy)	[213]
• 16S rRNA probe (dot blot hybridization of a directly isolated nucleic acid mixture)	
• Marker genes (quantified by colorimetry): *lacZ* (β-galactosidase, blue colonies), *gusA* (β-glucuronidase, indigo colonies), *xylE* (catechol 2,3-dioxygenase, yellow colonies), *tfdA* (2,4-dichlorophenoxyacetate, red colonies)	[214–216]
• *Lux* (luciferase) (bioluminescence quantified by charge-coupled device cameras or visualized *in planta*), GFP (green fluorescent protein detected *in situ* with confocal laser microscopy or epifluorescence microscopy), *inaZ* (ice-nucleation protein quantified by freezing assay)	[23,217–221]

(ii) The PGP traits exhibited by inoculant strains and their expression under plant rhizosphere influence need to be examined.

(iii) Physiological and molecular mechanisms in regulating PGP traits of the inoculant strain should be investigated.

(iv) The role of QS in bacteria–bacteria and plant–bacteria interactions and its influence need to be understood.

Acknowledgments

The authors are grateful to Professor Pichtel (USA) for his critical input and valuable suggestions in the preparation of this manuscript. We are also thankful to Mr Arshad, Ikram Ansari, Mohd Imran and Ms Maryam Zahin for their cooperation.

References

1 Atlas, R.M. and Bartha, R. (1991) *Microbial Ecology: Fundamental and Applications*, Addison-Wesley, Reading, MA.

2 Garbeva, P., van Veen, J.A. and van Elass, J.D. (2004) *Annual Review of Phytopathology*, **42**, 243–270.

3 Belimov, A.A., Kunakova, A.M., Kozhemiakov, A.P., Stepanole, V.V. and Yudkin, L.Y. (1998) Effect of bioassociative bacteria on barley grown in heavy metal contaminated soil. Proceedings of International Symposium on Agro-Environmental Issues and Future Strategies: Towards 21st Century, May 25–30, Faislabad, Pakistan.

4 Glick, B.R. (1995) *Canadian Journal of Microbiology*, **41**, 109–117.

5 Antoun, H. and Prevost, D. (2005) Ecology of plant growth promoting rhizobacteria in *PGPR: Biocontrol and Biofertilization* (ed. Z.A. Siddiqui), Springer, Dordrecht, The Netherlands, pp. 1–38.

6 Bashan, Y. and Holguin, G. (1998) *Soil Biology & Biochemistry*, **30**, 1225–1228.

7 Whipps, J.M. (2001) *Journal of Experimental Botany*, **52**, 487–511.

8 Ahmad, F., Ahmad, I. and Khan, M.S. (2006) *Microbiological Research*, in press, doi: 10.1016/j.micres.2006.04.001.

9 Tilak, K.V.B.R., Ranganayaki, N., Pal, K.K., De, R., Saxena, A.K., Nautiyal, C.S., Mittal, S., Tripathi, A.K. and Johri, B.N. (2005) *Current Science*, **89** (1), 136–150.

10 Glick, B.R. (2003) *Biotechnology Advances*, **21**, 383–393.

11 Van Loon, L.C. and Bakker, P.A.H.M. (2005) Induced systemic resistance as a mechanism of disease suppression by rhizobacteria in *PGPR: Biocontrol and Biofertilization* (ed. Z.A. Siddiqui), Springer, Dordrecht, The Netherlands, pp. 67–109.

12 Vessey, J.K. (2003) *Plant and Soil*, **255**, 571–586.

13 Dobbelaere, S., Vanderleyden, J. and Yakov, O. (2003) *Critical Reviews in Plant Sciences*, **22** (2), 107–149.

14 Pinton, R., Varanini, Z. and Nannipieri, P. (2001) *The Rhizosphere: Biodiversity and Organic Substances at the Soil–Plant Interface*, Marcel Dekker, New York.

15 Torsvik, V. and Ovreas, L. (2002) *Current Opinion in Microbiology*, **5**, 240–245.

16 Borneman, J. and Triplett, E.W. (1997) *Applied and Environmental Microbiology*, **63**, 2647–2653.

17 Ovreas, L. and Torsvik, V. (1998) *Microbial Ecology*, **36**, 303–315.

18 Marschner, H. (1995) *Mineral Nutrition of Higher Plants*, 2nd edn, Academic Press, London.

19 Sundin, P. (1990) Plant root exudates in interaction between plants and soil microorganisms. PhD Dissertation, Lund University, Sweden.

20 Dakora, F.D. and Philipps, D.A. (2002) *Plant and Soil*, 245, 35–47.

21 Lynch J.M. (ed.) (1990) *The Rhizosphere. Ecological and Applied Microbiology*, John Wiley & Sons, Ltd, West Sussex, UK.

22 van Elsas, J.D., Duarte, G.E., Rosado, A.S. and Smalla, K. (1998) *Journal of Microbiological Methods*, 32, 133–154.

23 Garland, J.L. (1996) *Soil Biology & Biochemistry*, 28, 213–221.

24 McCaig, A.E., Glover, L. and Prosser, J.I. (1999) *Applied and Environmental Microbiology*, 65, 1721–1730.

25 Heuer, H., Krsek, M., Baker, P., Smalla, K. and Wellington, E.M.H. (1997) *Applied and Environmental Microbiology*, 63, 3233–3241.

26 Muyzer, G. and Smalla, K. (1998) *Antonie van Leeuwenhoek*, 73, 127–141.

27 Massol-Deya, A.A., Odelson, D.A., Hickey, R.F. and Tiedje, J.M. (1995) Bacterial community fingerprinting of amplified 16S and 16–23S ribosomal RNA gene sequences and restriction endonuclease analysis (ARDRA) in *Molecular Microbial Ecology Manual* (eds A.D.L. Akkermans, J.D. van Elsas and F.J. de Bruijn), Kluwer Academic Publishers, Dordrecht.

28 Liu, W.T., Marsh, T.L., Cheng, H. and Forney, L.J. (1997) *Applied and Environmental Microbiology*, 63, 4516–4522.

29 Schmalenberger, A. and Tebbe, C.C. (2002) *FEMS Microbiology Ecology*, 40, 29–37.

30 Ranjard, L. and Richaume, A.S. (2001) *Research in Microbiology*, 152, 707–716.

31 Hill, S. (1992) Physiology of nitrogen fixation in free-living heterotrophs, in *Biological Nitrogen Fixation* (eds G. Stacey, R.H. Burris and H.J. Evans), Chapman & Hall, New York, pp. 87–134.

32 Cleveland, C.C., Townsend, A.R., Schimel, D.S., Fisher, H., Howarth, R.W., Hedin, L.O., Perakis, S.S., Latty, E.F., Von Fischer, J.C., Elseroad, A. and Wasson, M.F. (1999) *Biogeochemical Cycle*, 13, 623–645.

33 Piceno, Y.M. and Lovell, C.R. (2000) *Microbial Ecology*, 39, 32–40.

34 Hugenholtz, P., Goebel, B.M. and Pace, N.R. (1998) *Journal of Bacteriology*, 180, 4765–4774.

35 Pace, N.R. (1996) *ASM News*, 62, 463–470.

36 Sahgal, M. and Johri, B.N. (2003) *Current Science*, 84, 43–48.

37 Chabot, R., Antoun, H. and Cescas, M.P. (1996) *Plant and Soil*, 184, 311–321.

38 Antoun, H. Beauchamp, C.J. Goussard, N. Chabot, R. and Lalande, R. (1998) *Plant and Soil*, 204, 57–67.

39 Saxena, A.K. and Tilak, K.V.B.R. (1998) Free-living nitrogen fixers: its role in crop production in *Microbes for Health, Wealth and Sustainable Environment* (ed. A. K. Verma), Malhotra Publishing Company.

40 Rennie, R.J. (1981) *Canadian Journal of Microbiology*, 27, 8–14.

41 Paul, E.A. and Clark, F.E.N. (1989) *Soil Biology and Biochemistry*, 2nd edn, Academic Press, New York.

42 Barraquio, W.L., Segubre, E.M., Gonzalez, M.S., Verma, S.C., James, E.K., Ladha, J.K. and Tripathi, A.K. (2000) *The Quest for Nitrogen Fixation in Rice*, IRRI Los Banos, Philippines, pp. 93–118.

43 Haahtela, K., Helander, I., Nurmiaho-Lassila, E.L. and Sundman, V. (1983) *Canadian Journal of Microbiology*, 29 (8), 874–880.

44 Seldin, L., van Elsas, J.D. and Penids, E.G. C. (1984) *International Journal of Systematic Bacteriology*, 34, 451–456.

45 Baldani, V.L.D., Alvarez, M.A.d.-B., Baldani, J.I. and Döbereiner, J. (1986) *Plant and Soil*, 90, 35–46.

46 Tchan, Y.T.,((1984) Family II. Azotobacteraceae, in *Bergey's Manual of Systematic Bacteriology*, vol. 1 (eds N.R.

Krieg and J.G. Holt),Williams and
Wilkins, Baltimore, MD, p. 219.

47 Subba Rao, N.S. (1999) *Soil Microbiology*,
4th edn, Oxford and IBH publishing Co.
Pvt. Ltd, New Delhi, India.

48 Saikia, N. and Bezbaruah, B. (1995)
Indian Journal of Experimental Biology, 33,
571–575.

49 Somers, E. and Vanderleyden, J. (2004)
Critical Reviews in Microbiology, 30,
205–240.

50 Bashan, Y. and de Bahan, L.E. (2005)
Bacteria in *Encyclopedia of Soils in the
Environment*, vol. 1 (ed. D. Hillel),Elsevier,
UK, pp. 103–115.

51 Dobereiner, J. (1982) Prospects of
inoculation of grasses with *Azospirillum*
spp., in*Associative N₂ Fixation* (eds P.B.
Vose and A.P. Suschel), CRC Press, Boca
Raton, FL, pp. 1–9.

52 Kavimandan, S.K. Subba Rao, N.S. and
Mohrir, A.V. (1978) *Current Science*,
47, 96.

53 Tilak, K.V.B.R. and Murthy, B.N. (1981)
Current Science, 50, 496–498.

54 Fallik, E., Sarig, S. and Okon, Y. (1994)
Morphology and physiology of plant roots
associated with *Azospirillum*
in*Azospirillum–Plant Associations* (ed. Y.
Okon), CRC Press, Boca Raton, FL,
pp. 77–84.

55 Tripathi, A.K., Mishra, B.M. and Tripathi,
P. (1998) *Journal of Biosciences*, 23,
463–471.

56 Tripathi, A.K., Nagarajan, T., Verma, S.C.
and Le Rudulier, D. (2002) *Current
Microbiology*, 44, 363–367.

57 Saleena, L.M., Rangarajan, S. and Nair, S.
(2002) *Microbial Ecology*, 44 (3),
271–277.

58 Kadouri, D., Jurkevitch, E. and Okon, Y.
(2003) *Applied and Environmental
Microbiology*, 69, 3244–3250.

59 Cavalcante, V.A. and Dobereiner, J. (1988)
Plant and Soil, 108, 23–31.

60 Li, R.P. and Mac-Rae, I.C. (1991) *Soil
Biology & Biochemistry*, 23, 999–1002.

61 Sharma, J. (1997) Isolation and
characterization of *Acetobacter*

diazotrophicus associated with sugarcane.
MSc Dissertation, Aligarh Muslim
University, Aligarh.

62 Muthukumarasamy, R., Revathi, G. and
Lakshminarasiman, C. (1999) *Biology
and Fertility of Soils*, 29, 157–164.

63 Ahmad, I., Sharma, J. and Ahmad, F.
(2004) *Sugar Technology*, 6, 41–46.

64 Fuentes-Ramirez, I.H., Salgado, J.T.,
Ocampo, A.I.R. and Caballaro-Mellado, J.
(1993) *Plant and Soil*, 15, 145–150.

65 Ureta, A., Alvarez, B., Ramon, A., Vera,
M.A. and Martinez-Drets, G. (1995)
Plant and Soil, 172, 271–277.

66 Yamada, Y., Hoshino, K. and Ishikawa,
T. (1998) *International Journal of
Systematic Bacteriology*, 48, 3270–3280.

67 Cabellaro-Mellado, J. and Martinez-
Romero, E. (1994) *Applied and
Environmental Microbiology*, 60, 1532–1537.

68 Suman, A., Shasany, A.K., Singh, M.,
Shahi, H.N., Gaur, A. and Khanuja, S.P.S.
(2001) *World Journal of Microbiology and
Biotechnology*, 17, 39–45.

69 Reinhold, B., Hurek, T., Niemann, E.G.
and Fendrik, I. (1986) *Applied and
Environmental Microbiology*, 52, 520.

70 Reinhold-Hurek, B., Hurek, T., Gillis, M.,
Hoste, B., Vancanneyt, M., Kersters, K.
and De Ley, J. (1993) *International
Journal of Systematic Bacteriology*, 43,
574–584.

71 Coenye, T. and Vandamme, P. (2003)
Environmental Microbiology, 5, 719–729.

72 Gillis, M., Van, T.V. and Bradin, R., Goor,
M., Hebbar, P., Willems, A., Segers, P.,
Kersters, K., Heulin, T., and Fernandez,
M.P. (1995) *International Journal of
Systematic Bacteriology*, 45, 274–289.

73 Vandamme, P., Goris, J., Chen, W.-M.,
de Vos, P. and Willems, A. (2002)
Systematic and Applied Microbiology, 25,
507–512.

74 Kloepper, J.W., Lifshitz, R. and
Zablotowicz, R.M. (1989) *Trends in
Biotechnology*, 7, 39–43.

75 Othman, A., Amer, W., Fayez, M., Monib,
M. and Hegazi, N. (2003) *Archives of
Agronomy and Soil Science*, 49 (6), 683–705.

76 Madiagan, M.T. and Martinko, J.M. (2006) *Brook Biology of Microorganisms*, 11th edn, Pearson Prentice Hall, Pearson Education Inc., Upper Saddle River, NJ.

77 Mavingui, P. and Heulin, T. (1994) *Soil Biology & Biochemistry*, **26**, 801–803.

78 Lebuhn, M., Heulin, T. and Hartmann, A. (1997) *FEMS Microbiology Ecology*, **22**, 325–334.

79 Pires, M.N. and Seldin, L. (1997) *Antonie van Leeuwenhoek*, **71**, 195–200.

80 Timmusk, S. and Wagner, E.G.H. (1999) *Molecular Plant–Microbe Interactions*, **12**, 951–959.

81 Heyndrickx, M., Vauterin, L., Vandamme, P., Kersters, K. and De Vos, P. (1996) *Journal of Microbiological Methods*, **26**, 247–259.

82 Coelho, M.R.R., von der Weid, I., Zahner, V. and Seldin, L. (2003) *FEMS Microbiology Letters*, **222**, 243–250.

83 Döbereiner, J. and Day, J.M. (1976) Associative symbioses in tropical grasses: characterization of microorganisms and dinitrogen-fixing sites. Proceedings of 1st International Symposium on Nitrogen Fixation (eds W.E. Newton and C.J. Nyman), Washington State University Press, pp. 518–538.

84 Young, J.P.W. (1994) Phylogenetic classification of nitrogen-fixing organisms, in *Biological Nitrogen Fixation* (eds G. Stacey, R.H. Burris and H.J. Evans), Chapman & Hall, New York, pp. 43–86.

85 Boddey, R.M., da Silva, L.G., Reis, V.M., Alves, B.J.R. and Urquiaga, S. (1999) Assessment of bacterial nitrogen fixation in grass species in *Nitrogen Fixation in Bacteria: Molecular and cellular Biology* (ed. E.W. Triplett), Horizon Scientific Press, pp. 705–726.

86 James, E.K. (2000) *Field Crops Research*, **65**, 197–209.

87 Boddey, R.M. and Dobereiner, J. (1995) *Fertilizers Research*, **42**, 241–250.

88 Boddey, R.M., Urquiaga, S., Reis, V. and Döbereiner, J. (1991) *Plant and Soil*, **137**, 111–117.

89 Watanabe, I., Yoneyama, T., Padre, B. and Ladha, J.K. (1987) *Soil Science and Plant Nutrition*, **33**, 407–415.

90 Roger, P.A. and Ladha, J.K. (1992) *Plant and Soil*, **141**, 41–55.

91 Garcia de Salamone, I.E., Hynes, R.K. and Nelson, L.M. (2005) Role of cytokinins in plant growth promotion by rhizosphere bacteria, in *PGPR: Biocontrol and Biofertilization* (ed. Z.A. Siddiqui), Springer, Dordrecht, The Netherlands, pp. 197–216.

92 Garcia de Salamone, I.E. and Döbereiner, J. (1996) *Biology and Fertility of Soils*, **21**, 193–196.

93 Okon, Y. and Labandera-Gonzalez, C.A. (1994) *Trends in Biotechnology*, **3**, 223–228.

94 Eskew, D.L., Eaglesham, A.R.J. and App, A.A. (1981) *Plant Physiology*, **68**, 48–52.

95 Bashan, Y., Singh, M. and Levanony, H. (1989) *Canadian Journal of Botany*, **67**, 2429–2434.

96 Hurek, T., Reinhold-Hurek, B., Van Montagu, M. and Kellenberger, E. (1994) *Journal of Bacteriology*, **176**, 1913–1923.

97 Van de Broek, A., Michiels, J., Van Gool, A. and Vanderleyden, J. (1993) *Molecular Plant–Microbe Interactions*, **6**, 592–600.

98 Egener, T., Hurek, T. and Reinhold-Hurek, B. (1999) *Molecular Plant–Microbe Interactions*, **12**, 813–819.

99 Hurek, T., Handley, L., Reinhold-Hurek, B. and Piché, Y. (1998) Does *Azoarcus* sp. fix nitrogen with monocots? in *Biological Nitrogen Fixation for the 21st Century* (eds C. Elmerich, A. Kondorosi and W.E. Newton), Kluwer Academic Press, Dordrecht, The Netherlands, p. 407.

100 Hurek, T., Handley, L.L., Reinhold-Hurek, B. and Piché, Y. (2002) *Plant–Microbe Interactions*, **15**, 233–242.

101 Saikia, S.P., Jain, V. and Shrivastava, G.C. (2004) *Indian Journal of Agricultural Science*, **74** (4), 213–214.

102 Park, M., Kim, C., Yang, J., Lee, H., Shin, W., Kim, S. and Sa, T. (2005) *Microbiological Research*, **160**, 127–133.

103 Canbolat, M.Y., Bilen, S., Cakmaki, R., Sahin, F. and Aydin, A. (2006) *Biology and Fertility of Soils*, **42**, 350–357.

104 Arshad, M. and Frankenberger, W.T., Jr (1993) Microbial production of plant growth regulators, in *Soil Microbial Ecology* (ed. F.B. Metting, Jr), Marcel Dekker, Inc., New York, pp. 307–347.

105 Patten, C.L. and Glick, B.R. (1996) *Canadian Journal of Microbiology*, **42**, 207–220.

106 Gaur, A.C. (1990) *Phosphate Solubilizing Microorganisms as Biofertilizers*, Omega Scientific Publishers, New Delhi, p. 198.

107 de Freitas, J.R., Banerjee, M.R. and Germida, J.J. (1997) *Biology and Fertility of Soils*, **24**, 358–364.

108 Nautiyal, C.S. (1999) *FEMS Microbiology Letters*, **170**, 265–270.

109 Khan, M.S., Zaidi, A. and Wani, P.A. (2007) *Agronomy for Sustainable Development*, **27**, 29–43.

110 Scher, F.M. and Baker, R. (1982) *Phytopathology*, **72**, 1567–1573.

111 Goel, A.K., Sindhu, S.S. and Dadarwal, K.R. (2002) *Biology and Fertility of Soils*, **36**, 391–396.

112 O'Sullivan, D.J. and O'Gara, F. (1992) *Microbiological Reviews*, **56** (4), 662–676.

113 Fridlender, M., Inbar, J. and Chet, I. (1993) *Soil Biology & Biochemistry*, **25**, 1211–1221.

114 Renwick, A., Campbell, R. and Coe, S. (1991) *Plant Pathology*, **40**, 524–532.

115 Shanahan, P., O'Sullivan, D.J., Simpson, P., Glennon, J.D. and O'Gara, F. (1992) *Applied and Environmental Microbiology*, **58**, 353–358.

116 Flaishman, M.A., Eyal, Z.A., Zilberstein, A., Voisard, C. and Hass, D. (1996) *Molecular Plant–Microbe Interactions*, **9**, 642–645.

117 Doneche, B. and Marcantoni, G. *Comptes Rendus de L'Académie des Sciences Série III — Sciences de la Vie*, **314**, 279–283.

118 Pati, B.R., Sengupta, S. and Chandra, A.K. (1995) *Microbiological Research*, **150**, 121–127.

119 Martin, P., Glatzle, A., Kolb, W., Omay, H. and Schmidt, W. (1989) *Zeitschrift fur Pflanzenernährung und Bodenkunde*, **152**, 237–245.

120 Bastián, F., Cohern, A., Piccoli, P., Luna, V., Baraldi, R. and Bottini, R. (1998) *Plant Growth Regulation*, **24**, 7–11.

121 Tank, N. and Saraf, M. (2003) *Indian Journal of Microbiology*, **43** (1), 37–40.

122 Ash, C., Priest, F.G. and Collins, M.D. (1993) *Antonie von Leeuwenhoek*, **64**, 253–260.

123 Salantur, A., Ozturk, A., Akten, S., Sahin, F. and Donmez, F. (2005) *Plant and Soil*, **275**, 147–156.

124 Achouak, W., Normand, P. and Heulin, T. (1999) *International Journal of Systematic Bacteriology*, **49**, 961–967.

125 Minamisawa, K., Ogawa, K.-I., Fukuhara, H. and Koga, J. (1996) *Plant & Cell Physiology*, **37**, 449–453.

126 Atzorn, R., Crozier, A., Wheeler, C.T. and Sandberg, G. (1988) *Planta*, **175**, 532–538.

127 Barea, J.M. and Brown, M.E. (1974) *The Journal of Applied Bacteriology*, **37**, 583–593.

128 Azcón, R. and Barea, J.M. (1975) *Plant and Soil*, **43**, 609–619.

129 Gonzalez-Lopez, J., Salmeron, V., Martinez-Toledo, M.V., Ballesteros, F. and Ramos-Cormenzana, A. (1986) *Soil Biology & Biochemistry*, **18**, 119–120.

130 Janzen, R., Rood, S., Dormar, J. and McGill, W. (1992) *Soil Biology & Biochemistry*, **24**, 1061–1064.

131 Piccoli, P. and Bottini, R. (1994) *Symbiosis*, **17**, 229–236.

132 Piccoli, P., Masciarelli, O. and Bottini, R. (1996) *Symbiosis*, **21**, 263–274.

133 Gutierrez-Manero, F.J., Ramos, S.B., Probanza, A., Mehouachi, J., Tadeo, F.R. and Talon, M. (2001) *Physiologia Plantarum*, **111**, 206–211.

134 Nieto, K.F. and Frankenberger, W.T., Jr (1989) *Soil Biology & Biochemistry*, **21**, 967–972.

135 Horemans, S., De Koninck, K., Neuray, J., Hermans, R. and Vlassak, K. (1986) *Symbiosis*, **2**, 341–346.

136 Cacciari, I., Lippi, D., Pietrosanti, T. and Pietrosanti, W. (1989) *Plant and Soil*, **115**, 151–153.

137 Timmusk, S., Nicander, B., Granhall, U. and Tillberg, E. (1999) *Canadian Journal of Microbiology*, **39**, 610–615.

138 Upadhyaya, N.M., Letham, D.S., Parker, C.W., Hocart, C.H. and Dart, P.J. (1991) *Biochemistry International*, **24**, 123–130.

139 Kumar, V. and Narula, N. (1999) *Biology and Fertility of Soils*, **28**, 301–305.

140 Lin, W., Okon, Y. and Hardy, R.W.F. (1983) *Applied and Environmental Microbiology*, **45**, 1775–1779.

141 Kapulnik, Y., Gafny, R. and Okon, Y. (1985) *Canadian Journal of Botany*, **63**, 627–631.

142 Cleland, R.E. (1990) Auxin and cell elongation, in *Plant Hormones and Their Role in Plant Growth and Development* (ed. P.J. Davies), Kluwer Academic Publishers, Dordrecht, The Netherlands, pp. 132–148.

143 Hagen, G. (1990) The control of gene expression by auxin, in *Plant Hormones and Their Role in Plant Growth and Development* (ed. P.J. Davies), Kluwer Academic Publishers, Dordrecht, The Netherlands, pp. 149–163.

144 Jurkevitch, E., Hadar, Y. and Chen, Y. (1988) *Soil Science Society of America Journal*, **52**, 1032–1037.

145 Bar-Ness, E., Chen, Y., Hadar, Y., Marschner, H. and Römheld, V. (1991) *Plant and Soil*, **130**, 231–241.

146 Guerinot, M.L. (1994) *Annual Review of Microbiology*, **48**, 743–772.

147 Carson, K.C., Holliday, S., Glenn, A.R. and Dilworth, M.J. (1992) *Archives of Microbiology*, **157**, 264–271.

148 Nielsen, P. and Sørensen, J. (1997) *FEMS Microbiology Ecology*, **22**, 183–192.

149 Oliveira, R.G.B. and Drozdowicz, A. (1987) *Mikrobiolohichnyi Zhurnal*, **142**, 387–391.

150 Tapia-Hernandez, A., Mascarua-Esparza, M.A. and Caballero-Mellado, J. (1990) *Microbios*, **64**, 73–83.

151 Souto, G.I., Correa, O.S., Montecchia, M.S., Kerber, N.L., Pucher, N.L. and Bachur, M. (2004) *Journal of Applied Microbiology*, **97**, 1247–1256.

152 De Weger, L.A., Van der Bij, A.J., Dekkers, L.C., Simons, M., Wiffelman, C.A. and Lulenberg, B.J.J. (1995) *FEMS Microbiology Ecology*, **17**, 221.

153 Holflich, G., Wiehe, W. and Hecht-Buchholz, C. (1995) Rhizosphere colonization of different crops with growth promoting *Pseudomonas* and *Rhizobium* bacteria. *Microbiological Research*, **150**, 139–147.

154 Rodelas, B., González-López, J., Salmerón, V., Pozo, C. and Martinez-Toledo, M.V. (1996) *Symbiosis*, **21**, 175–186.

155 Burdman, S., Vedder, D., German, M., Itzigsohn, R., Kigel, J., Jurkevitch, E. and Okon, Y. (1998) Legume crop yield promotion by inoculation with *Azospirillum* in *Biological Nitrogen Fixation for the 21st Century* (eds C. Elmerich, A. Kondorosi and W.E. Newton), Kluwer Academic Publishers, Dordrecht, The Netherlands, pp. 609–612.

156 Burdman, S., Jurkevitch, E. and Okon, Y. (2000) Recent advances in the use of plant growth promoting rhizobacteria (PGPR) in agriculture in *Microbial Interactions in Agriculture and Forestry (Volume II)* (eds N.S. Subba Rao and Y.R. Dommergues), Science Publishers Inc., Plymouth, UK, pp. 229–250.

157 Ahmad, F. (2007) Diversity and potential bioprospection of certain plant growth promoting rhizobacteria. PhD Thesis, Aligarh Muslim University, Aligarh.

158 Volpin, H. and Kapulnik, Y. (1994) Interaction of *Azospirillum* with beneficial soil microorganisms in *Azospirillum/ Plant Associations* (ed. Y. Okon), CRC Press, Boca Raton, FL, pp. 111–118.

159 Okon, Y. and Itzigsohn, R. (1995) *Biotechnology Advances*, **13**, 415–424.

160 Yahalom, E., Okon, Y., Dovrat, A. and Czosnek, H. (1991) *Israel Journal of Botany*, **40**, 155–164.

161 Burdman, S., Kige, l.J. and Okon, Y. (1997) *Soil Biology & Biochemistry*, **29**, 923–929.

162 Rodelas, B., Gonzalez-Lopez, J., Martinez-Toledo, M.V., Pozo, C. and Salmeron, V. (1999) *Biology and Fertility of Soils*, **29**, 165–169.

163 Okon, Y. and Kapulnik, Y. (1986) *Plant and Soil*, **90**, 3–16.

164 Itzigsohn, R., Kapulnik, Y., Okon, Y. and Dovrat, A. (1993) *Canadian Journal of Microbiology*, **39**, 610–615.

165 Burdman, S., Volpin, H., Kigel, J., Kapulnik, Y. and Okon, Y. (1996) *Applied and Environmental Microbiology*, **62**, 3030–3033.

166 Syono, K., Newcomb, W. and Torrey, J.G. (1976) *Canadian Journal of Botany*, **54**, 2155–2162.

167 Gupta, A., Saxena, A.K., Gopal, M. and Tilak, K.V.B.R. (1998) *Journal of Scientific & Industrial Research*, **57**, 720–725.

168 Paul, S. and Verma, O.P. (1999) *Indian Journal of Microbiology*, **39**, 249–251.

169 Bai, Y., D'Aoust, F., Smith, D.L. and Driscoll, B.T. (2002) *Canadian Journal of Microbiology*, **48**, 230–238.

170 Bai, Y., Zhou, X. and Smith, D.L. (2003) *Crop Science*, **43**, 1774–1781.

171 Tilak, K.V.B.R., Ranganayaki, N. and Manoharachari, C. (2006) *European Journal of Soil Science*, **57**, 67–71.

172 Smith, S.E. and Read, D.J. (1997) *Mycorrhizal Symbiosis*, Academic Press, Sand Diego, CA.

173 Ness, R.L.L. and Vlek, P.L.G. (2000) *Soil Science Society of America Journal*, **64**, 949–955.

174 Frey-Klett, P., Curin, J.L., Pierrat, J.C. and Garbaye, J. (1999) *Soil Biology & Biochemistry*, **31**, 1555–1562.

175 Geric, B., Rupnik, M. and Kraigher, H. (2000) *Karst Phyton*, **40**, 65–70.

176 Albrecht, C., Geurts, R. and Bisseling, T. (1999) *The EMBO Journal*, **18** (2), 281–288.

177 Barea, J.M., Bonis, A.F. and Olivares (1983) *Journal of Soil Biology & Biochemistry*, **15**, 705–709.

178 Strzelczyk, E. and Leniarska, U. (1985) *Plant and Soil*, **86**, 387–394.

179 Petersen, D.J., Srinivasan, M. and Chanway, C.P. (1996) *FEMS Microbiology Letters*, **142**, 271–276.

180 Binaciotto, V., Andreotti, S., Baletrini, R., Bonfante, P. and Perotto, S. (2001) *European Journal of Histochemistry*, **45**, 39–49.

181 Binaciotto, V. and Bonfante, P. (2002) *Antonie van Leeuwenhoek*, **81**, 365–371.

182 Sood, S.G. (2003) *FEMS Microbiology Ecology*, **45**, 219–227.

183 Raimama, M.P., Albinob, U., Cruza, M.F., Lovatoa, G.M., Spagoa, F., Ferracina, T.P., Limaa, D.S., Goularta, T., Bernardia, C.M., Miyauchia, M., Nogueiraa, M.A. and Andradea, G. (2007) *Applied Soil Ecology*, **35** (1), 25–34.

184 Burd, G.I., Dixon, D.G. and Glick, B.R. (1998) *Applied and Environmental Microbiology*, **64**, 3663–3668.

185 Klee, H.J., Hayford, M.B., Kretzmer, K.A., Barry, G.F. and Kishore, G.M. (1991) *Plant Cell*, **3**, 1187–1193.

186 Glick, B.R., Patten, C.L., Holguin, G. and Penrose, D.M. (1999) *Biochemical and Genetic Mechanisms Used by Plant Growth Promoting Bacteria*, Imperial College Press, London.

187 Lifshitz, R., Kloepper, J.W., Kozlowski, M., Simonson, C., Carlson, J., Tipping, E. M. and Zaleska, I. (1987) *Canadian Journal of Microbiology*, **33**, 390–395.

188 Glick, B.R., Jacobsen, C.B., Schwarze, M. M.K. and Pasternak, J.J. (1994) *Canadian Journal of Microbiology*, **40**, 911–915.

189 Glick, B.R., Liu, C.P., Ghosh, S. and Dumbroff, E.B. (1997) *Soil Biology & Biochemistry*, **29**, 1233–1239.

190 Holguin, G. and Glick, B.R. (2001) *Microbial Ecology*, **41**, 281–288.

191 Grichko, V.P. and Glick, B.R. (2001) *Plant Physiology and Biochemistry*, **39**, 11–17.

192 Holguin, G. and Glick, B.R. (2000) Inoculation of tomato plants with *Azospirillum brasilense* transformed with the ACC deaminase gene from *Enterobacter cloacae* UW4. Auburn University web sitehttp://www.ag.

auburn.edu/argentina/pdfmanuscripts/
holguin.pdf (Accessed 08/09/2001).

193 Glick, B.R., Penrose, D.M. and Li, J.
(1998) *Journal of Theoretical Biology*, **190**,
63–68.

194 Arshad, M., Saleem, M. and Hussain, S.
(2007) Trends in Biotechnology Jun 15;
[Epub ahead of print] PMID: 17573137.

195 Sarig, S., Kapulnik, Y. and Okon, Y. (1986)
Plant and Soil, **90**, 335–342.

196 Razi, S.S. and Sen, S.P. (1996) *Biology and
Fertility of Soils*, **23**, 454–458.

197 Creus, C.M., Sueldo, R.J. and Barassi,
C.A. (1998) *Canadian Journal of Botany*,
76, 238–244.

198 Girma, F.S. and Krieg, D.R. (1992) *Plant
Physiology*, **99**, 577–582.

199 Alvarez, M.I., Sueldo, R.J. and Barassi,
C.A. (1996) *Cereal Research Communications*, **24**, 101–107.

200 Sies, H. (1991) *Oxidative Stress: Oxidants
and Antioxidants*, Academic Press, London.

201 Bowler, C., Van Montagu, M. and Inze, D.
(1992) *Plant Molecular Biology*, **43**, 83–116.

202 Elstner, E.E., Schempp, H., Preibisch, G.,
Hippeli, S. and Oswald, W. (1994)
Biological sources of free radicals in *Free
Radicals in the Environment, Medicine and
Toxicology*, Richelieu Press, London,
pp. 13–45.

203 Štajner, D., Kevreaan, S., Gašaić, O.,
Mimica-Dudić, N. and Zongli, H. (1997)
Biologia Plantarum, **39**, 441–445.

204 Štajner, D., Gašaić, O., Matković, B. and
Varga, Sz.I. (1995) *Agricultural Medicine*,
125, 267–273.

205 Fuqua, W.C., Winaus, S.C. and
Greenberg, E.P. (1994) *Journal of
Bacteriology*, **176**, 269–275.

206 Kleerebezem, M., Quadri, L.E., Kuipers,
O.P. and de Vos, W.M. (1997)
Molecular Microbiology, **24** (5), 895–904.

207 Lazazzera, B.A. and Grossman, A.D. (1998)
Trends in Microbiology, **6** (7), 288–294.

208 Von Bodman, S.B., Bauer, W.D. and
Coplin, D.L. (2003) *Annual Review of
Phytopathology*, **41**, 455–482.

209 Ziegler-Heitbrock, H.W., Frankenberger,
M. and Wedel, A. (1995) *Immunobiology*,
193, 217–223.

210 Wei, H.-L. and Zhang, L.-Q. (2006)
Antonie van Leeuwenhoek, **89** (2), 267–280.

211 Tsuchiya, K., Homma, Y., Komoto, Y. and
Susui, T. (1995) *Annals of the Phytopathological Society of Japan*, **61**, 318–324.

212 Paulitz, T.C. (2000) *European Journal of
Plant Pathology*, **106**, 410–413.

213 Amann, R., Ludwig, W. and Schleifer,
K.H. (1995) *Microbiological Reviews*, **59**,
153–169.

214 Bowen, G.D. and Rovira, A.D. (1999)
Advanced Agronomy, **66**, 1–102.

215 Wilson, K.J., Sessitsch, A. and
Akkermans, A. (1994) Molecular markers
as tools to study the ecology of
microorganisms in *Beyond the Biomass*
(eds K. Ritz, J. Dighton and K.E. Giller),
British Society of Soil Science, London,
pp. 149–156.

216 Wilson, K.J., Sessitsch, A., Corbo, J.C.,
Giller, K.E., Akkermans, A.D.L. and
Jefferson, R.A. (1995) *Microbiology*, **141**,
1691–1705.

217 Kloepper, J.W. and Beauchamp, C.J.
(1992) *Canadian Journal of Microbiology*,
38, 1219–1232.

218 Blackbum, N.T., Seech, A.G. and Trevors,
J.T. (1994) *Systematic and Applied
Microbiology*, **17**, 574–580.

219 Tombolini, R., Van Der Gaag, D.J.,
Gerhardson, B. and Jansson, J.K. (1999)
Applied and Environmental Microbiology,
65, 3674–3680.

220 Errampali, D., Leung, K., Cassidy, M.B.,
Kostrzynska, M., Blears, M., Lee, H.
and Trevors, J.T. (1999) *Journal
of Microbiological Methods*, 35–187.

221 Normander, B., Hendriksen, N.B. and
Nybroe, O. (1999) *Applied and
Environmental Microbiology*, **65** (10),
4646–4651.

222 Glandorf, D.M.C., Brand, I., Bakker, P.A.
H.M. and Schippers, B. (1992) *Plant and
Soil*, **147**, 135–142.

6

Molecular Mechanisms Underpinning Plant Colonization by a Plant Growth-Promoting Rhizobacterium

Christina D. Moon, Stephen R. Giddens, Xue-Xian Zhang and Robert W. Jackson

6.1
Introduction

Pseudomonads are versatile and ubiquitous inhabitants of terrestrial and aquatic ecosystems. They can be found in close association with animals and plants, where their ecological relationships span the continuum from antagonism (exemplified by the opportunistic human respiratory pathogen *Pseudomonas aeruginosa*) to mutualism (displayed in associations between plants and plant growth-promoting rhizobacteria (PGPR), such as strains of *P. putida* and *P. fluorescens*). PGPR encounter continually changing environmental conditions, where variation in factors such as temperature, water and nutrient availability, as well as competition with other environmental microbes, is typical. Thus, in order to persist in such habitats, PGPR must be able to rapidly respond and acclimatize to their changing environments.

P. *fluorescens* strain SBW25 has become a model plant-associated bacterium, where numerous studies have provided detailed insight into its biology, ecology and evolution [1–10]. SBW25 was originally isolated from the leaf surface of a field-grown sugar beet plant (*Beta vulgaris*) in Oxfordshire, UK [11]. It is an efficient colonizer of the plant environment, or phytosphere (Figure 6.1), where it thrives particularly well in the rhizosphere (root surfaces and closely associated soil) and displays biocontrol activity against soilborne *Pythium* pathogens [1,5]. Unlike PGPR that produce antimicrobial metabolites that are effective in disease control, SBW25 has not been observed to produce specific antifungal metabolites [1]. However, one of the factors identified as contributing to biocontrol activity against the damping-off disease agent *Pythium ultimum* is the competitive ability of SBW25 for utilizing carbon sources [1]. Thus, it is apparent that efficient phytosphere colonization is central to the ability of SBW25 to promote plant growth.

There is a general ongoing interest in understanding the genetic causes of ecological success (fitness). The acclimation and adaptation of SBW25 to the plant environment is a model system that has been adopted by various laboratories [5,7,12–15]. Ecological success is the net result of interactions between many diverse

Plant-Bacteria Interactions. Strategies and Techniques to Promote Plant Growth
Edited by Iqbal Ahmad, John Pichtel, and Shamsul Hayat
Copyright © 2008 WILEY-VCH Verlag GmbH & Co. KGaA, Weinheim
ISBN: 978-3-527-31901-5

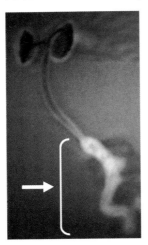

Figure 6.1 *Pseudomonas fluorescens* SBW25 colonization of plant surfaces. An *Arabidopsis thaliana* seedling was transplanted onto minimal nutrient agar overlaid with a dilute suspension of SBW25. Bacteria have accumulated around the seedling to form a biofilm, particularly surrounding the hypocotyl and root (bright white halo, arrowed). Image obtained under UV illumination by epifluorescence microscopy shown at four times magnification.

traits such as antimicrobial compound production, motility, nutrient acquisition and physical attachment to surfaces. The significance of each trait is dependent on the precise environmental conditions that the bacterium is exposed to. Thus, identifying the traits involved and unraveling their regulation and degree of contribution toward ecological success is a significant technical challenge. However, with the availability of whole-genome sequences and high-throughput technologies, experimental procedures that utilize genome-wide screens are highly feasible and have proven to be powerful tools allowing significant progress to be made toward realizing these goals.

One of the first steps toward identifying fitness-enhancing traits is to identify genes showing elevated levels of expression in the plant environment relative to a broader range of environments (including the laboratory) [7]. These genes are considered more likely to contribute to ecological performance in the plant environment. A genes-to-phenotype functional genomic approach for identifying plant-induced genes by using *in vivo* expression technology (IVET), a promoter trapping technique, has been performed [2,7]. This method was first used to identify plant-induced genes from *Xanthomonas campestris* [16] and was subsequently termed IVET [17]. It has since been used to identify niche-specific genes in many other microbes including animal pathogens [18,19]. IVET assays in SBW25 have been successfully developed [2,7,10]. This chapter provides an overview of the development and application of these assays and reports on the identity of SBW25 genes found to be active in the phyllosphere (above-ground plant surfaces) and the rhizosphere.

The identification of niche-specific genes opened up new opportunities to explore how SBW25 responds to the plant environment. Inducible gene expression occurs when the bacterium senses an environmental stimulus, which in turn triggers a cascade of regulatory events resulting ultimately in the expression of niche-specific genes; conversely, removal of the stimulus shuts down gene expression. Therefore, studies were undertaken to identify regulators that govern environment-specific gene expression and to determine the environmental niches in which the inducing factors are present. A suppressor analysis of IVET fusion strains that contain plant-inducible promoters provided insights into the regulatory pathways that direct the expression of plant-induced genes and uncovered examples of positive and negative regulators, and local and global networking.

Within the plant's environment, gene-inducing factors can be located in highly defined niches. To determine the precise locations where plant-inducible genes are expressed, and hence identify the locations of gene-inducing factors, a recombinase-based IVET (RIVET) system [20] has been adopted for use in SBW25 [21]. This system allows gene activity to be monitored on spatial and temporal scales [22] and is particularly suitable for use in complex environments such as the rhizosphere. The RIVET technique, like IVET, was developed as a promoter trap, though in many applications it has been used as a gene reporter [22].

In this chapter, we report recent progress toward understanding the molecular mechanisms underlying the adaptation of rhizobacteria to the plant environment. Throughout, we have used SBW25 as a model to show how the molecular tools and genomic approaches discussed above have been successfully utilized to reveal the complex interactions between bacteria and their environment.

6.2
Identification of Plant-Induced Genes of SBW25 Using IVET

The main principle employed in plant-based IVET screens is the bacterial biosynthesis of some essential growth factor (EGF) that is present only in negligible quantities in the environment of interest (Figure 6.2). For SBW25, two complementary IVET systems have been developed that utilize one of two different *egf* genes [2,7]. The first system was based on a *panB egf* reporter gene [7], which encodes a ketopantoate hydroxymethyltransferase. This enzyme catalyzes the first committed step in the biosynthetic pathway for the essential water-soluble B-group vitamin, pantothenate. The second system, providing a more stringent selection regime than *panB* (because pantothenate is at trace levels in the plant environment), was based on *dapB* as an *egf* reporter gene [2], which encodes 2,3′-dihydrodipicolinate reductase. This is required for the biosynthesis of lysine and diaminopimelate (DAP), a component of peptidoglycan found in bacterial cell walls, but not present in soil, water, plant and animal tissues. Both *panB* and *dapB* auxotrophic deletion mutants are conditional in that they can be rescued by the provision of exogenous pantothenate, or DAP and lysine, respectively. Accordingly, both mutants were severely compromised in their ability to colonize sugar beet seedlings [2,7].

To create a library of IVET strains of SBW25 for environmental screening, initially small fragments (3–5 kb) of SBW25 genomic DNA were ligated upstream of

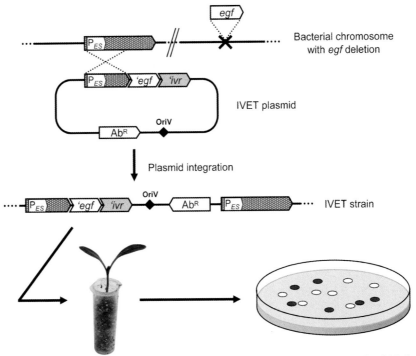

In situ selection: strains will only survive if
'*egf* is expressed (**P**_{ES} is active)

In vitro screen to find strains in which '*ivr* is
silent in the absence of environmental
signals (**P**_{ES} **is silent**)

Figure 6.2 The IVET principle. A chromosomal deletion of a gene encoding an 'essential growth factor' (*egf*) is carried out, rendering the host bacterium auxotrophic for the product (EGF) synthesized by the deleted gene. A library of IVET plasmids is created by ligating random genomic fragments immediately upstream of a promoterless copy of the *egf* gene ('*egf*). These constructs are integrated into the genome via single homologous crossover recombination at the cloned region, resulting in a collection of merodiploid IVET fusion strains. IVET strains are screened *in situ* (e.g. on sugar beet seedlings). Strains will only survive and grow if '*egf* is expressed due to the presence of a promoter (P_{ES}) upstream of '*egf* that is active in this environment. The strains recovered from the *in situ* screen are grown *in vitro* on media supplemented with EGF and screened using a second promoterless '*in vitro* reporter' gene ('*ivr*), such as '*lacZ*, to differentiate strains where the promoter driving expression of '*egf* is expressed constitutively (i.e. is not environment specific).

promoterless copies of '*panB*–'*lacZY* or '*dapB*–'*lacZY* present in the integration vectors pIVETP [7] and pIVETD [2], respectively. These libraries were conjugated into the corresponding *egf* mutant (SBW25 strains deleted for *panB* or *dapB*) and pools of clones were inoculated onto sugar beet seeds. Selection for plant-induced promoter fusions was achieved by harvesting bacteria from 2–3-week-old sugar beet seedlings (from phyllosphere, rhizosphere and/or bulk soil-associated populations). Growth of any clone will only ensue if there is an active promoter upstream of the '*egf* reporter gene to drive its expression, resulting in complementation of the *egf* deletion; silent genes do not express '*egf* and the number of these bacterial strains decreases. Harvested bacteria were subsequently screened *in vitro* for constitutive promoter activity by virtue of the '*lacZY* reporter, thus enabling the identification of gene fusions that are active in the plant environment and not in the culture. Positive IVET fusion activities were confirmed by inoculating IVET strains (purified mono-cultures) back onto the sugar beet seedlings and testing for colonization relative to wild-type SBW25. The IVET fusion points were identified by recovering integrated IVET constructs by conjugation into *Escherichia coli* [23] and sequencing across the junction of the reporter gene to the genomic fragment insertion. The availability of the SBW25 genome sequence (http://www.sanger.ac.uk/Projects/P_fluorescens/) has greatly facilitated the mapping of insertion points, providing a detailed context of the candidate plant-induced genes upstream of the reporter and insight into the promoters driving their activities.

In the preliminary IVET screening for rhizosphere-induced SBW25 genes employing the *panB* reporter, a panel of IVET clones representing ~10% of the coverage needed to comprehensively survey the SBW25 genome were screened, revealing the specific upregulation of 20 loci in the rhizosphere environment [7]. These results provided the first major insights into how SBW25 perceives and responds to this environment, as identified rhizosphere-induced genes had expected roles in nutrient acquisition, secretion, stress response and a number of other unidentified (novel) traits.

One of the most intriguing findings of this screening was the expression of a gene with 51% sequence identity to the plant pathogen *P. syringae* gene *hrcC*, which encodes a putative pore-forming outer membrane component of a type III protein secretion system (TTSS) [7]. Further analysis of this locus revealed a 20-kb cluster of TTSS-related genes (designated the Rsp cluster), which is closely related to the TTSS gene cluster of *P. syringae*, though SBW25 appears to lack a number of *P. syringae* TTSS gene homologues including an EBP gene, harpin gene and parts of the *hrpJ* cluster [6]. TTSSs are commonly associated with pathogenic bacteria and function in the delivery of 'virulence/modulating' proteins into host cells that may result in parasitism or elicitation of host defense responses. Screening of other *P. fluorescens* strains showed the gene cluster to be widely distributed in nonpathogens. Thus, the discovery that SBW25 possesses TTSS gene homologues led to further investigations into the functionality and ecological role of this locus in SBW25. While wild-type SBW25 does not cause disease symptoms in plants or elicit host defense responses in test host organisms, ectopic expression of a key sigma factor gene, *rspL*, that directs the expression of *rsp* genes, resulted in the elicitation of a hyper-sensitive response (HR) in *Nicotiana clevelandii*. Furthermore, expression of *rspL* in combination with the heterologous *P. syringae* avirulence protein, AvrB, resulted in a gene-for-gene specific HR reaction in *Arabidopsis thaliana* Col-0, albeit by using very high inoculum loads [6]. Although the SBW25 TTSS genes appear to function in protein secretion under contrived conditions, the ecological role and function of these genes in their natural environment remains to be elucidated. Evidence from plant colonization experiments with *rsp* mutants suggests that the *rsp* genes are involved in the active colonization of root surfaces and also affect growth *in vitro*.

These results suggest a different and more general role for the *rsp* genes in the function of SBW25 [4]. In another *P. fluorescens* strain, which has a full complement of TTSS structural genes, the TTSS genes have been implicated in antagonism of the plant pathogen *Py. ultimum* [24].

The SBW25 *panB* IVET screen also identified a rhizosphere-inducible gene, *hutT*, which is predicted to have a role in histidine uptake [7,21]. Subsequent functional analysis of *hutT* demonstrated that it is histidine-inducible and is required for the bacterium to grow on histidine as a sole carbon and nitrogen source [21]. Furthermore, the ecological significance of the *hut* locus was examined by competing a *hutT* deletion mutant (SBW25Δ*hutT*) against wild-type SBW25 during sugar beet seedling colonization. However, no impact on relative fitness was detected, suggesting that the ability to utilize plant-derived histidine is not required for competitive colonization *in planta*. This is likely because of the oligotrophic nature of the rhizosphere and the diverse nutritional capabilities of *P. fluorescens* [21].

The development of a *dapB*-based IVET system for SBW25 was based on the potentially weaker selection for rhizosphere-induced promoters using the *panB* system. Pantothenate is required in only small amounts to rescue auxotrophy and it is more generally available in the environment (than DAP) as it is produced freely by other microbes and plants [2]. The *dapB* system has proved to be highly rigorous and has been applied in a comprehensive screen of the entire SBW25 genome (the results of this work are to be published in the forthcoming SBW25 genome sequence paper), as well as a number of other plant- and soil-colonizing pseudomonads [25–27]. The *dapB* system has been utilized to examine quantitative expression of bacterial genes in the plant environment [4,10]. This IVET system has also been employed to gain insight into the genetic factors contributing to the successful persistence of a self-transmissible mercury resistant plasmid (pQBR103) in the plant-associated *Pseudomonas* community [10,28].

Almost 200 unique SBW25 IVET fusions have been identified by the *dapB* system (P.B. Rainey, personal communication), and the predicted functions of these can generally be classified into 11 broad functional categories [18,19]: motility and chemotaxis, nutrient scavenging, central metabolism, stress and detoxification, regulation, cell envelope, virulence and secretion, nucleic acid metabolism, transposition recombination and phage, unknown and cryptic. Investigations into a number of the identified loci have been, or are presently underway. The best characterized of these is the *wss* gene cluster (*wssA–wssJ*) that encodes the biosynthetic machinery for the production of an acetylated cellulose polymer [29,30]. This polymer is a major constituent of SBW25 biofilms and a key contributor to the large-spreading wrinkly spreader colony phenotype [8,29], where mutants with this phenotype overexpress the *wss* genes [30]. It is hypothesized that in the plant environment, the secreted acetylated cellulose polymer may act as a 'glue' to cause bacterial cells to adhere to each other and facilitate the spreading of cells across plant surfaces [2]. It was demonstrated that the *wss* locus is highly important to the ecological success of SBW25 during competitive colonization of the phyllosphere but plays a minor role in the rhizosphere and is not important for bulk soil [2] (Figure 6.3).

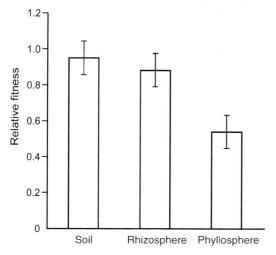

Figure 6.3 Fitness contribution of the *wss* locus during competitive colonization of sugar beet seedlings. The relative fitness of a *wssB* mutant compared to wild-type SBW25 in the soil, rhizosphere and phyllosphere showed that the *wssB* mutant was significantly less fit during phyllosphere colonization. Data were collected from three time points (1, 2 and 4 weeks) and are means and 95% confidence intervals of 11 replicates. Figure has been redrawn with permission [2].

Copper is an essential trace element, being an important cofactor for several enzymes involved in primary metabolism. However, it is toxic to bacteria in excess concentrations; thus, the maintenance of copper homeostasis is critical for survival. As measured on mannitol–glutamate–yeast extract medium, SBW25 is copper sensitive with a minimum inhibitory concentration of 500 µM [31].

Interestingly, the *dapB*-based promoter trap identified *cueA*, which encodes a P1-type ATPase with a predicted role in copper transport from the cytoplasm [31]. Further characterization of *cueA*, from both ecological and genetic perspectives, showed that *cueA* was induced by elevated concentrations of copper ions, and deletion of *cueA* resulted in a two-fold reduction of copper tolerance. Moreover, the deletion mutant strain (SBW25Δ*cueA*) was compromised in its ability to competitively colonize the roots of sugar beet and pea plants [31]. Taken together, the data indicate that copper ions accumulate on plant surfaces, which is consistent with earlier observations that copper accumulated in the roots of pea [32].

As in other IVET screens for bacterial genes upregulated in plant environments [33,34], genes involved in iron uptake were found to be upregulated in SBW25 [2]. These included a gene with high homology to a probable TonB-dependent ferric siderophore receptor gene from *P. aeruginosa* and a pyoverdine synthetase gene from *P. fluorescens*. Like copper, iron is an essential trace metal that is toxic in excess amounts. However, iron availability in the plant environment is generally limited, as under aerobic conditions and neutral pH, iron is usually found in the insoluble ferric (Fe^{3+}) form. Thus, systems to scavenge for this metal are highly important. Many pseudomonads produce the siderophore pyoverdine, which is excreted into the external environment, binds ferric iron and is then transported back into the cell via a dedicated receptor transport system. Investigations to determine the contribution of the pyoverdine synthetase gene, *pvdL*, to SBW25 fitness during early seedling development *in situ* revealed that pyoverdine biosynthesis contributed mainly to competitive fitness on the shoot (Moon, Zhang, Matthijs and Rainey, unpublished). However, the contributions to fitness from siderophore systems in other plant-colonizing bacteria have shown significant impact in the rhizosphere environment [35,36], though the ability to utilize exogenous siderophores is thought to have a bearing on the degree of the contribution to fitness [36]. *P. fluorescens* siderophores also appear to play a role in biocontrol of pathogenic fungi and oomycetes. Thio-quinolobactin and pyoverdine were recently shown to antagonize various oomycete and fungal pathogens [37]. Although pyoverdine has not been demonstrated experimentally to be an SBW25 antagonist of pathogens, the probable importance of pyoverdine production in plant growth promotion is supported by observations that SBW25 can reduce *Pythium* disease effects *in vivo* and stop *Pythium* growth on iron-free agar [5] (Jackson, unpublished), as well as the identification of *pvdL* as a plant-inducible gene.

In SBW25, IVET technologies have also been applied to attempt to uncover the fitness-enhancing traits of pQBR103, a 425 kb self-transmissible plasmid that confers mercury resistance, which was found in field-grown sugar beet-associated *Pseudomonas* populations [28,38]. Investigations into the cost of plasmid carriage showed that carriage was detrimental to SBW25 growth during early sugar beet development, but conferred an ecological advantage as the plants matured [13]. The screening of a *dapB*-based IVET library based on pQBR103 genomic fragments revealed 37 unique plant-inducible fusions; however, only three of these had orthologues in public DNA databases [28]. All of these showed similarity to genes encoding proteins with predicted helicase functions, though data suggest that they are not involved in the repair of UV-induced DNA damage [28]. An additional fusion was characterized that contained an unknown ORF adjacent to a functional oligoribonuclease (*orn*) gene, which was able to complement a *P. putida* KT2440 *orn* mutant. The *orn* gene was further found to be widely distributed among group I plasmids present in pseudomonads isolated from the same sugar beet fields as SBW25, suggesting that it is ecologically relevant [10]. However, the precise roles of each of these plant-inducible genes in the ecological success of their host bacteria remain unclear.

Almost one third of the IVET fusions identified in SBW25 screens have homology to hypothetical genes or have no homology to sequences in the databases. This is true of many IVET screens [18] and is largely a consequence of the growing knowledge gap between the difference in the rate of accumulation of genome sequencing data and the rate of experimental characterization of their biological functions. Another key general observation from IVET screens, including SBW25 screens, is the discovery of fusions that are orientated in the direction opposite to annotated genes [39]. It has been suggested that these 'cryptic fusions' may represent artifacts of the IVET system that do not truly function under natural conditions or that they may reflect the expression of regulatory RNA molecules or mRNA transcripts from

ORFs on the noncoding strand, a prediction supported by the observation that ORFs are often visible on the DNA strand opposite to predicted genes [39]. Alternatively, they may represent read-through from a strong promoter further upstream of the fusion. While the significance of cryptic fusions remains unclear, they do appear to be commonly identified from IVET screens [7,17,39,40], and it is of interest to understand their significance.

Overall, IVET screens have proven to be powerful tools in identifying genes that are expressed by SBW25 in response to the complex plant environment. This has provided an insight into revealing the genetic bases underlying SBW25's ecological success. In particular, in-depth studies into several plant-induced loci encoding traits in nutrition acquisition, stress response, physical attachment and potential eukaryotic signaling systems have provided insight into some of the stresses and obstacles that are encountered in the plant environment and the physiological responses of SBW25 to these. While the biological significance of each of the loci investigated has not always been apparent, investigations into their contributions to the ecological success of SBW25 have provided clues with regard to the necessity of these loci for plant colonization.

6.3
Regulatory Networks Controlling Plant-Induced Genes

Phenotypic acclimation is the reversible expression of one or more phenotypes in response to an environmental stimulus. The expression of various phenotypes is the result of cascades of interactions that are influenced by repressors and activators. IVET effectively identifies genes showing elevated levels of expression in a particular environment, such as the plant, and most of these niche-specific genes will be involved in bacterial acclimation to the environment to optimize fitness. Genes that are most likely identified by IVET include structural genes and activators, but not repressors of plant-induced genes, as these would be active *in vitro* but downregulated *in vivo*. Although IVET systems themselves are inadequate in identifying repressor genes because they are plant repressed, the auxotrophic basis of the IVET system is able to provide an efficient framework for identifying repressor genes.

To identify repressors of plant-induced genes in SBW25, a suppressor analysis of SBW25Δ*dapB* strains carrying plant-inducible IVET fusions was undertaken in a gene discovery method that is referred to as Suppressor-IVET or SPyVET [18] (Figure 6.4). Two transposons, MiniTn5Km and IS-Ω-Km/hah, were used to mutagenize IVET fusion strains. MiniTn5Km was used purely to identify repressors, but IS-Ω-Km/hah was used to identify both repressors and activators by virtue of an outward facing *npt* promoter located at one end of the transposon [41] (Giddens, unpublished). Mutants were plated at high density ($>10^5$ cfu cm^{-2}) on minimal medium containing kanamycin to select for the transposon, but lacking the essential growth factors DAP and lysine. The strict auxotrophy associated with the *dapB* deletion prevented the majority of mutants from growing, thus allowing a large number of mutants to be efficiently screened. Mutants that contained a transposon

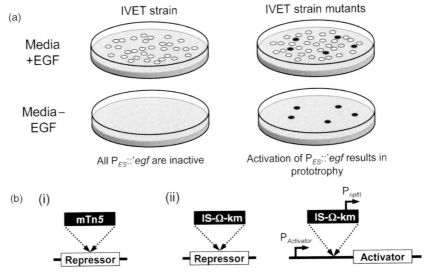

Figure 6.4 The SPyVET principle. (a) IVET strains will not grow *in vitro* in the absence of exogenous EGF owing to lack of 'P$_{ES}$::*egf* expression (see Figure 6.2), but regulatory mutants (suppressors) that result in P$_{ES}$ expression restore prototrophy enabling growth in the absence of exogenous EGF (Giddens, Jackson, Moon and Zhang, unpublished). (b) Two transposons were used to mutate IVET strains and disrupt expression of environment-specific regulators. (i) miniTn5 was used to solely disrupt repressors and (ii) IS-Ω-Km was used to disrupt repressors and induce environment-independent expression of neighboring activators by virtue of a constitutively expressed *nptII* promoter near one end of the transposon.

disruption of a repressor gene enable expression of the plant-inducible gene linked to the '*dapB*::*lacZ* reporter, thus allowing expression of '*dapB* and growth. Similarly, IS-Ω-Km/hah *nptII* promoter activity enabled expression of positive regulators that activate the plant-inducible gene fusion. By using arbitrary-primed PCR [41] and sequencing, it was possible to identify the insertion points of the transposons, which were then mapped to the SBW25 genome sequence.

Approximately 2–3 million mutants of each plant-inducible gene fusion were screened and a total of 16 regulators were identified for eight plant-inducible genes (Giddens, Jackson, Moon and Zhang, unpublished). Most of the activators identified were not isolated by IVET. This implies one of several possibilities. First, these activators are constitutively active and are themselves regulated post-transcriptionally. These would be recovered from an IVET screen through plants, but eliminated as 'housekeeping-type' gene fusions that are also active *in vitro*. For example, one gene, *algR*, was identified as a positive regulator of the plant-inducible *wss* gene cluster, and also identified as a positive activator of hydrogen peroxide resistance [42]. This trait is expressed *in vitro*, which suggests the regulator must be active both *in vitro* and *in vivo*. A second possibility is that the activators are too weakly expressed to rescue the *egf* mutant strain in an IVET screen; an example of this is the *rspL* activator, which has been shown to be expressed in the plant rhizosphere, but at relatively low levels [4]. Thirdly, the activators are transiently expressed in the plant environment and expression is not for a sufficient period of time to rescue the strain. Transient gene expression has been demonstrated in other systems [22].

Two plant-inducible genes (*wssE* and *cueA*) had more than three regulators controlling their expression compared with the other gene systems for which only one or two regulators were identified. In SBW25, *cueA* encodes a copper transporting P1-type ATPase [31]. To date, the majority of bacterial genes for copper homeostasis that have been characterized encode P1-type ATPases, and their expression is induced by high levels of extracellular copper [43]. Notably, the copper-exporting ATPase in Gram-negative bacteria, such as *E. coli* and *P. putida*, is regulated by a MerR-type activator (CueR) [44], whereas in the Gram-positive bacterium *Enterococcus hirae*, it is regulated by a repressor protein CopY and a copper chaperone CopZ [45]. SPyVET analyses identified CueR, a putative MerR-type activator, as a regulator of *cueA*. This was consistent with the previous work on *P. putida* that *cueA* is activated by CueR in a copper-responsive manner [46]. Additionally, SPyVET analysis suggested that CueA activated the expression of *cueA* and that CopZ negatively regulated *cueA*. It was also found that CueR and CopZ activated a plant-inducible permease locus of unknown function in SBW25 that was originally identified by IVET. These data suggest a role for the permease locus in copper homeostasis, which is currently under investigation.

The second plant-induced gene controlled by multiple regulatory inputs in SBW25 is *wssE*. The *wssE* gene is part of an operon (*wssABCDEFGHIJ*) that encodes a membrane-bound cellulose synthase complex. The production of cellulose is important for bacterial fitness in the sugar beet rhizosphere and particularly the phyllosphere [2]. The function of cellulose in plant colonization is unknown, but like many secreted extracellular polysaccharides (EPS), it is central to the formation of biofilms formed by SBW25. A total of seven regulators have been identified in controlling expression of *wssE*, which include repressors (AwsX, WspF, AlgZ (AmrZ) and FleQ) and activators (WspR, AwsR and AlgR). AlgR and AlgZ (AmrZ) have previously been implicated in controlling expression of another EPS, alginate [47]. In *P. aeruginosa*, AlgZ (AmrZ) is a transcriptional activator that acts on alginate genes, but in SBW25, it acts as a repressor, in this case of *wss* expression. Although SBW25 carries an intact alginate operon, the production of alginate has not been identified either *in vivo* or in biofilms. This indicates that cellulose is probably the more important EPS for plant colonization and that the regulators have been recruited to either control expression of both EPS gene systems or just the *wss* system, with alginate gene regulation becoming redundant.

FleQ is the master regulator of the flagellum biosynthesis genes [48] and this is the first time the regulator has been implicated in controlling the expression of EPS genes. However, the inverse relationship between EPS and flagellum gene expression is well documented [49]. Removing the bulk of the IS transposon from the *fleQ* mutant via Cre–*loxP* recombination [41] and subjecting the resulting unmarked *fleQ* mutant to another round of transposon mutagenesis led to the discovery that FleQ repressed the activity of AlgR. Since FleQ is an RpoN-interacting enhancer binding protein, the *algR* gene and upstream sequence were examined but RpoN binding sites were not identified; however, a putative RpoN promoter was identified upstream of *algZ*. It was proposed that FleQ activates expression of *algZ* and that AlgZ represses either *algR* or *wss*.

Two other pairs of regulators were identified that control expression of *wss*: *wspF* and *wspR*, and *awsX* and *awsR*. Mutations in *wspF* or *awsX* led to activation of *wss*, and consequent expression of the two genes in the mutants repressed *wss*. Through complementation and/or expression analysis, it was also found that AwsR and WspR activate *wss*. These genes encode proteins with GGDEF domains involved in the production of cyclic di-GMP, a bacterial signaling molecule. The WspR protein has been shown to cause a two-fold upregulation of *wss* expression [30]. There is no indication that these GGDEF-domain proteins have DNA-binding domains, which suggests that either they activate transcription *via* activation of an unknown regulator or the activation of the Wss enzymes leads to auto-activation. As yet, there is no indication why both the Wsp and Aws proteins would be involved in activation of the *wss* genes, but it possibly provides a means for SBW25 to sense different environments and induce *wss* expression appropriately. Both the Wsp and Aws systems appear to encode proteins that are membrane bound and could be important in sensing changes in the external environment.

Only one or two regulators have been identified for the other plant-inducible genes identified by IVET. One of these was a putative adhesin, AidA, which activates a putative nitrilase gene. Nitrilases catalyze the hydrolysis of nitrile compounds and may represent an important nitrogen source for bacterial growth in the plant environment. The mode of action of AidA in the regulation of the nitrilase gene, however, remains unclear at present.

The outcome of this study was the ability to correlate regulators with genes induced in the plant environment. However, it was also noted that some of the regulators and their functions either had been previously described or had predicted functions (*in silico*). This allowed these data to be considered in a broader context, at a systems level (in this case the plant acclimation system), by analyzing phenotypes associated with the regulators. It was deduced that FleQ is important for flagellum biosynthesis, for bacterial swimming and for the suppression of swarming motility, the FleQ and AlgR regulators control expression of hydrogen peroxide resistance, the Aws and Wsp systems influence colony morphology (probably due to cellulose synthesis), the *cueAR* system is important for copper resistance and the AidA regulator controls bacterial adhesion as well as nitrilase expression. A global functional model was formulated that links genes, quantitative gene expression and phenotypes (Figure 6.5). This forms a solid basis for examining how the plant acclimation system might change in response to differing environmental signals. It also forms a very important foundation for biologists examining other bacterial systems, who can identify overlaps and differences in the way bacteria acclimatize or adapt to different environments. Therefore, SPy-VET has become a valuable new method to utilize and comple-ment IVET screens and identify regulatory systems. This will be broadly applicable to different bacterial systems and for assessing the regulation of phenotypic acclimation.

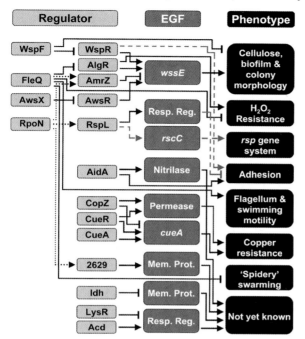

Figure 6.5 Regulatory networks revealed by SPyVET analysis of SBW25 illustrating the power of this approach to decipher complex regulatory interactions (Giddens, Jackson, Moon and Zhang, unpublished). Positive interactions are shown as arrows and negative interactions as blunt-end lines. Dotted lines represent hypothetical interactions and gray dashed lines represent interactions determined in studies prior to the SPyVET study (Giddens, Jackson, Moon and Zhang, unpublished).

6.4
Spatial and Temporal Patterns of Plant-Induced Gene Expression

To gain a better understanding of how SBW25 functions in the wild, there is a need to determine the distribution of gene-inducing signals in the environment of interest. One powerful technique that has been adopted for this purpose is RIVET, which was originally developed to identify *Vibrio cholerae* genes that were induced during infection of the mouse gastrointestinal tract [20,40] and later was used to successfully determine the spatial and temporal patterns of two critical virulence genes using the same mouse models [22]. The original RIVET strategy is based on a reporter gene, *tnpR*, which encodes a site-specific resolvase. When *tnpR* is expressed in a genome containing the artificial substrate cassette (*res1–tet–res1*) it catalyzes a recombination event between the two *res1* sequences, resulting in the excision of the *tet* gene, rendering the cell and its progeny tetracycline sensitive [20]. This RIVET system has been adapted for use in SBW25 (Figure 6.6a) and has been used to examine the expression of the plant-inducible histidine utilization (*hut*) genes *in situ* [21]. Here, the *res1–tet–res1* cassette was integrated into the SBW25 genome at an intergenic region upstream of the *wss* operon. A *hutT* fragment was fused to the *tnpR* reporter gene and integrated by single homologous recombination with the native *hutT* sequence in the genome. Thus, upon induction of the *hut* promoter by the presence of histidine, this would drive expression of *tnpR*, resulting in the production of the enzyme resolvase. Resolvase catalyzes a

(a) *hutT* RIVET activity

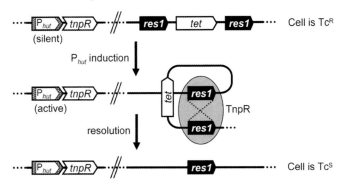

(b) *hutT* activity in response to histidine

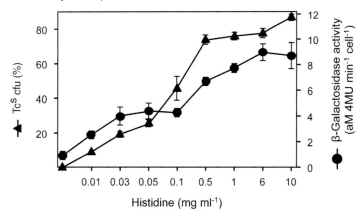

Figure 6.6 The RIVET principle. RIVET, as applied to SBW25 to report the activity of *hutT*, is shown [21]. (a) The promoter of interest, P*hut*, was ligated upstream to *tnpR*, which encodes resolvase. Upon P*hut* activation, *tnpR* is expressed and resolvase (TnpR) catalyzes a site-specific recombination between two *res1* sites flanking a Tc^R marker gene (*tet*). Thus, P*hut* expression results in the heritable loss of Tc^R. (b) P*hut* expression as measured by a RIVET reporter strain (circles), compared to a beta galactosidase reporter (triangles) after exposure to a range of concentrations of histidine (a known inducer of P*hut*) (data represent means and standard errors of three replicates). Part (b) has been redrawn with permission [21].

site-specific recombination between *res1* sites, resulting in the removal of the *tet* marker gene. Induced cells and their progeny are thereby rendered tetracycline susceptible.

The *hutT* RIVET reporter strain displayed detection sensitivity across a range of histidine concentrations (from 0.01 to 10 mg ml^{-1}) in culture, which was supported independently by a *lacZ* reporter (Figure 6.6b) [21]. Furthermore, the concentration of histidine in the plant environment was estimated using a *hutT* RIVET strain, based on which, it was extrapolated that free histidine was present in the rhizosphere and shoot environments at a concentration of 0.6 µg ml^{-1} [21].

To enhance the utility of the RIVET system in SBW25, modifications have been incorporated that report the induced cells with a visual reporter, green fluorescent protein (GFP), that eliminates the requirement for replica plating to determine which cells are Tc^S after loss of the *tet* gene (Figure 6.7). Here, the *gfp* gene is under the control of a LacI^q-repressible promoter ($P_{A1/O4/O3}$), and the constitutively expressed repressor gene (*lacI^q*) is inserted into the *res* cassette. When the subject gene is not induced, the *res1–tet–lacI^q–res1* cassette remains stably integrated within the chromosome and LacI^q is expressed resulting in *gfp* repression. However, if the gene linked to *tnpR* is expressed, then resolvase is produced, which catalyzes the resolution of the *res1–tet–lacI^q–res1* cassette. The loss of the *lacI^q* repressor gene lifts *gfp* repression and GFP is expressed. This system has been demonstrated to work in principle in SBW25 (Moon, Zhang and Rainey, unpublished). A tunable RIVET system for SBW25 has been developed, as described previously for *V. cholerae* [22], thereby allowing a range of promoter activities to be analyzed with the RIVET system. Furthermore, RIVET strains that report the activities of a variety of plant-induced genes identified by IVET have been constructed to investigate the spatial and temporal gene expression patterns of these genes *in situ* (Moon, Zhang and Rainey, unpublished).

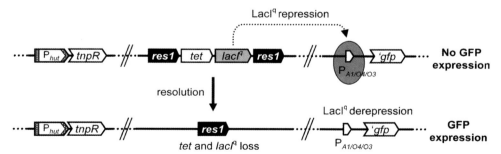

Figure 6.7 The RIVET-GFP principle. A modification to RIVET that enables induced cells to be detected visually, rather than by replica plating for loss of antibiotic resistance (Moon, Zhang and Rainey, unpublished). Expression of *gfp* is under control of a LacI^q-repressible promoter ($P_{A1/O4/O3}$), and a constitutively expressed *lacI^q* repressor gene is incorporated into the *res* cassette. Upon *tnpR* expression and resolution of the *res* cassette, loss of *lacI^q* results in derepression of *gfp* and expression of fluorescence.

6.5
Concluding Remarks and Future Perspectives

Significant progress has been made in unraveling the molecular mechanisms that underpin plant colonization by the model plant growth-promoting rhizobacterium *P. fluorescens* SBW25. This has been greatly advanced by the development and application of a number of advanced genetic tools for SBW25, including IVET, SPyVET and RIVET, in conjunction with the availability of the SBW25 genome sequence. Investigations to date have identified genes that are specifically induced in the plant environment, offering insights into the regulatory networks that control them. The contributions of plant-induced genes to bacterial fitness during plant colonization have also been investigated in many cases, and it has been found that not all plant-induced genes contribute to fitness. A foundation model for the regulatory networks influencing plant-induced genes has been established, and this has provided a valuable knowledge base with which to begin to understand the responses of SBW25 to the plant environment.

Future investigations into the genetic bases underlying plant colonization are likely to focus on teasing apart the complex regulatory networks that control both plant-induced genes and genes in general. The scale of these types of analyses is enormous and will require a highly in-depth and multidisciplinary approach. With functional genomics technologies becoming more readily accessible, such as transcriptomics, proteomics and metabolomics, the application of these to SBW25 will greatly enhance our understanding of its biology. However, to relate these data to behavior in the complex plant environment, the development and use of additional *in situ* based tools will almost certainly be required. These data will provide an important contribution to our understanding of the molecular bases underpinning plant colonization, and in particular, identification and characterization of the key genes and traits required for colonization will likely advance our understanding of field requirements for PGPR enabling enhanced performance of biocontrol strains.

Acknowledgments

SRG is funded by the Leverhulme Trust. We thank Paul B. Rainey for support and helpful discussion.

References

1 Ellis, R.J., Timms-Wilson, T.M. and Bailey, M.J. (2000) *Environmental Microbiology*, **2**, 274–284.

2 Gal, M., Preston, G.M., Massey, R.C., Spiers, A.J. and Rainey, P.B. (2003) *Molecular Ecology*, **12**, 3109–3121.

3 Goymer, P., Kahn, S.G., Malone, J.G., Gehrig, S.M., Spiers, A.J. and Rainey, P.B. (2006) *Genetics*, **173**, 515–526.

4 Jackson, R.W., Preston, G.M. and Rainey, P.B. (2005) *Journal of Bacteriology*, **187**, 8477–8488.

5 Naseby, D.C., Way, J.A., Bainton, N.J. and Lynch, J.M. (2001) *Journal of Applied Microbiology*, **90**, 421–429.

6 Preston, G.M., Bertrand, N. and Rainey, P. B. (2001) *Molecular Microbiology*, **41**, 999–1014.

7 Rainey, P.B. (1999) *Environmental Microbiology*, **1**, 243–257.

8 Rainey, P.B. and Travisano, M. (1998) *Nature*, **394**, 69–72.

9 Spiers, A.J. and Rainey, P.B. (2005) *Microbiology*, **151**, 2829–2839.

10 Zhang, X.X., Lilley, A.K., Bailey, M.J. and Rainey, P.B. (2004) *Microbiology*, **150**, 2889–2898.

11 Bailey, M.J., Lilley, A.K., Thompson, I.P., Rainey, P.B. and Ellis, R.J. (1995) *Molecular Ecology*, **4**, 755–763.

12 Humphris, S.N., Bengough, A.G., Griffiths, B.S., Kilham, K., Rodger, S., Stubbs, V., Valentine, T.A. and Young, I.M. (2005) *FEMS Microbiology Ecology*, **54**, 123–130.

13 Lilley, A.K. and Bailey, M.J. (1997) *Applied and Environmental Microbiology*, **63**, 1584–1587.

14 Turnbull, G.A., Morgan, J.A., Whipps, J.M. and Saunders, J.R. (2001) *FEMS Microbiology Ecology*, **36**, 21–31.

15 Unge, A. and Jansson, J. (2001) *Microbial Ecology*, **41**, 290–300.

16 Osbourn, A.E., Barber, C.E. and Daniels, M.J. (1987) *EMBO Journal*, **6**, 23–28.

17 Mahan, M.J., Slauch, J.M. and Mekalanos, J.J. (1993) *Science*, **259**, 686–688.

18 Jackson, R.W. and Giddens, S.R. (2006) *Infectious Disorders – Drug Targets*, **6**, 207–240.

19 Rediers, H., Rainey, P.B., Vanderleyden, J. and De Mot, R. (2005) *Microbiology and Molecular Biology Reviews*, **69**, 217–261.

20 Camilli, A., Beattie, D.T. and Mekalanos, J. J. (1994) *Proceedings of the National Academy of Sciences of the United States of America*, **91**, 2634–2638.

21 Zhang, X.X., George, A., Bailey, M.J. and Rainey, P.B. (2006) *Microbiology*, **152**, 1867–1875.

22 Lee, S.H., Hava, D.L., Waldor, M.K. and Camilli, A. (1999) *Cell*, **99**, 625–634.

23 Rainey, P.B., Heithoff, D.M. and Mahan, M.J. (1997) *Molecular & General Genetics*, **256**, 84–87.

24 Rezzonico, F., Binder, C., Defago, G. and Moenne-Loccoz, Y. (2005) *Molecular Plant–Microbe Interactions*, **18**, 991–1001.

25 Ramos-Gonzalez, M.I., Campos, M.J. and Ramos, J.L. (2005) *Journal of Bacteriology*, **187**, 4033–4041.

26 Rediers, H., Bonnecarrere, V., Rainey, P.B., Hamonts, K., Vanderleyden, J. and De Mot, R. (2003) *Applied and Environmental Microbiology*, **69**, 6864–6874.

27 Silby, M.W. and Levy, S.B. (2004) *Journal of Bacteriology*, **186**, 7411–7419.

28 Zhang, X.X., Lilley, A.K., Bailey, M.J. and Rainey, P.B. (2004) *FEMS Microbiology Ecology*, **51**, 9–17.

29 Spiers, A.J., Bohannon, J., Gehrig, S.M. and Rainey, P.B. (2003) *Molecular Microbiology*, **50**, 15–27.

30 Spiers, A.J., Kahn, S.G., Bohannon, J., Travisano, M. and Rainey, P.B. (2002) *Genetics*, **161**, 33–46.

31 Zhang, X.X. and Rainey, P.B. (2007) *Molecular Plant–Microbe Interactions*, **20**, 581–588.

32 Evans, K.M., Gatehouse, J.A., Lindsay, W. P., Shi, J., Tommey, A.M. and Robinson, N.J. (1992) *Plant Molecular Biology*, **20**, 1019–1028.

33 Yang, S., Perna, N.T., Cooksey, D.A., Okinaka, Y., Lindow, S.E., Ibekwe, A.M., Keen, N.T. and Yang, C.H. (2004) *Molecular Plant–Microbe Interactions*, **17**, 999–1008.

34 Zhao, Y., Blumer, S.E. and Sundin, G.W. (2005) *Journal of Bacteriology*, **187**, 8088–8103.

35 Loper, J.E. and Henkels, M.D. (1999) *Applied and Environmental Microbiology*, **65**, 5357–5363.

36 Mirleau, P., Delorme, S., Philippot, L., Meyer, J., Mazurier, S. and Lemanceau, P. (2000) *FEMS Microbiology Ecology*, **34**, 35–44.

37 Matthijs, S., Tehrani, K.A., Laus, G., Jackson, R.W., Cooper, R.M. and Cornelis,

P. (2007) *Environmental Microbiology*, **9**, 425–434.

38 Lilley, A.K., Bailey, M.J., Day, M.J. and Fry, J.C. (1996) *FEMS Microbiology Ecology*, **20**, 211–227.

39 Silby, M.W., Rainey, P.B. and Levy, S.B. (2004) *Microbiology*, **150**, 518–520.

40 Camilli, A. and Mekalanos, J.J. (1995) *Molecular Microbiology*, **18**, 671–683.

41 Manoil, C. (2000) *Methods in Enzymology*, **326**, 35–47.

42 Lizewski, S.E., Schurr, J.R., Jackson, D.W., Frisk, A., Carterson, A.J. and Schurr, M.J. (2004) *Journal of Bacteriology*, **186**, 5672–5684.

43 Rensing, C. and Grass, G. (2003) *FEMS Microbiology Reviews*, **27**, 197–213.

44 Stoyanov, J.V., Hobman, J.L. and Brown, N.L. (2001) *Molecular Microbiology*, **39**, 502–511.

45 Solioz, M. and Stoyanov, J.V. (2003) *FEMS Microbiology Reviews*, **27**, 183–195.

46 Adaikkalam, V. and Swarup, S. (2002) *Microbiology*, **148**, 2857–2867.

47 Baynham, P.J., Brown, A.L., Hall, L.L. and Wozniak, D.J. (1999) *Molecular Microbiology*, **33**, 1069–1080.

48 Dasgupta, N., Wolfgang, M.C., Goodman, A.L., Arora, S.K., Jyot, J., Lory, S. and Ramphal, R. (2003) *Molecular Microbiology*, **50**, 809–824.

49 Fredericks, C.E., Shibata, S., Aizawa, S., Reimann, S.A. and Wolfe, A.J. (2006) *Molecular Microbiology*, **61**, 734–747.

7
Quorum Sensing in Bacteria: Potential in Plant Health Protection

Iqbal Ahmad, Farrukh Aqil, Farah Ahmad, Maryam Zahin, and Javed Musarrat

7.1
Introduction

Quorum sensing (QS) is a widespread means for bacterial communities to rapidly and in coordination change genome expression pattern in response to environmental cues and population density. The term quorum sensing was first used in a review by Fuqua and Winans [1], which essentially reflected upon the minimum threshold level of individual cell mass required to initiate a concerted population response. Bacteria that use QS produce and secrete certain signaling compounds called autoinducers, or normally, N-acyl-homoserine lactone (AHL). The bacteria also have a receptor that can specifically detect the inducer. When the inducer binds the receptor, it activates transcription of certain genes, including those for autoinducers synthesis. When only a few other bacteria of the same kind are in the vicinity, diffusion reduces the concentration of the inducer in the surrounding medium to almost zero. So the bacteria produce small amounts only of the inducer. When a large number of bacteria of the same kind are in the vicinity, the inducer concentration crosses a threshold, whereupon greater amounts of the inducer are synthesized. This forms a positive feedback loop and the receptor becomes fully activated [2].

Many Gram-negative bacteria utilize AHL to coordinate expressions of virulence in response to the density of the surrounding bacterial population. Presently, several chemical classes of microbially derived signaling molecules have been identified. The most common signal molecule among Gram-negative bacteria is AHL. Molecules of AHL are produced by LuxI homologues, and constitute, in complex with LuxR homologues, a transcriptional regulator. AHL consists of a conserved homoserine lactone ring with a variable N-acyl chain. The predominant AHL variations induce the presence or absence of a keto or hydroxy group on the C_3 carbon atom as well as the length and saturation of this chain [1,3]. Bioassays and chemical methods (thin layer chromatography (TLC), chromatographic and spectroscopic methods) are routinely used for detection and characterization of signal molecules. Many types of QS systems have been characterized in different bacteria. Several bacterial

Plant-Bacteria Interactions. Strategies and Techniques to Promote Plant Growth
Edited by Iqbal Ahmad, John Pichtel, and Shamsul Hayat
Copyright © 2008 WILEY-VCH Verlag GmbH & Co. KGaA, Weinheim
ISBN: 978-3-527-31901-5

phenotypes essential for the successful establishment of symbiotic, pathogenic or commensal relationship with eukaryotic hosts, including motility, exopolysaccharide production, biofilm formation and toxin production are often regulated by QS [4,5]. Interestingly, production of quorum-sensing interfering (QSI) compounds by eukaryotic microorganisms has aroused immense interest among the researchers since such compounds can influence the bacterial signaling network positively or negatively. On the contrary, synthesis of structural homologues to the various QS signal molecules has resulted in the development of additional QSI compounds that could be used to control pathogenic bacteria. Further, the creation of transgenic plants that express bacterial QS genes is yet another strategy to interfere with bacterial behavior [6]. This chapter presents the concept of the acyl-HSL-based regulatory system (the Lux system), recent progress on bacterial traits under QS control, diversity, the detection and assay system for signal molecules and the role of anti-QS compounds in bacterial plant-disease control.

7.2
Acyl-HSL-Based Regulatory System: The Lux System

The first incidence of such a biological phenomenon came to light with the discovery of the luminescence produced by marine bacteria, *Vibrio fischeri* and *Vibrio harveyi*. These bacteria are nonluminescent in their free-living state in seawater (i.e. at low cell density). However, when grown in high cell densities in the laboratory, *V. fischeri* culture bioluminescence with a blue-green light. These bacteria develop symbiotic relationship with Japanese pinecone fish (*Monocentris japonica*) and squid species (*Euprymna scolopes*) [7].

For a long time, the bioluminescence expressed by *V. fischeri* remained a model system to study density-dependent expressions of gene function [8]. The molecular mechanism of bioluminescence regulation in *V. fischeri* MJI became known in 1983. A brief description of this system is illustrated in Figure 7.1.

Understanding the *lux* gene organization, regulation and function and molecular characterization of luminescence system of *V. fischeri* MJ1 became possible in 1983 through the cloning of a 9 kb fragment of its DNA that encodes all the functions required for autoinducible luminescence in the heterologous host *E. coli* [9]. The bioluminescence gene cluster of *V. fischeri* consists of eight *lux* genes (*luxA–E, luxG, luxI, luxR*) that are arranged in two bidirectionally transcribed operons separated by about 218 bp [9]. This structure is referred to as *lux* regulon. One unit contains the *luxR*, and the other unit, which is activated by *luxR* protein along with the auto-inducer, contains the *lux* ICDABEG operon [10]. The products of both *luxI* and *luxR* genes function as regulators of bioluminescence. The *luxI* gene is the only *V. fischeri* gene required for synthesis of the autoinducer 3-oxo-hexanoyl-homoserine lactone (3-oxo-C_6-HSL) or OOHL in *E. coli* [9]. The *luxA* and *luxB* genes encode subunits of the heterodimeric luciferase enzyme. Luciferase catalyzes the oxidation of an aldehyde and reduced flavin mononucleotide, and the products of this reaction are a long-chain fatty acid, water and flavin mononucleotide. Emission of blue-green light,

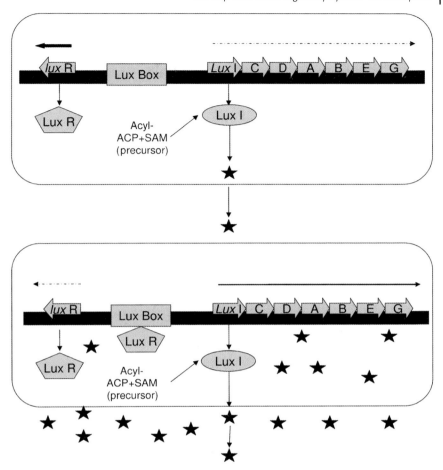

Figure 7.1 LuxI/LuxR quorum-sensing system circuit of *V. fisheri* [17].

with a maximum intensity at 490 nm, accompanying the oxidation reaction has led to this reaction being referred to as bioluminescence. Different luminescent bacteria may exhibit differences in the luminescence spectrum and the color of the emitted light due to differences in sensitizer proteins that cause shifts in wavelength [11]. While *luxC–E* encode products that form a multienzyme complex responsible for the synthesis of the aldehyde substrate utilized by luciferase [12], *luxG* encodes a probable flavin reductase [13] and is followed by a transcriptional termination site [14]. The bioluminescence induction involves an interaction between OOHL and the transcriptional regulator protein *luxR*. Once the autoinducer is bound to the N-terminal regulatory domain, multimer formation by *luxR* is enhanced and the C-terminal domain activates transcription from both the *lux* operons. The *lux* regulon is subjected to a tight regulation. Expression of *luxR* is regulated by two regulatory proteins *luxR* and CAP [15]. Induction of transcription from *lux* ICDA-BEG operon increases the cellular levels of mRNA transcripts required for both bioluminescence and OOHL molecules. The autoinduction mechanism is not initiated until a population has achieved a particular cell density. There are three components that are necessary to sense cell density: (i) signal, a *luxI* homologue; (ii) a means of recognizing the signal (a *luxR* homologue); and (iii) accumulation of the signal. Signal accumulation results either from an increase in cell numbering space with limited flow through or theoretically, by enclosing the cell in a smaller space [16,17].

7.3
QS and Bacterial Traits Underregulation

Quorum sensing among Gram-negative bacteria is widely known and is now better understood; however, a number of Gram-positive bacteria do possess QS systems. The nature of the signal molecule is different from that of the Gram-negative bacteria. For example, in *Staphylococcus aureus*, *LuxS* system utilizes the autoinducer AI-2. Gram-positive systems differ from Gram-negative AHL system in two main factors. (i) First, the signal substances are usually peptides that frequently have posttranscriptional modifications (ii) second, the signal substances often do not diffuse into the cell to bind to a cytoplasmic protein, but signal through binding to a membrane-located part of a two-component system. The cytoplasmic response regulator part of the two-component system then activates the target gene by binding to DNA [18]. Little information is available regarding QS virulence and pathogenicity traits in certain Gram-positive bacteria such as *S. aureus*, *S. epidermidis* and some *Streptococci* and *Bacilli* have been shown to exhibit QS control [18,19]. Other traits in Gram-positive bacteria such as competence development (*Streptococci*), sporulation (*B. subtilis*) and antibiotic biosynthesis (*Lactococcus lactis*) are also controlled by QS [20].

A variety of QS signal molecules are produced by various bacteria. Interestingly, among Gram-negative bacteria, the most common QS signal molecules produced belong to the family of *N*-acyl-homoserine lactones (Table 7.1).

Autoinduction was first described in the marine symbiotic bacterium *V. fischeri* (*Photobacterium*) [21], and AHLs were first identified in the Gram-negative marine bacteria, *V. fischeri* and *V. harveyi*, which play a central role in the regulation of bioluminescence [22,23]. Quorum sensing or cell-to-cell communication has now been discovered in a variety of Gram-negative and Gram-positive bacteria [3,24–26].

A considerable amount of data has been generated on QS and QS-control traits in bacteria, especially in Gram-negative bacteria. Regulation of many bacterial traits such as production of secondary metabolites and extracellular enzymes, biofilm formation, virulence and pathogenicity, bacterial cross-communication and the like have been suspected to arise from QS. The common traits of pathogenic and plant-associated bacteria regulated by QS are given in Table 7.1 and recent reports on the QS signal molecules and QS-regulated bacterial traits are summarized below.

Table 7.1 Plant-associated bacteria that produce autoinducer signal molecules in cell-to-cell communication.

Name of bacteria	Name of autoinducer molecule	Function	References
A. tumefaciens	OOHL	Conjugal transfer of Ti plasmids	[1,27–30]
B. japonicum USDA110	Unknown (structure under determination)	Growth inhibition, bacteriocin?	[31]
C. violaceum	HHL	Pigment production	[26]
Enterobacter agglomerans	OHHL Cyclo (Ala-Val)	NRa	[32,33]
E. carotovora	OHHL	Carbapenem biosynthesis	[34]
E. chrysanthemi	OHHL, HHL and DHL	Pectate lyases synthesis	[35,36]
Erwinia stewarti	OHHL	Capsular exopolysaccharide	[37]
P. aureofaciens	Unknown	Production of phenazine	[38]
P. fluorescens	HHHL, HOHL, HDHL and OOHL	Production of phenazine in some strains	[39–41]
Pseudomonas syringae	OOHL		[40,41]
P. aeruginosa	$3\text{-oxo-}C_{12}\text{-HSL}$	Multiple exoenzymes, biofilm formation, production of cyanide, lectin, pyocyanin, rhamnolipid and so on	[42]
R. solanacearum	OHL, HOOL, OHL, HHL volatile fatty acid methyl ester	NA	[41,43]
R. elti	Seven compounds	Nodulation	[44]
R. leguminosarum bv. viciae	N-(3-hydroxyl-7-cis-tetradecanoyl)-L-homoserine lactone(major). HHL, OHL and heptanoyl homoserine lactone	Rhizosphere genes, growth inhibition, plasmid transfer	[45–47]
S. liquefaciens	BHL, HHL	Swarming, protease	[48]
X. campestris	OOHL, low molecular mass diffusible factor	Extracellular enzymes and polysaccharides virulence determinants	[41,49–51]
V. fischeri	$3\text{-oxo-}C_{10}\text{-HSL}$	Bioluminescence	[47]
Streptomyces	γ-Butyrolactone	Antibiotic production	[52]
Streptococcus	Peptides	Competence, virulence	[53]
Rhodobacter sphaeroides	$7\text{-cis-}C_{14}\text{-HSL}$	Community escape	[54]
V. harveyi	$3\text{-hydroxy-}C_4\text{-HSL}$	Bioluminescence	[22]
Xenorhabdus nematophilus	$3\text{-hydroxy-}C_4\text{-HSL}$ or an antagonist $C_6\text{-HSL}$	Virulence, bacterial lipase	[55]
Nitrosomonas europaea	$3\text{-oxo-}C_6\text{-HSL}$	Emergence from lag phase	[56]
Myxococcus	Peptides, amino acids, others?	Development	[57]

(*Continued*)

Table 7.1 (*Continued*)

Name of bacteria	Name of autoinducer molecule	Function	References
Lactococcus and other lactic acid bacteria	Peptides	Bacteriocin production	[20]
Bacillus	Peptides	Competence development	[58]
Aeromonas hydrophila	C$_4$-HSL	Extracellular protease, biofilm formation	[59]
Aeromonas salmonicida	C$_4$-HSL	Extracellular protease	[59]

BHL: *N*-butyryl-L-homoserine lactone; HHL: *N*-hexanoyl-L-homoserine lactone; OHL: *N*-octanoyl-L-homoserine lactone; DHL: *N*-decanoyl-L-homoserine lactone; dDHL: *N*-dodecanoyl-L-homoserine lactone; TDHL: *N*-tetradecanoyl-L-homoserine lactone; OBHL: *N*-(3-oxo-butyryl)-L-homoserine lactone; OHHL: *N*-(3-oxo-hexanoyl)-L-homoserine lactone; OOHL: *N*-(3-oxo-octanoyl)-L-homoserine lactone; ODHL: *N*-(3-oxo-decanoyl)-L-homoserine lactone; OdDHL: *N*-(3-oxo-dodecanoyl)-L-homoserine lactone; OtDHL: *N*-(3-oxo-tetradecanoyl)-L-homoserine lactone; HBHL: *N*-(3-hydroxy-butyryl)-L-homoserine lactone; HHHL: *N*-(3-hydroxy-hexanoyl)-L-homoserine lactone; HOHL: *N*-(3-hydroxy-octanoyl)-L-homoserine lactone; HDHL: *N*-(3-hydroxy-decanoyl)-L-homoserine lactone; HdDHL: *N*-(3-hydroxy-dodecanoyl)-L-homoserine lactone; HtDHL: *N*-(3-hydroxy-tetradecanoyl)-L-homoserine lactone.

AHL-dependent QS has been extensively studied in *Pseudomonas aeruginosa* in which two hierarchically organized systems (the LasR/I and RhlR/I systems responsible for the production of and response to *N*-(3-oxododecanoyl)-AHL (3-oxo-C$_{12}$-AHL) and *N*-butyryl-AHL, respectively) regulate biofilm formation and virulence gene expression [60,61]. AHL-dependent regulation has been reported to be present in some strains of *P. fluorescens* and *Pseudomonas putida* [62,63].

In the above species of *Pseudomonas*, three global regulation of gene expression have been the subject of extensive investigation in recent years. Global regulator systems are involved in regulation of important traits such as virulence in *P. aeruginosa* and in plant bacterium interactions in beneficial strains of *P. putida* and *P. fluorescens*. These three systems are (i) the GacA–GacS two-component system, (ii) the stress and stationary phase sigma factor Rpos and (iii) the cell population density system [64].

The genus *Burkholderia* was defined by Yabuuchi *et al.* [65] to accommodate most of the former rRNA gpII *Pseudomonas* a widespread in different ecological niches [66]. Several *Burkholderia cepacia* (BCC) strains have attracted considerable interest from agricultural, medical and environmental scientists owing to their diverse physiological action in plant growth promotion and biocontrol purposes, and some are pathogenic to animals and humans [67,68].

The *luxI* of marine bacterium *V. fischeri* synthesizes *N*-(3-oxohexanoyl)-homoserine lactone (3-oxo-C$_6$-HSL), which regulates bioluminescence in a cell density-dependent manner, while *carI* of *Erwinia carotovora* also produces 3-oxo-C$_6$-HSL, which is responsible for the induction of plant cell wall degrading exoenzymes and the antibiotics, Carbapenem [32,69]. After analysis, Bassler *et al.* [70] indicated that at least two separate signal–response pathway converge to regulate the expression of luminescence in *V. harveyi*.

The second extracellular factor, Factor 2, present was *N*-butyryl-homoserine lactone (BHL), indicating that multiple QS systems can occur and interact with each other in a single bacterial species. One signal–response system is encoded by the luxL, luxM and luxN loci [71].

Swift *et al.* [25] investigated density-dependent multicellular behavior in prokaryotes such as in bioluminescence, sporulation, swarming, antibiotic biosynthesis, plasmid conjugal transfer and production of virulence determinants in animals, fish and plant pathogens. In the same year, Givskov *et al.* [72] reported swarming motility of *Serratia liquefaciens* to be QS controlled. Investigations by Wood and Pierson [73] led to the conclusion that the production of phenazine (Ph) antibiotics in *Pseudomonas aureofaciens* (Pau) 30–84 is regulated by the product of phzI, which is a member of the LuxI family *N*-acyl-homoserine lactone (*N*-acyl-HSL synthases). QS-regulated phenotypes including the swarming motility of *S. liquefaciens*, toxin production of *V. harveyi* and the bioluminescence of *V. fischeri* have been documented [72,74–76].

Pearson *et al.* [77] described that two QS systems (las and rhl) regulate virulence gene expression in *P. aeruginosa*. Rhamnolipid production has been reported to require both rhl system and rhlAB (encoding a rhamnosyl transferase). The las system directs the synthesis of the autoinducer *N*-(3-oxdodecanoyl)-homoserine lactone (PAl-1), which induces lasB responsible for the production of elastase. Wood *et al.* [78] reported that *P. aureofaciens* 30–84, which colonizes the wheat rhizosphere, produces three phenazine antibiotics to enhance its survival in competition with other organisms. Here, *N*-hexanoyl-homoserine lactone (HHL) serves as an inter-population signal molecule in the wheat rhizosphere; HHL is also required for phenazine expression *in situ*. Milton *et al.* [79] investigated that *V. anguillarum* may produce multiple AHL signal molecules to control virulence gene expression. Major AHL identified as *N*-3-oxodecanoyl-L-homoserine lactone (ODHL) synthesized by the gene *vanI* belonging to the *luxI* family of putative AHL synthases, *vanR* related to the luxR family of transcriptional activators. In the same year, Swift *et al.* [59] reported that in *Aeromonas* cell division may be linked to QS and that the major signal molecule *N*-butanoyl-L-homoserine lactone (BHL) is synthesized via both AhyI and AsaI. AhyR and BHL are both required for ahyI transcription. In addition, a minor AHL, *N*-hexanoyl-homoserine lactone, was identified.

Quorum sensing plays a role in orchestrating the expression of exoprotease, siderophores, exotoxins and several secondary metabolites and participates in the development of biofilms [80–83]. The *cviI* gene of the soil bacterium *Chromobacterium violaceum* encodes the enzyme for *N*-hexanoyl-L-homoserine lactone. C_6-HSL induces production of the purple pigment violacein as well as antifungal chitinase [26]. Chernin *et al.* [84] showed that *C. violaceum* produces a set of chitinolytic enzymes, whose production is regulated by HHL. In *C. violaceum* ATCC 31532 a number of phenotypic characteristics, including production of the purple pigment violacein, hydrogen cyanide and exoprotease are also known to be regulated by the endogenous AHL–HHL.

Surett *et al.* [85] reported the identification and analysis of the gene responsible for AI-2 production in *V. harveyi*, *S. typhimurium* and *E. coli* as *luxS* (V.h), *luxS* (S.t) and *luxS* (E.c), respectively; these are highly homologous to one another but not to any other identified gene, indicating that *luxS* genes define a new family of autoinducer-producing genes.

Freeman *et al.* [86] showed that *V. harveyi* control light production using two parallel QS systems. It produces two autoinducers, *AI-1* and *AI-2*, which are recognized by cognate membrane-bound two-component hybrid sensor kinases called *luxN* and *luxQ*, respectively. They have also showed that the *AI-1* and *LuxN* have a much greater effect on the level of *LuxO* phosphate and therefore on *Lux* expression than do *AI-2* and *LuxQ*. *LuxO* functions as an activator protein via interaction with alternative sigma factor sigma 54. *LuxO* together with sigma 54 activates the expression of negative regulator of luminescence; phenotypes other than *lux* are regulated by *LuxO* and *sigma 54* [87]. In the same year, Miyamoto *et al.* [88] observed that *LuxO* is involved in control of luminescence in *V. fischeri* but *luxO* was stimulated by *N*-acyl-HSL autoinducer, indicating that *luxO* is part of a second signal transduction system controlling luminescence.

McKnight *et al.* [89] revealed that a second type of intercellular signal is involved in *lasB* induction (elastase). This signal was identified as two heptyl-3-hydroxy-4-quinolone, designated as *Pseudomonas* quinolone signal (PQS). Its production and bioactivity depend on the las and rhl QS system, respectively. This signal is not involved in sensing cell density. Zhang and Pierson [90] reported that a second QS system, *CsaR–CsaI*, is involved in regulating biosynthesis of cell surface components in *P. aureofaciens* 30–84. Smith *et al.* [91] described that QS of *P. aeruginosa* contribute to its pathogenesis both by regulating expression of virulence factors (exoenzymes and toxins) and by inducing inflammation. 3-oxo-C_{12}-HSL activates T cells to produce the inflammatory cytokine gamma interferon [91]. It also induces cyclooxygenase 2 (Cox-2) expressions. *Sinorhizobium meliloti* required exopolysaccharides (EPSs) for efficient invasion of root nodules on the host plant: alfalfa. *S. meliloti ExpR* activates transcription of genes involved in *EPSII* production in a density-dependent manner [92].

von Bodman *et al.* [93] described how phytopathogenic bacteria have incorporated QS mechanisms into complex regulatory cascades that control genes for pathogenicity and colonization of host surface. QS involves the production of extracellular polysaccharide-degradative enzymes, antibiotics, siderophores and pigments as well as Hrp protein secretion, Ti plasmid transfer, mobility, biofilm formation and epiphytic fitness. Wagner *et al.* [94] investigated global gene expression patterns modulated by QS regulons. *Pseudomonas* quinolone signal is also an integral component of the circuitry and is required for the production of rhl-dependent exoproduct at the onset of stationary phase [95]. Twenty novel QS-regulated proteins were identified, many of which are involved in iron utilization, suggesting a link between QS and the iron regulatory system. *PhuP* and *HasR* are components of the two distinct heme uptake systems present in *P. aeruginosa*, both proteins are positively regulated by QS cascade.

McGrath *et al.* [96] reported that PQS production was dependent on the ratio of 3-oxo-C_{12}-HSL and C_4-HSL, suggesting a regulatory balance between the QS systems. Juhas *et al.* [97] identified the gene PA2591 as a major virulence regulator, *vqsR*, in the QS hierarchy. Sircili *et al.* [98] reported that QS activates that expression of the *lee* genes in EPEC (enteropathogenic *Escherichia coli*) with *QseA* activating transcription of ler, hence QS is involved in modulating the regulation of the EPEC virulence gene. QS regulates type III secretion (TTS) in *V. parahemolyticus*. At higher cell density (in the presence of autoinducers), QS represses TTS in *V. harveyi* and *V. parahaemolyticus* [99]. One of the possible QS-regulated phenotype is swarming: a flagella-driven movement of differentiated swarmer cells (hyperflagellated, elongated, multinucleated) by which bacteria can spread as a biofilm over a surface. The glycolipid or lipopeptide biosurfactants thereby produced function as wetting agents by reducing the surface tension. Surface migration is regulated by AHL and also other low-molecular-mass signal molecules (such as furanosyl borate diester AI-2) in biosurfactant production of different bacteria [100]. The first evidence for autoinduction in *E. amylovora* and a role of an AHL-type signal were reported. Further, two major plant virulence traits, production of extracellular polysaccharide (amylovora and levan) and tolerance to free oxygen radicals were controlled in a bacterial cell density dependent manner [101]. In *V. harveyi* *Hfq* together with sRNAs create an ultrasensitive regulatory switch that controls the critical transition into higher cell density, QS mode [102]. In *V. harveyi*, three-way coincidence detectors in the regulation of a variety of genes including those responsible for bioluminescence, type III secretion and metalloprotease (VVP) production [99]. Metalloprotease (VVP) production by *V. vulnificus* is known to be regulated by quorum sensing, supported by the fact that expression of *vvp* gene was closely related with expression of the *luxS* gene [104].

Chatterjee *et al.* [105] reported that *E. carotovora* subspecies possess two classes of QS signaling system, since AHL control the expression of various traits including extracellular enzyme/protein production and pathogenicity. The AHL response correlates with *expR*-mediated inhibition of exoprotein and secondary metabolite production. PQS production is negatively regulated by the *rhl* QS system and positively regulated by the *las* QS system. For more detailed information on QS in specific bacteria, see the excellent review articles on Vibrios (*V. harveyi*, *V. cholerae*, *V. fisheri*, *V. anghillarium*, *V. vulnificus*) by Milton [106]. It is concluded that AHL-dependent QS in *Vibrio fischeri* (*LuxI/R* system) is not found in all *Vibrios*. A more complex system is found in *V. harveyi*. Three parallel systems transmit signals via phosphorelase that converge on to one regulatory protein *LuxO*. Components of the three systems are found in Vibrios. However, the number and types of signal circuits found each strain indicates the diversity and complexity of the *Vibrio* QS systems [106].

7.4
QS in Certain Phytopathogenic Bacteria

7.4.1
E. carotovora

Erwinia are the only members of the family Enterobacteriaceae that are pathogenic to plants. Three major species among these are *E. amylovora*, *E. carotovora* and *E. herbicola*. They have gained economic significance due to the diseases they cause to various commercial crops such as onion, potato, carrot, celery, cucumber and pine apple. Some other species are also known to be pathogenic/opportunistic pathogens to insects and animals [103].

The soft rot *Erwinia* spp. includes *E. chrysanthemi*, *E. carotovora* sub sp. carotovora (Ecc), *E. atroseptica* (Eca) and *E. betavascolorum* (Ecb). These organisms characteristically produced an abundance of exoenzymes including pectin methyl esterase, pectate lyase, pectin lyase, polygalacturonase, cellulose and proteases [107,108]. These enzymes are virulence factors of the bacterium. Disruption of genes encoding above enzymes often leads to reduction of virulence in planta [109,110]. Other secondary metabolites widely studied in *Erwinia* sp. are the production and QS-based regulation of carbapenem, a β-lactam antibiotic. The cell density-dependent production of exoenzymes in Ecc is also dependent on the synthesis of OHHL by Car1, which encodes a luxI homologue in Ecc. Disruption of carI leads to a diminution of exoenzyme synthesis and a consequent reduction of virulence in planta [69,111]. It has been reported that CarR protein has the ability to sequester OHHL away from an additional LuxR homologue responsible for induction of exoenzymes synthesis. A second LuxR homologue *expR* (*eccR/rexR*) has been identified in ECC. Anderson *et al.* [112] deduced that control of exoenzyme production takes place through the input of many regulators, some of which interact with components of the QS system.

Other species of *Erwinia* have been found to produce acyl-HSLs; Eca, Ebc and *E. chrysanthemum* synthesize different acyl-HSLs (OHHL, HHL and *N*-decanoyl-L-homoserine lactone, DHL) [32,35,41].

7.4.2
R. solanacearum

R. solanacearum causes vascular wilt disease of many plants primarily due its ability to produce an acidic exopolysaccharide and plant cell wall degrading extracellular enzymes [113]. Expression of these virulence factors occurs in an apparent cell density dependent manner, with maximum expression at high cell densities [114]. The *lysR* type regulator, *PhcA*, is central to the complex regulation of EPS and extracellular enzymes, and hence, pathogenicity in *R. solanacearum*. PhcA activity is regulated by a two-component regulatory system, which, in turn, is responsive to the QS signal molecules, 3-hydroxy palmitic acid methyl ester (3OHPAME). Exogenous addition of 3OHPAME to cultures at low cell density does induce precocious production of EPS and enzymes [114–116] *PhcS* and *PhcR* that makeup the two-component system responsive to 3OHPAME. It has been shown that *PhcS* and *PhcR* act together to negatively regulate the expression of *PhcA*-regulated genes in the absence of 3OHPAME [117].

7.4.3
Xanthomonas campestris

Cell-to-cell communication in *X. campestris* (Xcc), the causative agent of the black rot of cruciferous plants, has been described. This organism also produces a variety of extracellular enzymes and EPS, which are the virulence factors of this bacterium. In Xcc 8004, production of extracellular enzymes and EPS has been shown to be subject to regulation by the *rpf* (regulation of pathogenicity factor) cluster comprising some nine genes (*rpfA-I*). It has indicated that Xcc appears to have a unique system for cell-to-cell signaling; it shares similarities with the system employed by *V. harveyi* and *R. solanacearum*. Most evident is that all three species appears to contain specialized two-component regulators to integrate and/or sense their respective QS signals [49,118–121].

7.4.4
Other Bacteria

More than 30 species of the genus *Burkholderia* are described, many of which are human pathogens [68]. It has been indicated that all *Burkholderia* sp. investigated so far employ QS systems that relay on AHL signal molecules to express certain phenotypic traits in a population density manner.

The role of the quorum-sensing system in the expression of a variety of traits was found in *Pseudomonas corrugate* that produce short-chain AHL quorum-sensing signal molecules. The main AHL produced was *N*-hexanoyl-L-homoserine lactone (C$_6$-AHL). These bacteria also possess an AHL quorum-sensing system designated PcoI/PcoR [122].

Pomianek and Semmelhack [123] have discovered effective modulators of the autoinducer-1 circuit for bacterial quorum sensing by the synthesis and evaluation of a small library of aryl-substituted acyl-homoserine lactone analogues. This series highlights the sensitivity to structure of the contrasting responses of agonism and antagonism of the natural signal and identifies an analogue that provokes the same response as the natural signal but at 10-fold lower concentration, a 'superagonist'.

AHL signal molecules are utilized by Gram-negative bacteria to regulate gene expression in a density-dependent manner, as for example, *S. liquefaciens* MG1 and *P. putida* IsoF colonize tomato roots, produce AHL in the rhizosphere and increase systemic resistance of tomato plants against the fungal leaf pathogen, *Alternaria alternata*. The AHL-negative mutant *S. liquefaciens* MG44 was less effective in reducing symptoms and *A. alternata* growth. Salicylic acid (SA) levels were increased in leaves when AHL-producing bacteria colonized the rhizosphere [131].

7.5
Quorum-Sensing Signal Molecules in Gram-Negative Bacteria

Three types of autoinducers are reported in literature (i) AHL, acyl-HSL or HSL) found in Gram-negative bacteria, autoinducer peptides (AIP) in Gram-positive bacteria and autoinducer-2 compounds (AI-2s), which are found in Gram-negative and Gram-positive bacteria. Some of the common signal molecules are listed in Table 7.2.

Table 7.2 Bioassay test strain used for the detection of autoinducer (AHL) and QS inhibition.

Reporter strain	Description	AHLs detected	References
E. coli VJS533 (pHV2001)	*V. fischeri* ES114 *lux* regulon with inactivated *LuxI* in pBR322; Ap^R	3-Oxo hexanoyl-HSL, hexanoyl-HSL, 3-oxooctanoyl-HSL, octanoyl-HSL	[129]
A. tumefaciens NTI (pJM749, pSVB33-23)	*traI::lacZ* and *traR* on separate plasmids; pTi is cured; Cb^R, Km^R	3-Oxo hexanoyl-HSLs, octanoyl-HSL and other acyl-HSLs	[130]
V. harveyi D1	Unknown mutation resulting in reduced autoinducer production	3-Hydroxy, butanoyl-HSL, 3-hydroxyl valeryl-HSL	[128]
E. coli MG-4 (pKDT17)	*LasB::lacZ* translational fusion and ptac-*rhl*R; Ap^R	2-Hydroxy,3-oxo and unsubstituted acyl-HSLs with side chain lengths of 8–14 carbons	[78,129]
C. violaceum CV0 blu	*cviI::*Tn5 *xylE* (inactivated *cviI*, an autoinducer synthase required for violacein production): Hg^R, Km^R, Cm^R	Hexanoyl-HSL, butanoyl-HSL, 3-oxohexanoyl-HSL, octanoyl-HSL, acyl-HSLs with longer side chains can be detected by screening for inhibition of hexanoyl-HSL-mediated violacein production	[59]
E. coli XL1 Blue (pECp61.5)	*rhlA::lacZ* translation fusion and ptac-*rhl*R in pSW205; Ap^R	Butanoyl-HSL, hexanoyl-HSL	[129]
R. solanacearum (p395B)	Inactivated *solI*, p395B contains *aidA::lacZ* fusion, Nx^R, Sp^R, Tc^R		[117]
A. tumefaciens A136	*traI–lacZ* fusion (pCF218) (pCF372), AHL biosensor		[24]
A. tumefaciens KYC6	Positive control for AHL assay	3-oxo-C$_8$-HSL overproducer	[24]
C. violaceum ATCC 12472	Type strain (QSI indicator strain)		ATCC
C. violaceum O26	MiniTn5 mutant of 31532, AHL biosensor		[26]
C. violaceum ATCC 31532	Nonpigmented, positive control for AHL assay	C$_6$-HSL production	ATCC
P. aeruginosa PAO1	Positive control for QSI	C$_4$-HSL and 3-oxo-C$_{12}$-HSL production	V Deretic
P. aeruginosa PDO 100	*RhlI* mutant	C$_4$-HSL$^-$	[132]
P. aeruginosa 30–84	*RhlI, lasI* mutant	C$_4$-HSL$^-$ and 3-oxo-C$_{12}$-HSL$^-$	[132]
P. aeruginosa PAO-MW1	Phenazine production, QSI indicator and type strain		[133]
Environmental bacterial isolates	QSI screening, Spring Lake, San Marcos, TX		[134]

Partially adapted from McCLean *et al.* [134]. Sources: ATCC, American Type Culture Collection; V. Deretic, Department Molecular Genetics and Microbiology, University of New Mexico Health Sciences Center, Albuquerque, NM; E.P. Greenberg, Department of Microbiology, University of Iowa, Iowa City, IA.

7.5.1
Bioassays for the Detection of Signal Molecules

Autoinducer signal molecules are produced at very low concentrations. Many bioassays and sensor systems have been developed that allow facile detection, characterization and quantitative analysis of microbial acyl-HSLs [40,117,124–128]. Most of the autoinducers are mutants that cannot synthesize their own AHL. So the wild-type phenotype is only expressed upon the addition of exogenous AHL.

Various bioassay strains developed have reporter genes including *lacZ*, *gfp*, *lux* and the production of an endogenous pigment. Some of the routinely used bioassays and their characteristic features are presented in Table 7.2. The details of these strains and their basis may be seen in more details from the literature published elsewhere [17,31].

Autoinducer sensors have generally been dependent on the use of lacZ reporter fusions in an *E. coli* or *A. tumefaciens* genetic background or on the induction or inhibition of the purple pigment, violacein in *C. violaceum* [135]. The inhibition of quorum sensing by *Bacillus* sp. in this system has been shown in Figure 7.2 as demonstrated in our laboratory [161].

An *A. tumefaciens* based AHL sensor pDC141E33 has been developed in which *lacZ* is fused to *traG*, that is regulated via the *luxR* homology *TraR* by incorporating 5-bromo-4-chloro-3-β-D-galactopyranoside (X-Gal) in the agar overlay. It is possible to visualize AHLs on TLC plates or on Petri dishes. It has been found that the pDC141E33 vector allows detection of the broadcast range of AHLs derivatives and shows the greatest sensitivity [31]. However, this bioreporter does not detect *N*-butanoyl-homoserine lactone, even at high concentrations [40].

Another frequently used bioreporter strain is based on the induction or inhibition of violacein production in *C. violaceum*. In this bacterium, production of pigment is regulated by HHL [26]. Strain CV026 is a violacein negative miniTn$_5$ mutants of *C. violaceum* in which pigment production can be restored by incubation with exogenous AHL.

AHL compound (C_{10}–C_{14}) *N*-acyl chains are unable to induce violacein production, but long-chain AHLs can be detected by their ability to inhibit violacein production when

Figure 7.2 Antiquorum-sensing activity of *Bacillus* sp. using *C. violaceum* ATCC 2147 [161].

an activating AHL (OHHL or HHL) is included into the assay medium [26]. In this case, a white halo on a purple background constitutes a positive result. However, this bioreporter cannot detect any of the 3-hydroxy-derivatives and lacks sensitivity to most of the 3-oxo derivatives [26,41]. However, this biosensor is activated by cyclic dipeptides [33].

Similarly, another strain that is used as bioreporter based on AHL-induced bacterial swarming is *S. liquefaciens* MGI. The *S. liquefaciem* MG44 strain is a SWrI::T45 mutant of MGI that cannot synthesize BHL or HHL and therefore requires an exogenous supply for serrawetin synthesis and swarming [136,137]. Many Gram-negative bacteria that employ QS systems produced multiple AHL molecules, for example *Rhizobium elti* produces at least seven AHLs [44]. It is presumed that these additional signals may be due to the presence of multiple QS systems or may be the products of a single AHL synthase.

Similarly, some bacteria may produce signals that are not detected by one of the reporters or they may produce molecules at levels below the threshold of sensitivity of the reporter [40]. Therefore, combinations of different bioreporters have been used to detect AHL-like activities.

7.5.2
Chemical Characterization of Signal Molecules

Autoinducers can be separated and purified by preparative reverse-phase high-pressure liquid chromatography. Normally 4–6 l of bacterial culture supernatant grown in chemically defined medium is extracted with dichloromethane, ethyl acetate or chloroform. The extract is evaporated on a rotary evaporator. The residue is then applied to a C_{18}-reverse-phase semi-preparation column and eluted with methanol gradient or an isocratic mobile phase of acetonitrile : water [31]. Fractions can be employed by using bioreporters and active fractions can be rechromatographed. Once a single active peak has been obtained, it can be further analyzed by analytical HPLC or subjected to identification techniques (Figure 7.3).

Partial characterization of autoinducers is normally carried out by TLC on C_{18}-reversed-phase plates with the sample (supernatant or extracts) and with different standards and after chromatography overlaid with a soft agar suspension of the indicators [26,40]. Using *C. violaceum* biosensor strain, *N*-acyl-homoserine lactones and cyclic peptides are detected [33].

TLC provides preliminary information on both the number and the structural groups of the compounds present in the supernatant or the extract fractions. The Rfs value can be compared with known standards.

Not only AHL detection but also QS inhibition by various compounds could be easily detected when TLC is coupled with other QSI indicator strains. Using *A. tumefaciens* bioreporter and CV026, bioassays are useful and rapid to test a large number of different microbial isolates [41,47]. Characterization can also be carried out by analytical HPLC using C_{18}-reversed-phase column.

Further identification of QS molecules have been carried out using spectroscopic techniques, such as mass spectrometry (MS), nuclear magnetic resonance spectroscopy (NMR) and infrared spectroscopy (IR).

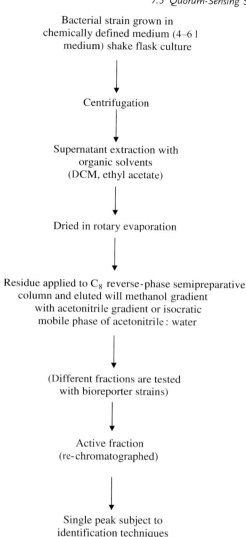

Bacterial strain grown in
chemically defined medium (4–6 l
medium) shake flask culture

Centrifugation

Supernatant extraction with
organic solvents
(DCM, ethyl acetate)

Dried in rotary evaporation

Residue applied to C_8 reverse-phase semipreparative
column and eluted will methanol gradient
with acetonitrile gradient or isocratic
mobile phase of acetonitrile : water

(Different fractions are tested
with bioreporter strains)

Active fraction
(re-chromatographed)

Single peak subject to
identification techniques

Figure 7.3 Schematic representation of the detection and characterization of quorum-sensing signal molecules from the bacterial cells.

Mass spectrometry detects even picomols of sample and can be coupled to HPLC or GC and various types ionization available such as electron impact (EI-MS), fast atom bombardment (FAB-MS) and chemical ionization, positive ion atmospheric pressure chemical ionization (APCI-MS) as described by various authors [17,31,47,135,139].

NMR is very useful to elucidate organic structures. The hydrogens and the carbons in an organic structure resonate at different chemicals shifts, depending on their environment and further appear as singlet, doublet, triplet and so on. Structans of many autoinducers was determined by NMR and combination with FAB-MS and MS with chemical ionization [32,46,140]. Similarly, IR spectrometry is useful for the identification of the functional groups in a molecule and can be used in combination with other techniques to precisely assign chemical structures.

Yang *et al.* [138] described a high-throughput detection method of QS molecules by colorimetry. The colorimetric assay is a modified version of the method of Goddu and coworkers described for the analysis of ester molecules [139]. This method can be used very quickly and easily to analyze the amount of lactone compounds and luctonase activity using $\approx$20–50 μl of sample in a 96-well plate. Its detection limit is 1 nmol of the lactone compounds and comparable with HPLC.

Fekete *et al.* [140] identified the rhizospheric bacterial (*Acidovorax* sp.) AHLs, CN-C$_3$-hydroxydecanoyl)-homoserine lactone. Combination of ultraperformance liquid chromatography (UPLC), ultrahigh-resolution mass spectrometry and *in situ* biosensors were used. The results were obtained by the analysis of bacterial QS molecules (HR-MS, Fourier transform ion cyclotron resonance mass spectrometry, nano-LC-MS) and by the aid of a biosensor. The results obtained from UPLC, FTICR-MS, nano-LC-MS and bioassays have been compared to attempt structural characterization of AHL without chemical synthesis of analytical standard.

7.6
Interfering Quorum Sensing: A Novel Mechanism for Plant Health Protection

It has now become apparent that AHLs are widely used for regulating diverse behavior in epiphytic, rhizosphere-inhabiting and pathogenic bacteria including phytopathogens, and that plants may produce their own metabolites that interfere with this signaling. Transgenic plants that produce high levels of AHLs or that can degrade bacteria-produced AHLs have been made. These plants have dramatically altered susceptibilities to infection by pathogenic *Erwinia* species. Further, such plants will prove to be useful tools in determining the roles of AHL-regulated density-dependent behavior in growth promotion and control of phytopathogenic bacteria [6]. Practically, there are three ways to interfere with QS mediated by AHLs: (i) blocking binding of AHLs with its receptor; (ii) competitive inhibition; and (iii) degradation of AHL.

There are three major groups of AHL biosynthetic enzymes:

(i) LUX 1 type, which appear to be most common.
(ii) AHL biosynthetic enzyme LUX M type, which has no significant homology with LUX 1 [141].
(iii) A third class of AHL synthase Hdts has been identified from the biocontrol strain (*P. fluorescens*, F113) [142].

It is presumed that possession of different AHL synthases may afford some protection from competitors or host species, developing inhibitory molecules that target the synthase.

As described earlier, there is a widespread occurrence of cell-to-cell signaling among different bacterial species. Many pathogenic bacteria, related to humans (*P. aeruginosa*, *Vibrio*, *S. aureus*) and plants, regulate virulence gene expression by quorum sensing. Also many organisms utilize the same species of molecule to regulate different phenotypes. Therefore, interspecies communication is likely to occur where different autoinducer-producing bacterial species inhabit a particular niche [82,143–145].

Plants have long been known to interact with symbiont bacteria though QS signaling, and plant pathogens use QS to colonize their hosts. A review of QS signaling in plant–pathogen symbiont interaction described some of the potential applications that could arise from their relationship [93]. Newton and Fray [146] focused more specifically on *E. carotovora* and *Agrobacterium* in their review of AHL expression and repression in the plant rhizosphere.

P. aeruginosa regulates several gene systems, including those required for the production of extracellular enzymes and toxins [139,147].

Many excellent articles are available on interfering QS among human pathogenic bacteria for developing novel anti-infective drug. Similarly, approaches are also applicable to control pathogenicity of phytopathogenic bacteria by disrupting QS system. Some of these attempts are described below.

Plant–microbe relationships with potential for pathogen control are described by other workers, such as *B. cepacia* in onion [148], *Ceratocystis ulmi* (adimorphic fungus that causes Dutch elm disease) [149], *S. meliloti* symbiont in legumes and P38 (pathogen) [150] and wine grape consortia [151].

Bacteria produce various types of AHL molecules that may result in bacterial crosstalk and may modulate the bacterial activity [78]. Some bacteria produce lactonases and acylases, which can disrupt cell-to-cell communication by hydrolyzing the cyclic ester or amide linkage of the QS molecule. Part of the biocontrol activity of *Bacillus thuringiensis* is through AHL lactonase [152]. Biocontrol efficacy against *E. carotovora* was reduced in *B. thuringiensis* mutated for AHL lactonase. Also strain of *E. coli*, *Bacillus fusiformis* lacking AHL lactonase showed similar lack of antipathogenic capability when cultivated on potatoes [153]. Intentionally, BT did not inhibit the growth of *E. carotovora*, but rather inhibited its virulence and ability to cause soft rot disease in potatoes. Molina *et al.* [154] demonstrated the effect of recombinant AHL lactonase in transforming strains incapable of biocontrol into biocontrol agents.

P. fluorescens was transformed with the *adiiA* gene encoding AHL lactonase under constitutive promoter. Another enzyme (porcin kidney acylase I) was shown to have AHL-degrading capability *in vitro* and the ability to inhibit growth of aquatic biofilms in an aquarium water sample [155].

Plants have also evolved a mechanism that enables them to detect and respond to acyl-HSL messaging systems to prevent or limit infections. Such interference could include the production of signal mimics, signal blocker or signal-degrading enzymes or production of compounds that block the activity of AHL-producing enzymes [6].

A complex blocking molecule is produced by the Australian marine alga *Delisea pulchra*. This produces halogenated furanones that have some structural similarity to AHL. It appears that *D. pulchra* uses these AHL blockers *in vivo* to inhibit bacterial cell swarming and attachment responses, thus preventing the buildup of bacterial biofilms on the algal surfaces [136,156].

Teplitski *et al.* [157] reported AHL inhibitory activities in exudates from pea seedlings. This observation was also confirmed in our laboratory (unpublished data). However, the nature of compounds has not yet been characterized. Similarly, plants such as carrots, garlic, habanen (chili) and water lily produced compounds that interfere with bacterial QS [158].

In addition, the waxy compound, propolis, produced by bees also contains QSI activity. Garlic extracts contain at least three different QSI compounds, one of which is a heterocyclic compound, containing four carbon and two sulfur atoms [159]. Other plants including crown vetch, soybean and tomato have also been found to be able to interfere with QS [157].

Recently, QS interference by aqueous and alcoholic extracts of medicinal plants/ weeds as well as certain essential oil have been reported from the United States and India [160–163].

Bacterial phenotypes controlled QS are frequently regulated by additional environment factors. In some cases, population-density signals can be modulated or overridden by factors such as oxygen tension, nutrient starvation, iron limitation or catabolite repression. It is possible that the plant-produced compounds indirectly alter the bacterial AHL response rather than do it directly, but even if this was the case, such compounds could prove to be important in determining the outcome of interactions between higher plants and pathogenic microbes [6]. Supplying transgenic plants with the ability to block or degrade AHL signals may provide another approach for engineering resistance to phytopathogenic bacteria such as *E. carotovora* [6].

One Gram-negative bacterium belonging to the species *Variovorax paradoxus* has been isolated from soil sample and degrades 3-oxo-C_6-HSL [164]. In this case, the enzyme has been identified but not yet cloned.2 Another bacterium, a Gram-positive *Bacillus* sp. has a N-AHSL hydrolase encoded by the *aiiA* gene [165]. The *aiiA* gene has been cloned and introduced into tobacco to generate transgenic plants that exhibit increased resistance toward *E. carotovora*, whose pathogenicity is dependent on QS-regulated production of enzymes macerating the plant cell wall [166]. Other strains of *Bacillus* sp. have been now found to have *aiiA* gene [152,167].

Induction of systemic resistance in tomato by *N*-acyl-L-homoserine lactone producing rhizosphere bacteria has been demonstrated by Schuhengger *et al.* [131]. They observed that *S. liquefaciens* MG1 and *P. putida* IsoF colonize tomato roots, produce AHL in the rhizosphere and increase systemic resistance of tomato plants against the fungal leaf pathogen *Alternaria alternata*. The AHL-negative mutant *S. liquefaciens* MG44 was less effective in reducing symptoms and *A. alternata* growth as compared to the wild type. Salycyclic acid (SA) levels increased in leaves when AHL-producing bacteria colonized the rhizosphere. Macroarray and Northern blot analyses revealed that AHL molecules systematically induce SA and ethylene-dependent defense genes. Thus, AHL molecules play a role in the biocontrol activity of rhizobacteria through the induction of systemic resistance to pathogens.

This appears to be a very promising approach toward preventing AHL signaling in plant-associated/pathogenic bacteria compared to the GM plant approach.

7.7
Conclusion

It is now clear that QS is a widespread gene-regulatory mechanism among Gram-negative bacteria. However, QS in Gram-positive bacteria has now been explored in few cases. In plant-associated bacteria, including pathogenic ones, a variety of traits are under QS control. However, many traits are still to be reconfirmed. Extensive information on the chemical structure and function of QS molecules among Gram-negative bacteria is now available. The signal molecules exhibit structural diversity and most common signal molecules are AHLs. Interestingly, QS has been identified as a novel target to influence the bacterial virulence and pathogenicity. The natural and synthetic compounds having QS-interfering properties have been identified. It is conceivable that quorum-sensing inhibition may represent a natural, widespread, antimicrobial strategy utilized by plants and other organisms with significant impact on biofilm formation. The QS may be targeted in different way. The creation of transgenic plants that express bacterial QS genes is yet another interesting strategy to interfere with bacterial behavior and disease control. Plants are now known to harbor anti-QS activity/metabolites that could disrupt the QS-controlled pathogenicity of bacteria and manipulate plant–microbe interactions to obtain improved crop production. More fundamental research on this mechanism and the presence of multiple QS systems and their interaction with each other in a single bacterial species remains to be conducted. Now, with the advanced understanding of QS systems operating in various microorganisms and methods for characterization of QS molecules and existence of bacteria-to-bacteria and bacteria-to-plant interactions, we are able to target QS-regulated functions by either (i) degrading or inhibiting QS signals, signal–cell receptors.

However, many questions remain to be solved as an in-depth knowledge of the AHL signaling system, which is common in a number of important plant-associated bacteria, is needed. It is possible that gross disruption of AHL-based communication in the rhizosphere may adversely affect the colonization or behavior of a number of important growth-promoting or biocontrol species [90]. Many more interesting phenomena between mixed microbial communities and their interactions with plants are to be explored.

Acknowledgments

We are thankful to Professor Robert J.C. McLean (USA) for his encouragement and support to work on QS and Professor John Pitchel (USA) and Dr S. Hayat (AMU, Aligarh) for critical input and preparation of this manuscript. Finally, the cooperation received by students, especially by Mohd Imran and Miss Sameena Hasan, (AMU) are thankfully acknowledged.

References

1 Fuqua, W.C. and Winans, S.C. (1994) *Journal of Bacteriology*, **176**, 2796–3806.

2 Whitehead, N.A., Barnard, A.M.L., Slater, B.H., Simpson, N.J.L. and Salmond, G.P.C. (2001) *FEMS Microbiology Reviews*, **25**, 365–404.

3 Salmond, G.P., Bycroft, B.W., Stewart, G.S. and Williams, P. (1995) *Molecular Microbiology*, **16**, 615–624.

4 Kendall, M.M. and Sperandio, V. (2007) *Current Opinion in Gastroenterology*, **23** (1), 10–15.

5 Gonzalez, J.E. and Keshavan, N.D. (2006) Messing with bacterial quorum sensing. *Microbiology and Molecular Biology Reviews*, **70**, 859–875.

6 Fray, R.G. (2002) *Annals of Botany*, **89**, 245–253.

7 Visick, K.L. and McFall-Ngai, M.J. (2000) *Journal of Bacteriology*, **182**, 1779–1787.

8 Hastings, J.W. and Greenberg, E.P. (1999) *Journal of Bacteriology*, **181**, 2667–2668.

9 Engebrecht, J., Nealson, K. and Silverman, M. (1983) *Cell*, **32**, 773–781.

10 Engerbrecht, J. and Silverman, M. (1987) *Nucleic Acids Research*, **15**, 10455–10467.

11 Meighen, E.A. (1992) *Encyclopedia of Microbiology* (ed. J. Lederberg), Academic Press, pp. 309–319.

12 Boylan, M., Miyamoto, C., Wall, L., Graham, A. and Meighen, E.A. (1989) *Photochemistry and Photobiology*, **49**, 681–688.

13 Zenno, S. and Saigo, K. (1994) *Journal of Bacteriology*, **176**, 3544–3551.

14 Swartzman, E., Kapoor, S., Graham, A.E. and Meighen, E.A. (1990) *Journal of Bacteriology*, **172**, 6797–6802.

15 Shadel, G.S. and Baldwin, T.O. (1992) *The Journal of Biological Chemistry*, **267**, 7690–7695.

16 Shadel, G.S. and Baldwin, T.O. (1991) *Journal of Bacteriology*, **173**, 568–574.

17 Gera, C. and Srivastava, S. (2006) *Current Science*, **90** (5), 666–677.

18 Otto, M. (2004) *FEMS Microbiology Letters*, **241**, 135–141.

19 Balaban, N., Giacometti, A., Cirioni, O., Gov, Y., Ghiselli, R., Mocchegiani, F., Viticchi, C., Del Prete, M.S., Saba, V., Scalise, G. and Dell' Acqua, G. (2003) *The Journal of Infectious Diseases*, **187**, 625–630.

20 Kleerebezem, M., de Vos, W.M. and Kuipers, O.P. (1999) *Cell–Cell Signalling in Bacteria* (eds G.M. Dunny and S.C. Winans), ASM Press, Washington, pp. 159–174.

21 Nealson, K.H., Platt, T. and Hastings, J.W. (1970) *Journal of Bacteriology*, **104**, 313–322.

22 Cao, J.-G. and Meighen, E.A. (1989) *The Journal of Biological Chemistry*, **264**, 21670–21676.

23 Eberhard, A., Burlingame, A.L., Eberhard, C., Kenyon, G.L., Nealson, K.H. and Oppenheimer, N.J. (1981) *Biochemistry*, **163**, 2444–2449.

24 Fuqua, C., Winans, S.C. and Greenberg, E.P. (1996) *Annual Review of Microbiology*, **50**, 727–751.

25 Swift, S., Throup, J.P., Salmond, G.P.C., Williams, P. and Stewart, G.S.A.B. (1996) *Trends in Biochemical Sciences*, **21**, 214–219.

26 McClean, K.H., Winson, M.K., Fish, L., Taylor, A., Chhabra, S.R., Camara, M., Daykin, M., Lamb, J.H., Swift, S., Bycroft, B.W., Stewart, G.S.A.B. and Williams, P. (1997) *Microbiology*, **143**, 3703–3711.

27 Zhang, L. and Kerr, A. (1991) *Journal of Bacteriology*, **173**, 1867–1872.

28 Zhang, L., Murphy, P.J., Kerr, A. and Tate, M.E. (1993) *Nature*, **362**, 446–448.

29 Piper, K.R., von Bodman, B.S. and Farrand, S.K. (1993) *Nature*, **362**, 448–450.

30 Hwang, I., Pei-Li, L., Zhang, L.H., Piper, K.R., Cook, D.M., Tate, M.E. and Farrand, S.K. (1994) *Proceedings of the National*

Academy of Sciences of the United States of America, **91**, 4639–4643.

31 Brelles-Marino, G. and Bedmar, E.J. (2001) *Journal of Biotechnology*, **91**, 197–209.

32 Bainton, N.J., Stead, P., Chhabra, S.R., Bycroft, B.W., Salmond, G.P.C., Stewart, G.S.A.B. and Williams, S.P. (1992) *Biochemical Journal*, **288**, 997–1004.

33 Holden, M.T., Ram-Chhabra, S., de Nys, R., Stead, P., Bainton, N.J., Hill, P.J., Manefield, M., Kumar, N., Labatte, M., England, D., Rice, S., Givskov, M., Salmond, G.P., Stewart, G.S., Bycroft, B. W., Kjelleberg, S. and Williams, P. (1999) *Molecular Microbiology*, **33**, 1254–1266.

34 Bainton, N.J., Bycroft, B.W., Chhabra, S. R., Stead, P., Gledhill, L., Hill, P.J., Rees, C.E.D., Winson, M.K., Salmond, G.P.C., Stewart, G.S.A.B. and Williams, P. (1992) *Gene*, **116**, 87–91.

35 Nasser, W., Bouillant, M.L., Salmond, G. and Reverchon, S. (1998) *Molecular Microbiology*, **29**, 1391–1405.

36 Reverchon, S.L., Bouillant, M.L., Salmond, G. and Nasser, W. (1998) *Molecular Microbiology*, **29**, 1407–1418.

37 von Bodman, S.B. and Farrand, S.K. (1995) *Journal of Bacteriology*, **177**, 5000–5008.

38 Pierson, L.R., Keppenne, V.D. and Wood, D.W. (1994) *Journal of Bacteriology*, **176**, 3966–3974.

39 Mavrodi, D.V., Ksenzenko, V.N., Bonsall, R.F., Cook, R.J., Boronin, A.M. and Thomashow, L.S. (1998) *Journal of Bacteriology*, **180**, 2541–2548.

40 Shaw, P.D., Ping, G., Daly, S.L., Cha, C., Cronan, J.E., Jr, Rinehart, K.L. and Farrand, S.K. (1997) *Proceedings of the National Academy of Sciences of the United States of America*, **94**, 6036–6041.

41 Cha, C., Gao, P., Chen, Y.C., Shaw, P.D. and Farrand, S.K. (1998) *Molecular Plant–Microbe Interactions*, **11**, 1119–1129.

42 Pesci, E.C., Pearson, J.P., Seed, P.C. and Iglewski, B.H. (1997) *Journal of Bacteriology*, **179**, 3127–3132.

43 Flavier, A.B., Clough, S.J., Schell, M.A. and Denny, T.P. (1997) *Molecular Microbiology*, **26**, 251–259.

44 Rosemeyer, V., Michiels, J., Verreth, C. and Vanderleyden, J. (1998) *Journal of Bacteriology*, **180**, 815–821.

45 Gray, K.M., Pearson, J.P., Downie, J.A., Boboye, B.E.A. and Greenberg, E.P. (1996) *Journal of Bacteriology*, **178**, 372–376.

46 Schripsema, J., de Rudder, K.E.E., van Vliet, T.B., Lankhorst, P.P., de Vroom, E., Kijne, J.W. and van Brussel, A.A.N. (1996) *Journal of Bacteriology*, **178**, 366–371.

47 Lithgow, J.K., Wilkinson, A., Hardman, A., Rodelas, B., Wisniewski-Dye, F., Williams, P. and Downie, J.A. (2000) *Molecular Microbiology*, **37**, 81–97.

48 Eberl, L., Molin, S. and Givskov, M. (1999) *Journal of Bacteriology*, **181**, 1703–1712.

49 Barber, C.E., Tang, J.L., Feng, J.X., Pan, M.Q., Wilson, T.J.G., Slater, H., Dow, J. M., Williams, P. and Daniels, M.J. (1997) *Molecular Microbiology*, **24**, 555–566.

50 Chun, W., Cui, J. and Poplawsky, A.R. (1997) *Physiological and Molecular Plant Pathology*, **51**, 1–14.

51 Poplawsky, A.R., Chun, W., Slater, H., Daniels, M.J. and Dow, M. (1998) *Molecular Plant–Microbe Interactions*, **11**, 68–70.

52 Horinouchi, S. (1999) *Cell–Cell Signalling in Bacteria* (9th eds, Dunny, G.M. and Winas, S.C.), ASM Press, Washington, pp. 193–207.

53 Lazazzera, B.A. and Grossman, A.D. (1998) *Trends in Microbiology*, 7, 288–294.

54 Puskas, A., Greenberg, E.P., Kaplan, S. and Schaefer, A.L. (1997) *Journal of Bacteriology*, **179**, 7530–7537.

55 Dunphy, G., Miyamoto, C. and Meighen, E. (1997) *Journal of Bacteriology*, **179**, 5288–5291.

56 Batchelor, S.E., Cooper, M., Chhabra, S. R., Glover, L.A., Stewart, G.S.A.B., Williams, P. and Prosser, J.I. (1997) *Applied and Environmental Microbiology*, **63**, 2281–2286.

57 Plamann, L. and Kaplan, H.B. (1999) *Cell–Cell Signalling in Bacteria* (eds G.M. Dunny and S.C. Winans), ASM Press, Washington, pp. 67–82.

58 Dunny, G.M. and Leonard, B.A. (1997) *Annual Review of Microbiology*, **51**, 527–564.

59 Swift, S., Karlyshev, A.V., Fish, L., Durant, E.L., Winson, M.K., Chhabra, S.R., Williams, P., Macintyre, S. and Stewart, G.S.A.B. (1997) *Journal of Bacteriology*, **179**, 5271–5281.

60 Schuster, M., Lostroh, C.P., Ogi, T. and Greenberg, E.P. (2003) *Journal of Bacteriology*, **185**, 2066–2079.

61 Smith, R.S. and Iglewski, B.H. (2003) *Current Opinion in Microbiology*, **6**, 56–60.

62 El-Sayed, A.K., Hothersall, J. and Thomas, C.M. (2001) *Microbiology*, **147**, 2127–2139.

63 Steidle, A., Allesen-Holm, M., Riedel, K., Berg, G., Givskov, M., Molin, S. and Eberl, L. (2002) *Applied and Environmental Microbiology*, **68**, 6371–6382.

64 Bertani, I. and Venturi, V. (2004) *Applied and Environmental Microbiology*, **70** (9) 5493–5502.

65 Yabuuchi, E., Kosako, Y., Oyaizu, H., Yano, I., Hotta, H., Hashimoto, Y., Ezaki, T. and Arakawa, M. (1992) *Microbiology and Immunology*, **36**, 1251–1275.

66 Coenye, T. and Vandamme, P. (2003) *Environmental Microbiology*, **5**, 719–729.

67 Mahenthiralingam, E., Urban, T.A. and Goldberg, J.B. (2005) *Nature Reviews Microbiology*, **3**, 144–156.

68 Eberl, L. (2006) *International Journal of Medical Microbiology*, **296**, 103–110.

69 Jones, S., Yu, B., Bainton, N.J., Birdsall, M., Bycroft, B.W., Chhabra, S.R., Cox, A.J.R., Golby, P., Reeves, P.J., Stephens, S., Winson, M.K., Salmond, G.P.C., Stewart, G.S.A.B. and Williams, P. (1993) *EMBO Journal*, **12**, 2477–2482.

70 Bassler, B.L., Wright, M. and Silverman, M.R. (1994) *Molecular Microbiology*, **13**, 273–286.

71 Pearson, J.P., Passador, L., Iglewski, B.H. and Greenberg, E.P. (1995) *Proceedings of the National Academy of Sciences of the United States of America*, **92**, 1490–1494.

72 Givskov, M., de Nys, R., Manefield, M., Gram, L., Maximillen, R., Ebrel, L., Molin, S., Steinberg, P.D. and Kjelleberg, S.

73 Wood, D.W. and Pierson, L.S., III (1996) *Gene*, **168**, 49–53.

74 Kjelleberg, S., Steinberg, P., Givskov, M., Gram, L., Manefield, M. and de Nys, R. (1997) *Aquatic Microbial Ecology*, **13**, 85–93.

75 Manefield, M., Harris, L., Rice, S.A., de Nys, R. and Kjelleberg, S. (2000) *Applied and Environmental Microbiology*, **66**, 2079–2084.

76 Rasmussen, T.B., Manefield, M., Andersen, J.B., Eberl, L., Anthoni, U., Christophersen, C., Steinberg, P., Kjelleberg, S. and Givskov, M. (2000) *Microbiology*, **146**, 3237–3244.

77 Pearson, J.P., Pesci, E.C. and Iglewski, B.H. (1997) *Journal of Bacteriology*, **179**, 5756–5767.

78 Wood, D.W., Gong, F., Daykin, M.M., Williams, P. and Pierson, L.S., III (1997) *Journal of Bacteriology*, **179**, 7663–7670.

79 Milton, D.L., Hardman, A., Cámara, M., Chhabra, S.R., Bycroft, B.W., Stewart, G.S.A.B. and Williams, P. (1997) *Journal of Bacteriology*, **179**, 3004–3012.

80 Davies, D.G., Parsek, M.R., Pearson, J.P., Iglewski, B.H., Costerton, J.W. and Greenberg, E.P. (1998) *Science*, **280**, 295–298.

81 Hentzer, M., Wu, H., Anderson, J.B., Riedel, K., Rasmusen, T.B., Bagge, N., Kumar, N., Schemberi, M.A., Song, Z., Kristoffersen, P., Manfield, H.N. and Givskov, M. (2003) *EMBO Journal*, **22**, 3803–3815.

82 Passador, L., Tucker, K.D., Guertin, K.R., Journet, M.P. and Kende, A.S. (1996) *Journal of Bacteriology*, **178**, 5995.

83 Winson, M.K., Camara, M., Latifi, A., Foglino, M., Chhabra, S.R., Daykin, M., Bally, M., Chapon, V., Salmond, G.P.C., Bycroft, B.W., Lazdunski, A., Stewart, G.S.A.B. and Williams, P. (1995) *Proceedings of the National Academy of Sciences of the United States of America*, **92**, 9427–9431.

84 Chernin, L.S., Winson, M.K., Thompson, J.M., Haran, S., Bycroft, B.W., Chet, I.,

(1996) *Journal of Bacteriology*, **178**, 6618–6622.

Williams, P. and Stewart, G.S.A.B. (1998) *Journal of Bacteriology*, **180**, 4435–4441.

85 Surette, M.G., Miller, M.B. and Bassler, B. L. (1999) *Proceedings of the National Academy of Sciences of the United States of America*, **96**, 1639–1644.

86 Freeman, J.A., Lilley, B.N. and Bassler, B.L. (2000) *Molecular Microbiology*, **35**, 139–149.

87 Lilley, B.N. and Bassler, B.L. (2000) *Molecular Microbiology*, **36**, 940–954.

88 Miyamoto, C.M., Lin, Y.H. and Meighen, E.A. (2000) *Molecular Microbiology*, **36**, 594–607.

89 McKnight, S.L., Iglewski, B.H. and Pesci, E.C. (2000) *Journal of Bacteriology*, **182**, 2702–2708.

90 Zhang, Z. and Pierson, L.S., III (2001) *Applied and Environmental Microbiology*, **67**, 4305–4315.

91 Smith, R.S., Harris, S.G., Phipps, R. and Iglewski, B. (2002) *Journal of Bacteriology*, **184** (4), 1132–1139.

92 Teplitski, M.I.B.S. (2002) Quorum sensing in *Sinorhizobium meliloti* and effect of plant signals on bacterial quorum sensing, PhD Thesis, Ohio State University.

93 von Bodman, S.B., Bauer, W.D. and Coplin, D.L. (2003) *Annual Review of Phytopathology*, **41**, 455–482.

94 Wagner, V.E., Bushnell, D., Passador, L., Brooks, A.I. and Iglewski, B.H. (2003) *Journal of Bacteriology*, **185**, 2080–2095.

95 Diggle, S.P., Winzer, K., Chhabra, S.R., Worrall, K.E., Camara, M. and Williams, P. (2003) *Molecular Microbiology*, **50**, 29–43.

96 McGrath, S., Wade, D.S. and Pesci, E.C. (2004) *FEMS Microbiology Letters*, **230** (1), 27–34.

97 Juhas, M., Wiehlmann, L., Huber, B., Jordan, D., Lauber, J., Salunkhe, P., Limpert, A.S., von Gotz, F., Steinmetz, I., Eberl, L. and Tummler, B. (2004) *Microbiology*, **150** (4), 831–841.

98 Sircili, M.P., Walters, M., Trabulsi, L.R. and Sperandio, V. (2004) *Infection and Immunity*, **72**, 2329–2337.

99 Henke, J.M. and Bassler, B.L. (2004) *Journal of Bacteriology*, **186**, 6902–6914.

100 Daniels, R., Vanderleyden, J. and Michiels, J. (2004) *FEMS Microbiology Reviews*, **28**, 261–289.

101 Martinelli, D., Grossmann, G., Sequin, U., Brandl, H. and Bachofen, R. (2004) *BMC Microbiology*, **4**, 25.

102 Lenz, D.H., Mok, K.C., Lilley, B.N., Kulkarni, R.V., Wingreen, N.S. and Bassler, B.L. (2004) *Cell*, **118**, 69–82.

103 Perombelon, M.C.M. and Kelman, A. (1980) *Annual Review of Pathology*, **18**, 361–387.

104 Kawase, T., Miyoshi, S., Sultan, Z. and Shinoda, S. (2004) *FEMS Microbiology Letters*, **240**, 55–59.

105 Chatterjee, A., Cui, Y., Liu, Y., Dumenyo, C.K. and Chatterjee, A.K. (1995) *Applied and Environmental Microbiology*, **61**, 1959–1967.

106 Milton, D.L. (2006) *International Journal of Medical Microbiology*, **296**, 61–71.

107 Py, B., Barras, F., Harris, S., Robson, N. and Salmond, G.P.C. (1998) *Methods in Microbiology*, **27**, 158–168.

108 Thomson, N.R., Thomas, J.D. and Salmond, G.P.C. (1999) *Methods in Microbiology*, **29**, 347–426.

109 Mae, A., Heikinheimo, R. and Palva, E.T. (1995) *Molecular & General Genetics*, **247**, 17–26.

110 Marits, R., Koiv, V., Laasik, E. and Mae, A. (1999) *Microbiology*, **145**, 1959–1966.

111 Pirhonen, M., Flego, D., Heikinheimo, R. and Palva, E.T. (1993) *EMBO Journal*, **12**, 2467–2476.

112 Andersson, R.A., Eriksson, A.R.B., Heikinheimo, R., Mae, A., Pirhonen, M., Koiv, V., Hyytiainen, H., Tuikkala, A. and Palva, E.T. (2000) *Molecular Plant–Microbe Interactions*, **13**, 384–393.

113 Schell, M.A. (1996) To be or not to be: how *Pseudomonas solanacearum* decides whether or not to express virulence genes. *European Journal of Plant Pathology*, **102**, 459–469.

114 Clough, S.J., Flavier, A.B., Schell, M.A. and Denny, T.P. (1997) *Applied and Environmental Microbiology*, **63**, 844–850.

115 Clough, S.J., Lee, K.E., Schell, M.A. and Denny, T.P. (1997) *Journal of Bacteriology*, **179**, 3639–3648.

116 Brumbley, S.M., Carney, B.F. and Denny, T.P. (1993) *Journal of Bacteriology*, **175**, 5477–5487.

117 Flavier, A.B., Ganova-Raeva, L.M., Schell, M.A. and Denny, T.P. (1997) *Journal of Bacteriology*, **179**, 7089–7097.

118 Dow, J.M., Feng, J.X., Barber, C.E., Tang, J.L. and Daniels, M.J. (2000) *Microbiology*, **146**, 885–891.

119 Tang, J.L., Liu, Y.N., Barber, C.E., Dow, J. M., Wootton, J.C. and Daniels, M.J. (1991) *Molecular & General Genetics*, **226**, 409–417.

120 Wilson, T.J.G., Bertrand, N., Tang, J.L., Feng, J.X., Pan, M.Q., Barber, C.E., Dow, J.M. and Daniels, M.J. (1998) *Molecular Microbiology*, **28**, 961–970.

121 Slater, H., Alvarez-Morales, A., Barber, C. E., Daniels, M.J. and Dow, J.M. (2000) *Molecular Microbiology*, **38**, 986–1003.

122 Licciardello, G., Bertani, I., Steindler, L., Bella, P., Venturi, V. and Catara, V. (2007) FEMS Microbiology Ecology [E-Pub ahead of printing].

123 Pomianek, M.E. and Semmelhack, M.F. (2007) *ACS Chemical Biology*, **2** (5), 293–295.

124 Winson, M.K., Swift, S., Fish, L., Throup, J.P., Jorgensen, F., Chhabra, S.R., Bycroft, B.W., Williams, P. and Stewart, G.a.S.B. (1998) *FEMS Microbiology Letters*, **163**, 185–192.

125 Andersen, J.B., Heydom, A., Hentzer, M., Eberl, L., Geisenberger, O., Christensen, B.B., Molin, S. and Givskov, M. (2001) *Applied and Environmental Microbiology*, **67**, 575–585.

126 Zhu, J., Chai, Y., Zhong, Z., Li, S. and Winans, S.C. (2003) *Applied and Environmental Microbiology*, **69**, 6949–6953.

127 Blosser, R.S. and Gray, K.M. (2000) *Journal of Microbiological Methods*, **40**, 47–55.

128 Cao, J.G., Wei, Z.Y. and Meighen, E.A. (1995) *Biochemical Journal*, **312**, 439–444.

129 Brint, J.M. and Ohman, D.E. (1995) *Journal of Bacteriology*, **177**, 7155–7163.

130 Chhabra, S.R., Stead, P., Bainton, N.J., Salmond, G.P.C., Stewart, G.S.A.B., Williams, P. and Bycroft, B.W. (1993) *Journal of Antibiotics*, **46**, 441–454.

131 Schuhegger, R., Ihring, A., Gantner, S., Bahnweg, G., Knappe, C., Vogg, G., Hutzler, P., Schmid, M., Van-Breusegem, F., Eberl, L., Hartmann, A. and Langebartels, C. (2006) *Plant Cell & Environment*, **29** (5), 909–918.

132 Whiteley, M., Lee, K.M. and Greenberg, E.P. (1999) *Proceedings of the National Academy of Sciences of the United States of America*, **96**, 13904–13909.

133 Pierson, L.S., III and Thomashow, L.S. (1992) *Molecular Plant–Microbe Interactions*, **5**, 330–339.

134 McCLean, R.J.C., Leland, S., Pierson, L. S., III and and Fuqua, C. (2004) *Journal of Microbiological Methods*, **58**, 351–360.

135 Camara, M., Daykin, M. and Chhabra, S. R. (1998) *Methods in Microbiology*, **27**, 319–330.

136 Givskov, M., Eberl, L. and Molin, S. (1997) *FEMS Microbiology Letters*, **148**, 115–122.

137 Lindum, P.W., Anthoni, U., Christophersen, C., Eberl, L., Molin, S. and Givskov, M. (1998) *Journal of Bacteriology*, **180**, 6384–6388.

138 Yang, Y.-H., Lee, T.-H., Kim, J.H., Kim, E. J., Joo, H.-S., Lee, C.-S. and Kim, B.-G. (2006) *Analytical Biochemistry*, **356**, 297–299.

139 Goddu, R.F., Leblanc, N.F. and Wright, C.M. (1995) *Anal. Chem.*, **27**, 1251–1255.

140 Fekete, A., Frommberger, M., Rothballer, M., Li, X., Englmann, M., Fekete, J., Hartmann, A., Eberl, L. and Schmitt-Kopplin, P. (2007) *Analytical and Bioanalytical Chemistry*, **387** (2), 455–467.

141 Hanzelka, B.L., Parsek, M.R., Val, D.L., Dunlap, P.V., Cronan, J.E., Jr and Greenberg, E.P. (1999) *Journal of Bacteriology*, **181**, 5766–5770.

142 Laue, B.E., Jiang, Y., Chhabra, S.R., Jacob, S., Stewart, G.S.A.B., Hardman, A.,

Downie, J.A., O'Gara, F. and Williams, P. (2000) *Microbiology*, **146**, 2469–2480.

143 Parsek, M.R., Val, D.L., Hanzelka, B.L., Cronan, J.E., Jr and Greenberg, E.P. (1999) *Proceedings of the National Academy of Sciences of the United States of America*, **96**, 4360.

144 Eberhard, A., Widrig, C.A., McBath, P. and Schineller, J.B. (1986) *Archives of Microbiology*, **146**, 35.

145 Kline, T., Bowman, J., Iglewski, B.H., de Kievit, T., Kakai, Y. and Passador, L. (1999) *Bioorganic & Medicinal Chemistry Letters*, **9**, 3447.

146 Newton, J.A. and Fray, R.G. (2004) *Cellular Microbiology*, **6**, 213–224.

147 Ochsner, U.A. and Reiser, J. (1995) *Proceedings of the National Academy of Sciences of the United States of America*, **92**, 6424–6428.

148 Aguilar, C., Bertani, I. and Venturi, V. (2003) *Applied and Environmental Microbiology*, **69**, 1739–1747.

149 Hornby, J.M., Jacobitz-Kizzier, S.M., McNeel, D.J., Jensen, E.C., Treves, D.S. and Nickerson, K.W. (2004) *Applied and Environmental Microbiology*, **70**, 1356–1359.

150 Mathesius, U., Mulders, S., Gao, M.S., Teplitski, M., Caetano-Anolles, G., Rolfe, B.G. and Bauer, W.D. (2003) *Proceedings of the National Academy of Sciences of the United States of America*, **100**, 1444–1449.

151 Fleet, G.H. (2003) *International Journal of Food Microbiology*, **86**, 11–22.

152 Dong, Y.H., Gusti, A.R., Zhang, Q., Xu, J.L. and Zhang, L.H. (2002) *Applied and Environmental Microbiology*, **68**, 1754–1759.

153 Dong, Y.H., Zhang, X.F., Xu, J.L. and Zhang, L.H. (2004) *Applied and Environmental Microbiology*, **70**, 954–960.

154 Molina, L., Constantinescu, F., Michel, L., Reimmann, C., Duffy, B. and Defago, G. (2003) *FEMS Microbiology Ecology*, **45**, 71–81.

155 Xu, F., Byun, T., Dussen, H.J. and Duke, K.R. (2003) *Journal of Biotechnology*, **101**, 89–96.

156 Gram, L., de-Nys, R., Maximilien, R., Givskov, M., Steinberg, P. and Kjelleberg, S. (1996) *Applied and Environmental Microbiology*, **62**, 4284–4287.

157 Teplistski, M. Robinson, J.B. and Wolfgang, D.B. (2000) *The American Phytopathological Society*, **13**, 637–648.

158 Rasmussen, T.B., Bjarnsholt, T., Skindersoe, M.E., Hentzer, M., Kristoffersen, P., Kote, M., Nielsen, J., Eberl, L. and Givskov, M. (2005) *Journal of Bacteriology*, **187**, 1799–1814.

159 Persson, T., Hansen, T., Rasmussen, T., Skinderso, M., Givskov, M. and Nielsen, J. (2005) *Organic and Biomolecular Chemistry*, **3**, 253.

160 Adonizio, A.L., Downum, K., Bennett, B.C. and Mathee, K. (2006) *Journal of Ethnopharmacology*, .

161 Hasan, S. (2006) Quorum sensing inhibition and antimicrobial properties of certain medicinal plants and natural products, MSc Dissertation, Aligarh Muslim University.

162 Ahmad, I., Hasan, S. and Zahin, M. (2007) Paper presented in 9th Indian Agriculture Scientists and Farmers' Congress held at Bioved Research and Communication Centre, Allahabad, India, 29-30, Jan., 2007.

163 Rasmussen, T.B., Bjarnsholt, T., Skindoersoe, M.E., Hentzer, M., Kristoffersen, P., Kote M., Nielson, J., Eberl, L. and Givskov, M., (2005) *J. Bacteriology*, **187** (5), 799–814.

164 Leadbetter, J.R. and Greenberg, E.P. (2000) *Journal of Bacteriology*, **182**, 6921–6926.

165 Dong, Y.H., Xu, J.L., Li, X.Z. and Zhang, L.H. (2000) *Proceedings of the National Academy of Sciences of the United States of America*, **97**, 3526–3531.

166 Dong, Y.H., Wang, L.H., Xu, J.L., Zhang, H.B. and Zhang, L.H. (2001) *Nature*, **411** (6839), 813–817.

167 Lee, S.J., Park, S.Y., Lee, J.J., Yum, D.Y., Koo, B.T. and Lee, J.K. (2002) *Applied and Environmental Microbiology*, **68**, 3919–3924.

8

Pseudomonas aurantiaca SR1: Plant Growth Promoting Traits, Secondary Metabolites and Crop Inoculation Response

Marisa Rovera, Evelin Carlier, Carolina Pasluosta, Germán Avanzini, Javier Andrés, and Susana Rosas

8.1
Plant Growth Promoting Rhizobacteria: General Considerations

Natural agricultural ecosystems depend directly on microorganisms present in the soil and soil rhizosphere that lead to increase in crop yield. Beneficial rhizosphere microorganisms are important determinants of plant health and soil fertility since they participate in many key ecosystem processes such as those involved in the biological control of plant pathogens, nutrient cycling and seedling establishment [1]. However, the natural role of rhizospheric microorganisms has been marginalized because of conventional farming practices such as tillage and high inputs of inorganic fertilizers and pesticides [2].

The recent progress in our understanding of the biological interactions occurring in the rhizosphere and of the practical requirements for inoculant formulation and delivery should increase the technology's reliability in the field and facilitate its commercial development.

Plant growth promoting rhizobacteria (PGPR) were first defined by Kloepper and Schroth [3] as soil bacteria that colonize the plant roots after they are inoculated onto seeds and enhance plant growth [3]. The following actions are implicit in the colonization process: ability to survive inoculation onto seed, to multiply in the spermosphere (region surrounding the seed) in response to seed exudates, to attach to the root surface and to colonize the developing root system [4].

PGPR enhance plant growth by direct and indirect means, but the specific mechanisms involved have not yet been well explained [4,5]. Direct mechanisms of plant growth promotion by PGPR can be demonstrated in the absence of plant pathogens or other rhizospheric microorganisms, whereas indirect mechanisms involve the ability of PGPR to reduce the deleterious effects of plant pathogens on crop yield. PGPR have been reported to enhance plant growth directly by means of a variety of mechanisms, including fixation of atmospheric nitrogen that is transferred to the plant, production of siderophores that chelate iron and make it available

Plant-Bacteria Interactions. Strategies and Techniques to Promote Plant Growth
Edited by Iqbal Ahmad, John Pichtel, and Shamsul Hayat
Copyright © 2008 WILEY-VCH Verlag GmbH & Co. KGaA, Weinheim
ISBN: 978-3-527-31901-5

to the plant root, solubilization of minerals such as phosphorus and synthesis of phytohormones [5]. In the presence of PGPR, direct enhancement of mineral uptake owing to increases in specific ion fluxes at the root surface has also been reported [6,7]. In addition, PGPR enhance plant growth indirectly by suppressing phytopathogens through a variety of mechanisms, including the production of antibiotics [8–10], iron-sequestering compounds [11], extracellular lytic enzymes [12], other secondary metabolites such as hydrogen cyanide (HCN) [13] and induction of systemic resistance [14] or competition for physical space and nutrients [15].

8.2
Secondary Metabolites Produced by *Pseudomonas*

Certain fluorescent pseudomonads isolated from the soil promote plant growth by producing metabolites that inhibit bacteria and fungi deleterious to plants [8,9,12,16–18]. Some of these disease-suppressing antibiotic compounds have been characterized chemically and include phenazine-1-carboxylic acid, pyrrolnitrin, pyoluteorin and 2,4-diacetylphloroglucinol (DAPG) [10,19–21]. The results obtained by both application of molecular techniques and direct isolation have demonstrated unequivocally that these antibiotics are produced in the spermosphere and rhizosphere and are very important for suppressing soilborne plant pathogens [22–24].

The broad-spectrum antibiotic DAPG has wide antifungal, antibacterial, antihelminthic, nematicidal and phytotoxic activity [19,25,26]. DAPG produced by fluorescent pseudomonads is referred to as a major determinant in the biocontrol activity of PGPR in numerous studies [19,22,27–30]. However, studies on the production of antifungal metabolites by *Pseudomonas aurantiaca* strains are scarce [31,32]; at present, a few research groups are studying DAPG production by this species.

P. aurantiaca SR1 was isolated from soybean rhizosphere by our research group in Río Cuarto, Córdoba, Argentina. This strain produces an orange pigment associated with a strong *in vitro* inhibiting capacity against different pathogenic fungi such as *Macrophomina phaseolina*, *Rhizoctonia solani*, *Pythium* spp., *Sclerotinia sclerotiorum*, *Sclerotium rolfsii*, *Fusarium* spp. and *Alternaria* spp. [33]. The antifungal compound is secreted by the bacterium when culture media such as tryptic soy agar, nutrient agar or media supplemented with triptone or peptone are used. Among the tested carbon sources, mannitol and saccharose were found to induce pigment production, while glucose acted as a repressor [34].

We demonstrated that *P. aurantiaca* SR1 produces DAPG and it is not only generated under *in vitro* conditions but also under a rhizosphere environment of treated crops. This compound was isolated by chromatography and chemically characterized by spectroscopy (absorption, FTIR, mass and ^{1}H NMR) studies [35]. Chromatographic studies using TLC and HPLC showed the presence of a compound with medium polarity. The Rf and retention time values for the active fraction corresponded with the values found using a standard sample of DAPG. The IR studies showed the presence of hydroxyl and carbonyl groups and, in addition, showed that there was similarity with the DAPG IR spectrum. The results of the gas chromatography–mass spectroscopy indicated the presence of an ion molecular peak of m/z 210. The molecular weight and fragmentation pattern were on the expected lines for DAPG. Also, the ^{1}H NMR spectrum showed the presence of aromatic hydrogen and a singlet signal corresponding to aliphatic hydrogens. These results are indicative of an aromatic ring substituted by an acetyl group, which is characteristic and coincides with that of DAPG. The identification reveals that DAPG is the active compound produced by *P. aurantiaca* SR1. It is important to emphasize that this is the first study of its kind on this bacterium in Argentina.

Many studies indicate that DAPG is one of the most important isolated antibiotics [36,37]. This compound is widely distributed in antagonistic *P. fluorescens* strains that occur in natural disease-suppressive soil [38].

Bonsall *et al.* have reported DAPG isolation from soil and broth cultures by means of reversed-phase high-performance liquid chromatography [20]. Shanahan *et al.* published a reversed-phase liquid chromatographic analysis of DAPG in culture and in soil [22]. Later, Shanahan *et al.* employed a gradient LC assay for determining monoacetylphloroglucinol (MAPG) and DAPG in growth culture media [39].

In another study, we have demonstrated by liquid chromatography–tandem mass spectrometry that *P. aurantiaca* SR1 produces indol-3-acetic acid (IAA) (11 $\mu g\,ml^{-1}$) [40]. These results can explain a possible mechanism for which *P. aurantiaca* SR1 promotes plant growth. IAA is one of the physiologically most active auxins and is a common product of L-tryptophan metabolism in several microorganisms including PGPR [41]. Promotion of root growth is one of the major markers by which the beneficial effect of plant growth promoting bacteria is measured [5,42,43].

P. aurantiaca SR1 produces siderophores and HCN and moderately solubilizes phosphate. Furthermore, it possesses the capacity to colonize the root systems of different crops, maintaining an appropriate population in the rhizosphere area and in internal structures of the plants. It behaves as an endophyte in wheat and soybean [44].

The biosynthesis of IAA is correlated with DAPG production. When the production of DAPG is suppressed, for example, owing to exposition of the bacterial cultivation to permanent light, IAA presence is not detected.

8.3
Coinoculation Greenhouse Assays in Alfalfa (*Medicago sativa L.*)

P. aurantiaca SR1 in coinoculation with *Sinorhizobium meliloti* strain 3DOh13 was studied to determine its effects on nodulation and growth of alfalfa (*M. sativa L.*). Alfalfa is the most important forage legume in the semiarid Argentinean Pampas because of the quality nutrients that it provides [45]. Therefore, the effect that this plant has on soil fertility is important, as well as the contribution of its root system to the improvement and conservation of soil structure [46].

In coinoculation studies with PGPR and *Rhizobium/Bradyrhizobium* spp., an increase in the root and shoot weight, plant vigor, nitrogen fixation and grain yield has been shown in various legumes such as common bean [47], green gram [48], pea

Table 8.1 Effect of coinoculation with *P. aurantiaca SR1* on shoot
and root length in an alfalfa cultivar.

Treatment	Shoot length (cm)	Root length (cm)
Coinoculation	8.2000 ± 0.29^a	14.7800 ± 0.47^a
P. aurantiaca SR1	6.2000 ± 0.73^b	11.8200 ± 0.67^b
S. meliloti 3DOh13	7.0800 ± 0.42^a	13.2800 ± 0.52^b
Control N_2	5.3800 ± 0.51^c	9.7200 ± 0.43^c
Control	3.4000 ± 0.30^c	7.5400 ± 0.89^c

Mean $\pm$ SE, means with different letters ^{a,b,c} in the same column differ significantly at $P<0.05$ (Bonferroni test). Plants were grown for 45 days in the greenhouse.

[50] and soybean [50]. Knight and Langston-Unkefer found that inoculation of nodulating alfalfa roots by means of a toxin-releasing *Pseudomonas syringae* pv. *tabaci* significantly increased plant growth, nitrogenase activity, nodule number, total nodule weight and nitrogen yield under controlled growth conditions [51].

In our studies, both *P. aurantiaca* SR1 and *S. meliloti* strain 3DOh13 were cultured on tryptic soy broth (TSB) medium at $28 \pm 1\,°C$. The optical cell densities at 600 nm (OD_{600}) were 0.22 and 0.36, which corresponded to approximately 4.5×10^8 and $6.8 \times 10^8\,CFU\,ml^{-1}$ of *P. aurantiaca* SR1 and *S. meliloti* 3DOh13, respectively.

The inoculant was prepared by mixing *P. aurantiaca* SR1 and *S. meliloti* strain 3DOh13 in a 1 : 1 ratio (vol/vol). The optical cell density at 600 nm (OD_{600}) was 0.25, which corresponded to cell numbers of *S. meliloti* strain 3DOh13 of approximately 6.6×10^8 and $6.3 \times 10^8\,CFU\,ml^{-1}$ of *P. aurantiaca* SR1. One gram of sterilized seeds was inoculated with the mixed bacterial suspension, and populations of bacteria on inoculated seeds were $10^5\,CFU/seed$.

The greenhouse conditions were as follows: air temperature of $28 \pm 2\,°C$ and additional illumination of $220\,\mu E\,m^{-2}\,seg^{-1}$ for a photoperiod of 16 : 8 h (day : night).

In bacterial coexistence assays, *P. aurantiaca* SR1 did not exercise any inhibiting effect on the growth of *S. meliloti*. *P. aurantiaca* SR1 coinoculated with *S. meliloti* stimulated the length and root shoot growth at 45 days after sowing (Table 8.1).

Additionally, the coinoculation with *S. meliloti* strain 3DOh13 and *P. aurantiaca* SR1 resulted in a significant increase in the fresh and dry shoots and root weight (Table 8.2) and in the number of nodules.

Even though there is a possibility of great variability in field results, if a positive effect of a PGPR is seen on a specific crop in greenhouse studies, there is a strong likelihood that such effect will carry through to field conditions [52].

8.4
Field Experiments with *P. aurantiaca SR1* in Wheat (*Triticum aestivum L.*)

PGPR strains promote wheat growth because of their ability to transform root exudates into phytohormones that are absorbed by the roots, thus allowing nitrogen fertilizer application to reduce [53–55]. This conversion into substances by the root

Table 8.2 Effects of *P. aurantiaca* SR1 on plant growth parameters
(shoot and root, fresh and dry weight) in an alfalfa cultivar.

Treatment	Shoot fresh weight (g)	Shoot dry weight (g)	Root fresh weight (g)	Root dry weight (g)
Coinoculation	0.045 ± 0.002^a	0.029 ± 0.06^a	0.079 ± 0.03^a	0.014 ± 0.007^a
P. aurantiaca SR1	0.031 ± 0.004^b	0.019 ± 0.06^b	0.025 ± 0.01^b	0.003 ± 0.004^c
S. meliloti 3DOh13	0.039 ± 0.004^a	0.026 ± 0.11^a	0.024 ± 0.01^b	0.009 ± 0.003^b
Control N_2	0.022 ± 0.002^c	0.005 ± 0.004^c	0.007 ± 0.001^c	0.005 ± 0.008^b
Control	0.018 ± 0.001^c	0.004 ± 0.006^c	0.005 ± 0.001^c	0.004 ± 0.004^c

Mean $\pm$ SE, means with different letters [a,b,c] in the same column differ significantly at $P < 0.05$ (Bonferroni test). Plants were grown for 45 days in the greenhouse.

that promote plant growth is a biochemical property of the PGPR [56]. This explains the increase in the fresh and dry weight of wheat inoculated with rhizobacteria and the increase in the capacity of radical absorption with lower doses of nitrogen fertilizer in gramineous plants [57]. This helps avoid the excess use of urea, which otherwise contaminates surface and ground water. The experiment in wheat with *P. aurantiaca* SR1, formulated as an inoculant by BIAGRO S.A. Laboratory (Buenos Aires, Argentina), was conduced in a fully randomized block design with seven replicate blocks for each treatment. Blocks measured $7.2 \, \text{m}^2$ (1.20 m wide and 6 m long) and were 0.20 m apart. Six treatments were established: untreated control, seeds inoculated with *P. aurantiaca* SR1, soil fertilized with nitrogen and phosphorus (dose 100% and 50%) and combination of both doses of fertilizers with *P. aurantiaca* SR1.

At stage V5 (30 days after sowing), emergence, length of shoots and roots, volume of root, fresh and dry mass of shoots and roots were recorded. The parameters of the yield components evaluated were number of spikes per plant, number of grains per spike, weight of one thousand grains and yield (kg ha^{-1}).

After 30 days of seeding, the positive effect of the inoculation of wheat with *P. aurantiaca* SR1 was observed when an increase in emergence, shoot and root length and radical dry weight occurred (Table 8.3). Inoculation would reduce the fertilizer use by 50%, as the yield parameters evaluated with the 50% fertilizer dose gave values similar to the ones obtained with the 100% dose without inoculation. *P. aurantiaca* SR1 increased wheat yield by 21% (kg ha^{-1}) and 16% in number of grains per spike (Table 8.4).

Numerous studies refer to the effects of inoculation with PGPR. Field experiments showed that treatment with *Pseudomonas* and *Bacillus* strains increased seedling emergence in wheat [58]. *Pseudomonas* species are able to grow in sufficient quantities on the roots of winter wheat [59]. Inoculation at the time of planting with *Pseudomonas chlororaphis* strain 2E3 increased the emergence of spring wheat by 8 and 6% at two different sites in northern Utah [60]. Yield increases in wheat by PGPR inoculation varied from 18 to 22% in Passo Fundo and from 27 to 28% in Pato Branco, Brazil [61].

Table 8.3 Response of wheat (*T. aestivum* L.) inoculated with *P. aurantiaca* SR1.

		Stage V5						
					Fresh weight		Dry weight	
Treatments	Emergence (plants $\times$ m^2)	Length shoot (cm)	Length root (cm)	Volume of root (cm^3)	Shoot (mg)	Root (mg)	Shoot (mg)	Root (mg)
P. aurantiaca SR1	534.86^{ab}	16.94^{ab}	160.4^{ab}	1.77^{ab}	0.97^a	0.33^a	223.43^a	190^a
Fertility P + N doses 100%	357.71^c	13.78^c	87.15^c	1.29^{bc}	1.01^a	0.25^a	217.71^a	128.86^{ab}
Fertility P + N 100% + *P. aurantiaca* SR1	571.43^a	14.33^{bc}	181.49^a	2.09^a	1.19^a	0.31^a	252.29^a	183.86^a
Fertility P + N 50%	384.57^{bc}	16.02^{abc}	159.59^{ab}	1.20^{bc}	1.39^a	0.16^a	228.14^a	109.57^{ab}
Fertility P + N 50% + *P. aurantiaca* SR1	540^{ab}	14.89^{abc}	150.16^{ab}	1.66^{bc}	1.20^a	0.26^a	262.14^a	156^{ab}
Untreated control	542^{abc}	16.83^{abc}	101.89^c	0.91^c	0.94^a	0.12^a	198.43^a	80.29^b

Means with different letters ^{a,b,c} in the same column differ significantly at $P < 0.05$ (Bonferroni test).

Table 8.4 Yield in wheat (*T. aestivum* L.) inoculated with *P. aurantiaca* SR1.

	Harvest			
Treatments	Yield (kg ha^{-1})	Weight of 1000 grains	Number of spikes per plant	Number of grains per spike
P. aurantiaca SR1	4933.57^a	39.98^a	3.30^a	37.05^a
Fertility P + N 100%	4215.43^{ab}	39.06^a	3.20^a	34.30^{ab}
Fertility P + N 100% + *P. aurantiaca* SR1	4790.14^{ab}	38.85^a	3.10^a	36.40^a
Fertility P + N 50%	4156.86^{ab}	38.27^a	2.90^a	34.60^{ab}
Fertility P + N 50% + *P. aurantiaca* SR1	4402.29^{ab}	37.34^a	3.30^a	35.60^{ab}
Untreated control	3895.00b	38.99^a	2.80^a	31.80^b

Means with different letters ^{a,b,} in the same column differ significantly at $P < 0.05$ (Bonferroni test).

8.5
Conclusions

We have demonstrated that *P. aurantiaca* SR1 produces compounds that stimulate the growth of both wheat and alfalfa cultivars. It increased the number and size of nodules in alfalfa roots compared to *S. meliloti* inoculated alone, suggesting that the nitrogen fixation is enhanced. Also, this strain promotes wheat growth and the inoculation would reduce the use of nitrogen fertilizers by 50%.

The understanding of the mechanisms involved in antibiosis adds another beneficial property of *P. aurantiaca* SR1 upon plant growth; thus, it may serve as an ideal microorganism to be used for enhancing crop yields through its biocontrol and PGPR effects.

Is important to state that *P. aurantiaca* SR1 has not been reported by other Argentina research groups and that it has been formulated by the industry as an inoculant for its application in different countries.

Acknowledgments

We are grateful to Secretaría de Ciencia y Técnica of Universidad Nacional de Río Cuarto and Agencia Nacional de Promoción Científica y Tecnológica (ANPCyT).

References

1 Jeffries, P., Gianinazzi, S., Perotto, S., Turnau, K. and Barea, J.M. (2003) *Biology and Fertility of Soil*, **37**, 1–16.

2 Mäder, P., Fliessbach, A., Dubois, D., Gunst, L., Fried, P. and Niggli, U. (2002) *Science*, **296**, 1694–1697.

3 Kloepper, J.W. and Schroth, M.N. (1978) Proceedings of the 4th International Conference on Plant Pathogenic Bacteria, vol. 2, Station de Pathologie Vegetale et Phytobacteriologie, INRA, Angers, France, pp. 879–882.

4 Kloepper, J.W. (1993) *Soil Microbial Ecology: Applications in Agricultural and Environmental Management* (ed. F.B. Metting, Jr), Marcel Dekker Inc., New York, USA, 255–274.

5 Glick, B.R. (1995) *Canadian Journal of Microbiology*, **41**, 109–117.

6 Bashan, Y. and Levanony, H. (1991) *Plant and Soil*, **137**, 99–103.

7 Bertrand, H., Plassard, C., Pinochet, X., Toraine, B., Normand, P. and Cleyet-Marel, J.C. (2000) *Canadian Journal of Microbiology*, **46**, 229–236.

8 Hill, D.S., Stein, R.I., Torkewitz, N.R., Morse, A.M., Howell, C.R., Pachlatko, J.P., Becker, J.O. and Ligon, J.M. (1994) *Applied and Environmental Microbiology*, **60**, 78–85.

9 Rodriguez, F. and Pfender, W.F. (1997) *Phytopathology*, **87**, 614–621.

10 Thomashow, L.S., Weller, D.M., Bonsall, R.F. and Pierson, L.S., III (1990) *Applied and Environmental Microbiology*, **56**, 908–912.

11 Loper, J.E. and Buyer, J.S. (1991) *Molecular Plant–Microbe Interactions*, **4**, 5–13.

12 Fridlender, M., Inbar, J. and Chet, I. (1993) *Soil Biology & Biochemistry*, **25**, 1211–1221.

13 Voisard, C., Keel, C., Hass, D. and Defago, G. (1989) *EMBO Journal*, **8**, 351–358.

14 Bull, C.T., Weller, D.M. and Thomashow, L.S. (1991) *Phytopathology*, **81**, 954–959.

15 Persello-Cartieaux, F., Nussaume, L. and Robaglia, C. (2003) *Plant Cell & Environment*, **26**, 189–199.

16 Keel, C., Wirthner, P.H., Oberhansii, T.H., Voisard, C., Burger, P., Hass, D. and Défago, G. (1990) *Symbiosis*, **9**, 327–341.

17 Schippers, B., Bakker, W. and Bakker, P.A. M. (1987) *Annual Review of Phytopathology*, **25**, 339–358.

18 Maurhofer, M., Keel, C., Schnider, U., Voisard, C., Haas, D. and Défago, G. (1991) *Phytopathology*, **82**, 190–195.

19 Keel, C., Snider, U., Maurhofer, M., Voisard, C., Laville, J., Burger, P., Wirthner, P.H., Hass, D. and Défago, G. (1992) *Molecular Plant–Microbe Interactions*, **5**, 4–13.

20 Bonsall, R.F., Weller, D.M. and Thomashow, L.S. (1997) *Applied and Environmental Microbiology*, **63**, 951–955.

21 Nielsen, M.N., Sørensen, J., Fels, J. and Pedersen, H.C. (1998) *Applied and Environmental Microbiology*, **64**, 3563–3569.

22 Shanahan, P., O'Sullivan, D., Simpson, P., Glennon, J. and O'Gara, F. (1992) *Applied and Environmental Microbiology*, **58**, 353–358.

23 Kraus, J. and Loper, J.E. (1995) *Applied and Environmental Microbiology*, **61**, 849–854.

24 Ross, I.L., Alami, Y., Harvey, P.R., Achouak, W. and Ryder, M.H. (2000) *Applied and Environmental Microbiology*, **66**, 1609–1616.

25 Dowling, D.N. and O'Gara, F. (1994) *Trends in Biotechnology*, **12**, 133–144.

26 Cronin, D., Moënne-Loccoz, Y., Fenion, A., Dunne, C., Dowling, D.N. and O'Gara, F. (1997) *Applied and Environmental Microbiology*, **63**, 1357–1361.

27 Vincent, M.N., Harrison, L.A., Brackin, J. M., Kovacevich, P.A., Mukerji, P., Weller, D.M. and Pierson, E.A. (1991) *Applied and Environmental Microbiology*, **57**, 2928–2934.

28 Harrison, L.A., Letendre, L., Kovacevich, P., Pierson, E.A. and Weller, D.M. (1993) *Soil Biology & Biochemistry*, **25**, 215–221.

29 Fenton, A.M., Stephens, P.M., Crowley, J., O'Callaghan, M. and O'Gara, F. (1992) *Applied and Environmental Microbiology*, **58**, 3873–3878.

30 Validov, S., Mavrodi, O., De La Fuente, L., Boronin, A., Weller, D., Thomashow, L. and Mavrodi, D. (2005) *FEMS Microbiology Letters*, **242**, 249–256.

31 Esipov, S.E., Adanin, V.M., Baskunov, B.P., Kiprianova, E.A. and Garagulia, A.D. (1975) *Antibiotiki*, **20**, 1077–1081.

32 Nowak-Thompson, B., Hammer, P.E., Hill, D.S., Stafford, J., Torkewitz, N., Gaffney, T.D., Lam, S.T., Molnár, I. and Ligon, J.M. (2003) *Journal of Bacteriology*, **185** (3), 860–869.

33 Rosas, S., Altamirano, F., Schroder, E. and Correa, N. (2001) *Phyton-International Journal of Experimental Botany*, 203–209.

34 Rovera, M., Cabrera, S., Rosas, S. and Correa, N. (2000) *Nitrogen Fixation: From Molecules to Crop Productivity* (eds F.O. Pedrosa, M. Hungría,M.G. Yates and W.E. Newton),Kluwer Academic Publishers, p. 602.

35 Rovera, M., Correa, M., Reta, M., Andres, J., Rosas, S. and Correa, N. (2000) Chemical identification of antifungal metabolites produced by *Pseudomonas aurantiaca*. Proceedings of the 5th International PGPR Workshop web page http://www.ag.auburn.edu/argentina/pdfmanuscripts/rovera.pdf.

36 Raaijmakers, J.M., Bonsall, R.F. and Weller, D.M. (1999) *Phytopathology*, **89**, 470–475.

37 Siddiqui, I.A. and Shaukat, S.S. (2003) *Soil Biology & Biochemistry*, **32**, 1615–1623.

38 Weller, D.M., Raaijmakers, J.M., Gardener, B.B.M. and Thomashow, L.S. (2002) *Annual Review of Phytopathology*, **40**, 309–348.

39 Shanahan, P., Glennon, J.D., Crowley, J.J., Donnelly, D.F. and O'Gara, F. (1993) *Analytica Chimica Acta*, **272**, 271–277.

40 Prinsen, E., van Dongen, W., Esmans, E. and van Onckelen, H.J. (1997) *Mass Spectrometry*, **32**, 12–22.

41 Vladimir, K., Chebotar, A. and Shoichiro Akao, A. (2001) *Biology and Fertility of Soil*, **34**, 427–432.

42 Costacurta, A. and Vanderleyden, J. (1995) *Critical Reviews in Microbiology*, **21**, 1–18.

43 Patten, Ch.L. and Glick, B.R. (1996) *Canadian Journal of Microbiology*, **42**, 207–220.

44 Rosas, S., Rovera, M., Andrés, J.A., Pastor, N.A., Guiñaz,ú L.B., Carlier, E. and Correa, N. (2005) Proceeding Prospects and Applications for Plant Associated Microbes. 1st International Conference on Plant–Microbe Interactions: Endophytes and Biocontrol Agents (eds S. Sorvari and O. Toldo) Lapland, Finland, pp. 91–99.

45 Viglizzo, E.F. (1995) *La alfalfa en la Argentina* (eds E.H. Hijano and A. Navarro), Instituto Nacional de Tecnología Agropecuaria (INTA), Buenos Aires, Argentina, pp. 260–272.

46 Vance, C. (1997) *Biological Fixation of Nitrogen for Ecology and Sustainable Agriculture* (eds A. Legocki and A. Pühler), Springer, Berlin, pp. 179–186.

47 Grimes, H.D. and Mount, M.S. (1984) *Soil Biology & Biochemistry*, **16**, 27–30.

48 Sindhu, S.S., Grupta, S.K. and Dadarwal, K.R. (1999) *Biology and Fertility of Soil*, **29**, 62–68.

49 Bolton, H. Ellio, L.F., Jr, Turco, R.F. and Kennedy, A.C. (1990) *Plant and Soil*, **123**, 121–124.

50 Dashti, N., Zhang, F., Hynes, R. and Smith, D.L. (1998) *Plant and Soil*, **200**, 205–213.

51 Knight, T.J. and Langston-Unkefer, P.J. (1988) *Science*, **241**, 951–994.

52 De Freitas, J.R. and Germida, J.J. (1992) *Soil Biology & Biochemistry*, **24**, 1127–1135.

53 Valdivia-Urdiales, B., Fernández-Brondo, J.M. and Sánchez-Yáñez, J.M. (1999) *Revista Latinoamericana de Microbiologia*, **41**, 214–217.

54 Baldani, V.L.D., Baldani, J.I. and Dobereiner, J. (2000) *Biology and Fertility of Soil*, **30**, 485–491.

55 Bashan, Y. and Levanony, H. (1988) *Azospirillum IV* (ed. K1ingmüller W.), Springer-Verlag, Berlin, pp. 166–173.

56 Junior, F.D.B., Reis, V.M., Urquiaga, S. and Dobereiner, J. (2000) *Plant and Soil*, **21**, 153–159.

57 Withmore, A.P. (2000) *Journal of Agricultural Science*, **48**, 115–122.

58 van Elsas, J.D., Dijkstra, A.F., Govaert, J.M. and van Veen, J. (1986) *FEMS Microbiology Letters*, **38** (3), 151–160.

59 De Freitas, J.R. and Germida, J.J. (1990) *Canadian Journal of Microbiology*, **36**, 265–272.

60 Kropp, B.R., Thomas, E., Pounder, J.I. and Anderson, A.J. (1996) *Biology and Fertility of Soil*, **23**, 200–206.

61 Da Luz, W.C. (2001) *Fitopatologia Brasileira*, **26** (3), 597–600.

9

Rice–Rhizobia Association: Evolution of an Alternate Niche of Beneficial Plant–Bacteria Association

Ravi P.N. Mishra, Ramesh K. Singh, Hemant K. Jaiswal, Manoj K. Singh, Youssef G. Yanni, and Frank B. Dazzo

9.1
Introduction

The vision of self-fertilizing crops contributed to the euphoria created by the emergence of biotechnology and the Green Revolution. In his 1970 Nobel Peace Prize lecture, Norman Borlaug highlighted the need to extend the range of symbioses to include nitrogen-fixing bacteria, such as rhizobia, and cereals to sustain the Green Revolution. He went on to acknowledge that even though high-yielding dwarf rice and wheat varieties were the catalysts that had ignited the Green Revolution, chemical fertilizers were the fuel that gave it thrust [11]. Since then, there have been extensive discussions on the prospects of establishing such novel symbiotic systems including research plans for their implementation. However, it is now clear that the energy required for the reduction of nitrogen to ammonia in nitrogen fixation is not greater than that required for the production of ammonia by reduction of nitrate, the main form of nitrogen assimilated by plants. Consequently, cereals such as rice would not likely suffer any significant energy penalty if they were supporting nitrogen fixation [12].

A huge amount of natural gas is consumed in the synthesis of nitrogenous fertilizer as anhydrous ammonia. In addition, this industrial process produces carbon dioxide, the main cause of greenhouse effect and global warming. The industrial production of nitrogenous fertilizer is also expensive, and in developing countries, the additional costs often exceed the means of low-income farmers, limiting the yield potential of their crops. Once chemical fertilizers are applied, additional residual problems can arise. Roughly one third of the nitrogen fertilizer applied is actually used by the crop. The nonassimilated nitrogen may result in nitrate (NO_3^-) contamination of groundwater [13], posing a serious health hazard. In addition, excess nitrogen can also lead to soil acidification [14] and increased denitrification resulting in higher emission of nitrous oxide (N_2O), another potent greenhouse gas that may exacerbate global warming [15]. Therefore, cropping

Plant-Bacteria Interactions. Strategies and Techniques to Promote Plant Growth
Edited by Iqbal Ahmad, John Pichtel, and Shamsul Hayat
Copyright © 2008 WILEY-VCH Verlag GmbH & Co. KGaA, Weinheim
ISBN: 978-3-527-31901-5

systems requiring large additions of nitrogen fertilizer are nonsustainable systems because they deplete nonrenewable natural resources and can intensify health hazards and environmental pollution [16]. This problem could be mitigated if rice and other cereals were able to establish more intimate associations with plant growth promoting (PGP) microorganisms that can efficiently provide these crops (either directly or indirectly) with nonpolluting sources of plant nutrients (especially nitrogen), thereby reducing the plant's dependence on large chemical fertilizer inputs to achieve high crop yields.

Rhizobia are symbiotic bacteria that belong to a versatile and physiologically robust group of nitrogen-fixing microorganisms, some of which can induce root or stem nodules on leguminous plants. Rhizobia have great environmental and agricultural importance because their symbioses with legumes are responsible for a high proportion of the atmospheric nitrogen fixed biologically on earth. Rhizobia currently consists of 61 species belonging to 13 different genera, namely *Rhizobium*, *Mesorhizobium*, *Ensifer* (formally *Sinorhizobium*), *Bradyrhizobium*, *Azorhizobium*, *Allorhizobium*, *Methylobacterium*, *Burkholderia*, *Cupriavidus* (formally *Wautersia*/*Ralstonia*), *Devosia*, *Herbaspirillum*, *Ochrobactrum* and *Phyllobacterium*. The taxonomy of rhizobia is in constant flux, and its current status can be assessed at http://www.rhizobia.co.nz/taxonomy/rhizobia.html and http://edzna.ccg.unam.mx/rhizobial-taxonomy/.

Rice (*Oryza sativa* L.) is one of the most important food crops of the planet and serves as the staple diet for most people living in developing countries [2]. It is estimated that 8 billion people will populate the earth by the year 2020, and the expected 4.8 billion rice consumers will need 760 million tons of rice annually [3]. This means that rice production must be increased by 2% per year to meet future demand [4–6]. Such increased production will require twice the amount of nitrogen input compared to what is currently applied to the crop as fertilizer, which is neither economically feasible nor environmentally desirable [7–9]. Therefore, an alternative approach is to increase the contribution of biologically fixed nitrogen to rice by utilizing the natural association between nonlegumes and associative [4,10] and symbiotic (rhizobial) diazotrophs [1,7].

In the mid-1990s, a multinational collaborative effort was begun between several agricultural research institutes to explore the possible existence of natural, intimate plant growth promoting associations between cereals and rhizobia. In the present chapter, we summarize various initiatives undertaken to discover and exploit this novel plant–bacteria association.

9.2
Landmark Discovery of the Natural Rhizobia–Rice Association

During the last few years, there has been an increased interest in exploring the possibility of extending the beneficial interaction between rice and nitrogen-fixing bacteria. This line of investigation came into focus in 1992, when the IRRI hosted an international workshop to evaluate the current knowledge on the potential for nodulation and nitrogen fixation in rice associated with symbiotic bacteria [17]. The

New Frontier Project (1994) was developed by the IRRI to coordinate worldwide collaborative efforts among research centers to explore natural rice–bacteria associations to reduce the dependence of rice on synthetic mineral nitrogen resources. The long-term objective of that project was to enable self-fertilization of rice plants. The working group of the project concluded that exploratory research is primarily needed to assess the feasibility of nitrogen fixation in rice by an international multidisciplinary group. In fact, this research was considered to be scientifically risky because it was entering unfamiliar territory, but likely to have an enormous impact on agricultural productivity if successful. One of the research directions recommended at that workshop was to determine if rhizobia naturally colonize the interior of rice roots when this cereal grows in rotation with a legume crop, and if so, to assess the potential impact of this novel plant–microbe association on rice production. This idea is derived from the general concept that roots of healthy plants grown in natural soil eventually develop a continuum of root-associated microorganisms extending from the rhizosphere to the rhizoplane and even deeper into the epidermis, cortex, endodermis and vascular system [18–21]. Typically, the presence of these beneficial microorganisms within roots does not induce any disease symptom. These microorganisms are described as endophytes or internal root colonists since they can intimately colonize the interior of living plant tissue [22,23]. In the mid-1990s, a multinational collaborative project was initiated to search for natural, intimate associations between rhizobia and rice (*Oryza sativa* L.), assess their impact on plant growth and exploit those combinations that can enhance grain yield with less dependence on inputs of chemical nitrogen fertilizers.

An important development in the exploration of rice–endophytic rhizobia associations took place when a natural plant growth promoting association was discovered between rice and *Rhizobium leguminosarum* bv. *trifolii*, which is the root-nodule nitrogen-fixing symbiont of berseem clover (*Trifolium alexandrinum* L.) commonly cultivated in the Nile delta region of Egypt. This intimate association is believed to have evolved as a result of rice being rotated successfully with berseem clover for between 700 and 1400 years [1].

The guiding hypothesis was that the natural endophytic associations between rhizobia and cereal roots would most likely occur where cereals are successfully rotated with a legume crop that could enhance the soil population of the corresponding rhizobial symbionts. Such natural *Rhizobium*–cereal associations would be perpetuated if they were mutually beneficial. If this hypothesis was correct, the cereal roots growing at these sites should harbor, along with other microbes, a high population density of endophytic rhizobia that are already adapted to be highly competitive for colonization of the interior habitats of crop roots, being protected from stiff competition with other soil–rhizosphere microorganisms under field conditions. This is where endophytic rhizobia are strategically located, because a more rapid and intimate metabolic exchange is possible within host plant tissues rather than just on their epidermal surface. An ideal location to test this hypothesis was in the Egyptian Nile delta where rice has been rotated with the forage legume, Egyptian berseem clover (*T. alexandrinum* L.) since antiquity. In this area, japonica and (more recently) indica and hybrid rice cultivars are cultivated by transplantation in irrigated lowlands. Currently, about 60–70% of the 500 000 ha of land area used for rice production in Egypt is engaged in rice–clover rotation. Berseem clover's high yield, protein content and symbiotic nitrogen-fixing capacity enhance its use as a forage and green manure plant in this region. An interesting enigma for this successful farming system is that the clover rotation with rice can replace 25–33% of the recommended amount of nitrogen fertilizer needed for optimal rice production. However, nitrogen balance data indicate that this benefit of rotation with clover cannot be explained solely by the increase in available soil nitrogen created by mineralization of the biologically fixed, nitrogen-rich clover crop residues. So the question asked was whether there is a natural endophytic *Rhizobium*–rice association that has evolved, which contributes to this added benefit of clover–rice rotation. The answer is yes indeed. Studies have indicated that the well-known clover root-nodule occupant *R. leguminosarum* bv. *trifolii* does indeed participate in such an association with rice, independent of root nodule formation and biological nitrogen fixation [1,7]. Further studies showed that inoculation of some varieties of rice with certain strains of endophytic rhizobia can significantly improve their vegetative growth, grain productivity and agronomic fertilizer nitrogen use efficiency. Rhizobia can help to produce higher rice grain yield with less dependence on inorganic fertilizer inputs, which is fully consistent with sustainable agriculture.

Subsequently, Chaintreuil *et al.* [24] investigated the natural existence of endophytic photosynthetic *Bradyrhizobium* (an endosymbiont of *Aeschynomene indica* and *A. sensitiva*) within the roots of the wetland wild rice *Oryza breviligulata* from Senegal and Guinea. In Africa, the wetland rice *O. breviligulata*, which is the ancestor of the African cultivated rice *O. glaberrima*, has been harvested and consumed in the Sahel and Sudan regions for more than 10 000 years. *O. breviligulata* grows spontaneously in temporary ponds, wetland plains and river deltas of the semiarid and semihumid regions of Africa. This primitive rice species is frequently found growing in association with several aquatic legumes belonging to the genera *Aeschynomene* and *Sesbania* [25–27]. Among these aquatic species, *A. indica* and *A. sensitiva* form stem nodules with photosynthetic *Bradyrhizobium* strains [28], which occur as endophytes in *O. breviligulata* nodal roots.

Following the same working hypothesis, the possible existence of endophytic rhizobia from India, a major rice producer of the world. Several diverse strains of endophytic rhizobia has also been explored in India, identified as *R. leguminosarum* bv. *phaseoli* and *Burkholderia cepacia* were isolated from rice roots collected at different locations in India [29,30]. Greenhouse and field experiments revealed that these legume-nodulating isolates have a great potential to promote rice growth and productivity.

9.3
Confirmation of Natural Endophytic Association of Rhizobia with Rice

Several laboratory, greenhouse and field experiments were performed using an ecological approach to detect, enumerate and isolate rhizobial endophytes from surface-sterilized roots of field-grown rice and wheat [1,7,30,31]. Rice plants were sampled at different field sites during two rotations with berseem clover in the Nile delta. The first field sampling site was from vegetative regrowth of 'ratoon' rice that remained after rice harvest at the end of the previous growing season, intermingled among clover plants in their current rotation. The second sampling was from four different sites in flooded fields of transplanted rice, during the next rice-growing season. Field-sampled roots were promptly taken to the laboratory, washed free of soil, cut at the stem base, blotted, weighed and surface-sterilized sufficiently with penetrating sodium hypochlorite solution until viable rhizoplane organisms could no longer be cultivated. These surface-sterilized roots were washed and macerated in sterile diluent and five replicates of each decimal dilution were inoculated directly on axenic roots of berseem clover seedlings grown on nitrogen-free Fahraeus agar slopes in enclosed tube cultures. Nodulated plants were scored after 1 month of cultivation.

This experimental design [1] took advantage of the strong positive selection provided by the clover 'trap' host so as to select for the numerically dominant 'rice-adapted' clover-nodulating rhizobia present among the diversity of other natural rice endophytes that survived surface sterilization of the rice roots. It also provided an easy route to isolate the dominant strains of endophytic rhizobia within the clover root nodules that ultimately developed on plants inoculated with the highest dilutions of rice macerates in the MPN series. The results from all sample sites provided solid confirmation of the original guiding hypothesis that clover-nodulating rhizobia intimately colonize the rice root interior in these fields of the Egyptian Nile delta [1,7]. The population size of clover-nodulating rhizobia was 2–3 logs higher inside the roots of the ratooned rice that was growing among the clover plants than in the transplanted rice in flooded fields (Figure 9.1). These results suggested that rice root interiors provide more favorable growth conditions for rhizobia when cultivated in close proximity to clover in the drained, more aerobic

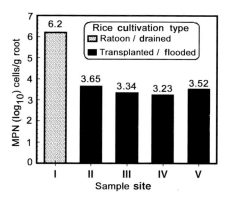

Figure 9.1 Most probable numbers of endophyte populations of *R. leguminosarum* bv. *trifolii* in rice roots cultivated in the Egyptian Nile delta. From Yanni *et al.* [7] and reprinted with permission from CSIRO Publishing (http://www.publish.csiro.au/journals/fpb).

soils rather than in monoculture within flooded soils, highlighting a long-term benefit of rice–legume rotation in promoting this intimate plant–microbe association.

Standard microbiological techniques were used to isolate into pure culture these rhizobial nodule occupants representing the numerically dominant endophytes of rice roots, verify that they were authentic rhizobia by testing their ability to nodulate berseem clover in gnotobiotic culture and evaluate their nitrogen-fixing activities on their natural clover host. These symbiotic performance tests (plus 16S rDNA sequencing) confirmed that the rice-adapted isolates were authentic strains of R. leguminosarum bv. trifolii capable of nodulating berseem clover under axenic conditions and that both effective and ineffective rhizobial isolates were included in the culture collection [1,7].

To fulfill Koch's postulates, these endophytic rhizobial strains were cultured on roots of rice plants under microbiologically controlled conditions, and then reisolated from surface-sterilized roots, 32 days after inoculation. Strain identification tests using plasmid profiling and BOX-PCR genomic fingerprinting showed that these reisolates were the same as the original inoculant strains, fulfilling Koch's postulates and confirming that they can form intimate endophytic associations with rice roots without requiring the assistance of other soil microorganisms [1]. Using a similar experimental approach, the existence of endophytic, bean-nodulating rhizobia in Indian cultivated rice was also verified [30].

We have used various molecular approaches such as 16S rDNA PCR-RFLP, BOX-PCR, plasmid profiling and SDS-PAGE to reveal the genomic diversity of 'rice-borne' rhizobial isolates. These studies helped to define the breadth of this ecological niche for rhizobia and guided our selection of isolates that can represent the genomic diversity in various studies of this association. It also indicated that our culture collections of rice-adapted rhizobia contained different strain genotypes that vary in their ability to evoke growth responses and that diverse populations of rhizobia colonize rice root interiors in different agroecosystems of Egypt and India [7,31].

9.4
Association of Rhizobia with Other Cereals Like Wheat, Sorghum, Maize and Canola

Since this discovery of a third ecological niche for Rhizobium (Figure 9.2), we have created an international network of collaborators to expand the intrinsic scientific merit of this project in both basic and applied directions of beneficial plant–microbe interactions. As an outcome, many tests of the generality of endophytic, plant growth promoting rhizobia within cereals have indicated that this type of association is widespread worldwide rather than being restricted to a particular crop (rice) and place (Nile delta). Other natural associations of endophytic plant growth promoting rhizobia within field-grown roots of wheat, barley, sorghum, canola, millet, rice and maize rotated with legumes have now been described in Canada, Mexico, Morocco, South Africa, Venezuela, China, India and elsewhere [32–36] (Y. Jing, personal communication). Thus, despite some initial reservations about this novel finding

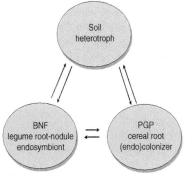

Figure 9.2 Widespread natural occurrence of three ecological
niches for *Rhizobium* in legume–cereal rotations. From Yanni
et al. [7] and reprinted with permission from CSIRO Publishing
(http://www.publish.csiro.au/journals/fpb).

within the scientific community, there is now no longer any scientific basis on which
to doubt the existence and potential benefits of this plant–microbe association. In
more recent work, a variety of legumes (berseem clover, alfalfa, soybean, lentil, faba
bean and bean) normally cultivated in rotation with wheat as trap hosts has been
used in an attempt to better reveal the species/biovar diversity of the numerically
dominant rhizobial endophytes of field-grown wheat in the Nile delta. The result
of that study was quite interesting in that the clover symbiont, *R. leguminosarum*
bv. *trifolii*, is the dominant *Rhizobium* endophyte within wheat roots in fields of
the Nile delta, while none of the other rhizobia represented by the other legume
cross-inoculation groups occupied this ecological niche in this same habitat [37].

9.5
Mechanism of Interaction of Rhizobia with Rice Plants

9.5.1
Mode of Entry and Site of Endophytic Colonization in Rice

The establishment of the *Rhizobium*–legume symbiosis and its formation of effective (i.e. nitrogen fixing) root nodules require a coordinated temporal and spatial
expression of both plant and bacterial genes [38]. A highly specialized and intricately
evolved interaction between these soil microorganisms and legume plants requires
the functions of the *nod* genes/Nod factors. Primary infection in most *Rhizobium*–
legume symbioses involves a coordinated development of wall-bound infection
threads within host target cells (most often root hairs) [39]. In contrast, rhizobial
interaction with rice and other cereals is *nod* gene/Nod factor independent and does
not involve the formation of infection threads [40]. A primary mode of rhizobial

invasion of rice roots is through 'cracks' in the epidermis and fissures created during emergence of lateral roots [41,42]. Further, several workers have reported that there may be three main portals for rhizobial entry in rice and other nonlegume roots, which include cracks created during emergence of the lateral root, lysed root hairs and cracks created between undamaged epidermal cells on the root surface [7,16,42–45]. The main site of rice epidermal root colonization by rhizobia is a collar ring surrounding the emergence of lateral roots (Figure 9.3a) [40–42]. The main site of internal root colonization is the intercellular space within root tissue, but rhizobia were also observed on the root surface, at root tips, in lateral root cracks and even in the cortex and vascular system [40,42–46]. More recently, it has been shown that rhizobia are able to achieve ascending migration from within roots, through the stem and into lower leaves of rice [41] Figure 9.3. There, they are able to grow to high local densities of up to 10^{10} bacteria per cm^3 of rice plant tissue [41]. Similar results were obtained for the invasion of wheat roots by *Azorhizobium caulinodans* when the cultures were supplemented with the flavonoid naringenin [45,47].

In the case of photosynthetic *Bradyrhizobium* colonization of rice roots, the bacteria colonize the root surface where numerous lateral roots emerge and produce fissures in the root epidermis and underlying cortex. These fissures are sites of intercellular bacterial proliferation where bacteria invade the fissure via disjoining epidermal cells, forming packets of proliferating bacteria in-between [24]. Interestingly, in *O. breviligulata*, photosynthetic bradyrhizobia are also present intracellularly in cortical cells. However, unlike in *Aeschynomene*, the number of invaded cells remained limited in *O. breviligulata*, and no division of these infected host cells was observed. The intracellular invasion could thus be the ultimate stage of rice infection by *Bradyrhizobium*. Nevertheless, the infection process in *O. breviligulata* by some photosynthetic bradyrhizobia is very similar to the first stages of infection in the leguminous plant *Aeschynomene*. In both, *O. breviligulata* and *A. sensitiva*, expression of *nod* genes is not necessary for the first steps of infection involving primitive 'crack entry' and direct intercellular invasion [24]. Most recently, the 'crack entry invasion' and intercellular colonization (within cortical cells, stele and aerenchyma) of rice by *Methylobacterium* sp. and *Burkholderia vietnamiensis* strains have been verified [48,49].

Several technologies, based on reporter gene assay, fluorescence confocal microscopy and scanning and transmission electron microscopy have been used to visualize and evaluate intercellular bacterial colonization, entry, spread and the establishment of internal colonization in this bacteria–plant association [40–42,50]. It has been seen that after inoculation of rice, rhizobia proliferate at the lateral root emergence site (Figure 9.3a). It is very likely that some type of signal communication/nutritional stimulation occurs in order to account for the preferential attraction of rhizobial cells toward root cracks and colonization of those sites. Rhizobial entry through the root epidermis is thought to be facilitated by cell wall degrading enzymes such as cellulase and pectinase that are produced by the rhizobial endophytes [7]. The current model is that rhizobial endophytes use those cell-bound enzymes to hydrolyze glycosidic bonds in adhesive polymers between epidermal cells, thereby allowing them ingress into and dissemination within cereal host roots [7].

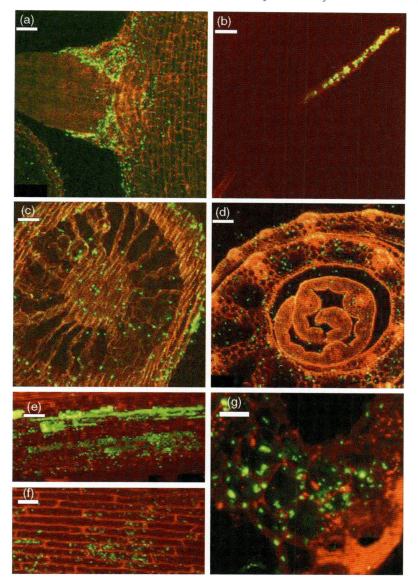

Figure 9.3 Confocal laser scanning micrographs of *gfp*-tagged cells of wild-type *Sinorhizobium meliloti* 1021 colonized within healthy rice tissues. (a) lateral root emergence; (b) lysed root hair; (c) cross section of the tap root; (d and g) cross sections of the leaf sheath above the stem base; (e and f) within leaves. Bar scales are 50 μm in (a)–(e) and 20 μm in (f)–(g). From Chi *et al.* [41] and reprinted with permission from the American Society for Microbiology Press (http://aem.asm.org/).

Strains of *A. caulinodans*, *Bradyrhizobium* and *Rhizobium* were genetically modified with the *lacZ* reporter gene to study their mode of invasion and extent of colonization in rice roots [40]. To study the timing and route of entry into rice tissues further, rhizobia have also been tagged with DNA sequences encoding the green fluorescent protein (Gfp) that imparts a green autofluorescence in the bacteria. This gene expression enables a nondestructive assay to be used to locate the bacteria, quantify their local abundance and follow their association with the root of young rice seedlings at single-cell resolution using epifluorescence microscopy [41,42]. After inoculation with *R. leguminosarum* bv. *trifolii* strain ANU843, the bacterial cells were observed along the root grooves and in microcolonies on the root surfaces of the primary main root. Gradually, this colonization progressed into lateral root cracks, the epidermis and finally deeper within intercellular spaces of the root cortex. The bacteria form long lines of fluorescent microcolonies inside lateral roots. Furthermore, no specific morphological changes of roots such as root hair curling, formation of infection threads or root nodule primordial were observed on the rice roots inoculated with rhizobial strains [42].

Quantitative microscopy is being used to evaluate spatial patterns of rhizobial colonization on rice roots to better understand how the rhizobia–rice association develops. This work involves scanning electron microscopy of rice roots inoculated with a reference biofertilizer strain of endophytic *R. leguminosarum* bv. *trifolii* analyzed at single-cell resolution using CMEIAS (Center for Microbial Ecology Image Analysis System) image analysis software developed for these studies in computer-assisted microscopy. New measurement features have been developed in CMEIAS, for example the Cluster Index [50], to extract information from digital images of microbial cells on the root surface needed to compute plotless, plot-based and geostatistical analyses that describe and mathematically model spatial patterns of their root surface colonization. This includes geostatistically defendable interpolation of their distribution and dispersion, even within areas of the root that are not sampled [39,51–53]. Typical scanning electron micrographs depicting various morphological features of the colonization of the Sakha 102 variety of rice roots by the rhizobial strain E11 used in these studies are illustrated in Figure 9.4a–e. These images reveal that these bacteria (1) attach in both supine and polar orientations, preferentially to the nonroot hair epidermis, in contrast to their preferential attachment in polar orientation to root hairs of their host legume, (2) commonly colonize small crevices at junctions between epidermal cells (white arrows in Figure 9.4a–c), suggesting this route to be a portal of entry into the root and (3) produce eroded pits on the rice root epidermis (Figure 9.4d and e). Similarly eroded plant structures are produced in the *Rhizobium*–white clover symbiosis by plant cell wall degrading enzymes bound to the bacterial cell surface, and these pits represent incomplete attempts of bacterial penetration that had only progressed through isotropic, noncrystalline outer layers of the plant cell wall [54]. Consistent with these results, activity gel electrophoresis indicated that the rice-adapted rhizobia produced a cell-bound CMcellulase [7]. This enzyme likely participates in the invasion and dissemination of the rhizobial endophyte within host roots.

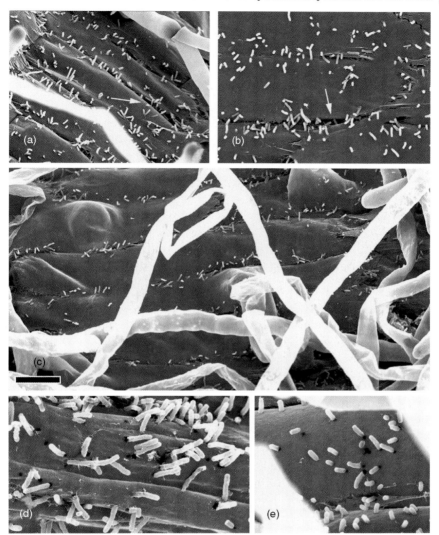

Figure 9.4 (a–e) Scanning electron micrographs of the rice epidermal root surface colonized by an endophyte strain of rice-adapted rhizobia. From Yanni et al. [7] and reprinted with permission from CSIRO Publishing (http://www.publish.csiro.au/journals/fpb).

This type of information derived from microscopy enhances the understanding of root colonization by inoculant strains, the dynamic aspects of rhizobial dispersion on the root and how different inoculant delivery systems will ultimately impact successful application of biofertilizer inoculants.

9.5.2
Systemic Movement of Rhizobial Endophytes from Rice Root to Leaf Tip

A plant assay was developed to analyze the relationship between the ability of *Rhizobium* strains to affect seedling growth and their ability to survive and multiply within rice tissues [54]. As the environment provided by the rice leaf can be easily used to study internal colonization by bacteria, this assay measured the multiplication, movement and compatibility of *Rhizobium* strains within rice tissues. In addition, it enabled the use of various bacterial strains as biological 'probes' of any induced responses or preformed systems of plant responses in the rice plants. In this bioassay, *gfp*-tagged bacterial cells were pressure infiltrated into sections of rice seedling leaves, and viable cell counts were recorded every 2 days up to 15 days. These assays were based on the current bacterial plant pathology leaf assays, which demonstrate that different bacterial cells could only grow and multiply within leaf tissues if they contain particular bacterial genes that are associated with specific nutrient uptake systems. Bacteria with mutation in these genes, or which do not contain these genes, do not grow in plant tissues. The results indicated that the *Rhizobium* strains could survive and multiply within the rice plant during this test period.

Further examination of the infection, dissemination and colonization of healthy rice plant tissues by different species of *gfp*-tagged rhizobia was done using laser scanning confocal microscopy [41]. Those studies indicated a dynamic infection process beginning with surface colonization of the rhizoplane (especially at lateral root emergence), followed by endophytic colonization within roots and then ascending endophytic migration into the stem base, leaf sheath and leaves where they developed high local populations (Figure 9.3). *In situ* CMEIAS image analysis indicated local endophytic population densities reaching levels as high as 9×10^{10} rhizobia per cm^3 of infected host tissues, whereas plating experiments indicated rapid, transient or persistent growth depending on the rhizobial strain and rice tissue examined. Considered collectively, the results indicate that this endophytic plant–bacterium association is far more inclusive, invasive and dynamic than previously thought, including dissemination in both below-ground and above-ground tissues and enhancement of growth physiology by several rhizobial species, therefore heightening its interest and potential value as a biofertilizer strategy for sustainable agriculture to produce the world's most important cereal crops [41].

9.5.3
Genetic Predisposition of Rice–Rhizobia Association

Isoflavonoids are derived from naringenin, a flavonone intermediate that is ubiquitous in plants, and they play a critical role in plant development and host defense responses. Isoflavonoids secreted by legumes also play an important role in promoting the formation of nitrogen-fixing nodules by symbiotic rhizobia. In these plants, the key enzyme that redirects phenylpropanoid pathway intermediates from flavonoids to isoflavonoids is cytochrome P450 monooxygenase, isoflavone synthase

(IFS). Rice does not naturally produce isoflavones, but similar kinds of signal molecules do exist in rice and other nonlegumes [55]. Some studies have analyzed rice seedlings for possible signal molecules that might interact with rhizobial cells [55]. Extract made from seedlings of rice cultivars were tested with the reporter strain ANU*gus* (PMD1), which contains the NGR 234 *nodD* gene and an inducible promoter. A higher level of signals similar to legumes was produced, which could induce the *nodD* gene [55]. In a recent effort to develop a rice variety possessing the ability to induce nodulation (*nod*) genes in rhizobia, the IFS gene from soybean was incorporated into rice cultivar Murasaki R86 under the control of the 35S promoter [56]. The presence of IFS in transgenic rice was confirmed by PCR and Southern blot analysis. Analyses of the 35S–IFS transgenic lines demonstrated that the expression of the IFS gene led to the production of the isoflavone genistein in rice tissues. These results showed that the soybean IFS gene-expressed enzyme is active in the R86 rice plant and that the naringenin intermediate of the anthocyanin pathway is available as a substrate for the introduced foreign enzyme. The genistein produced in rice cells was present in a glycoside form, indicating that endogenous glycosyltransferases were capable of recognizing genistein as a substrate. Studies with rhizobia demonstrated that the expression of IFS conferred rice plants with the ability to produce flavonoids that are able to induce *nod* gene expression, albeit to varied degrees, in different rhizobia [56]. Thus, the possibilities of establishing a more effective type of *Rhizobium*–nonlegume interaction are potentially available in rice because rice roots contain many of the plant compounds that can stimulate rhizobia.

9.6
Importance of Endophytic Rhizobia–Rice Association in Agroecosystems

9.6.1
Plant Growth Promotion by *Rhizobium* Endophytes

Early studies on endophytic colonization of rice by rhizobia indicated that some strains promoted the shoot and root growth of certain rice varieties in gnotobiotic culture [1]. Later, more extensive tests established the range of growth responses of japonica, indica and hybrid rice varieties from Egypt, Philippines, United States, India and Australia when these cultivars were inoculated with various rice-adapted rhizobia. The results indicated that the diverse rhizobial endophytes evoked a full spectrum of growth responses in rice (positive, neutral and sometimes even negative), often exhibiting a high level of strain–variety specificity [1,7,29,31,42,57–60]. On the positive side, a chronology of PGP^+ responses of rice to rhizobia manifested as increased seedling vigor (faster seed germination followed by increased root elongation, shoot height, leaf area, chlorophyll content, photosynthetic capacity, root length, branching and biomass). This effect carries over into increased yield and nitrogen content of the straw and grain at maturity. Similar results were obtained when wheat was inoculated with diverse genotypes of endophytic wheat-adapted rhizobia in gnotobiotic plant bioassays; there is high strain–variety specificity in rhizobial promotion of wheat growth.

Several field inoculation trials have been conducted to assess the agronomic potential of these *Rhizobium*–cereal associations under field conditions, with the long-term goal of identifying, developing and implementing superior biofertilizer inoculants that can promote rice and wheat productivity in real-world cropping systems while reducing their dependence on nitrogen fertilizer inputs. A direct agronomic approach was adopted to address the importance of continued evaluation of the various strain genotypes in our diverse collections of cereal-adapted rhizobia under experimental field conditions [1,7]. This meant first acquiring useful information from laboratory PGP bioassays. This information was then applied to design and implement small field trials at the Sakha Agricultural Research Station in the Kafr El-Sheikh area of the Nile delta, Egypt. Information from these results was then utilized to conduct scaled-up experiments on large farmers' fields in neighboring areas of the Nile delta where cereal–legume rotations are used, so our results could be compared to real on-farm baselines in grain production.

Experiments included nitrogen fertilizer applications at three rates: one third, two third and the full recommended rate previously assessed without inoculation with nitrogen fixer(s). The inocula for rice and wheat were used as indicated in Tables 9.1 and 9.2. The field trials were conducted in 20-m^2 subplots or sub-subplots with four replications. The various trials were supplemented with calcium superphosphate before tillage, with potassium sulfate added 1 month after wheat sowing or rice transplantation. Appropriate broad-spectrum herbicide(s) were applied to control the major narrow- and broadleaf weeds.

In total, we conducted 24 different field inoculation trials using selected endophytic strains of rhizobia with rice and wheat in the Nile delta. So far, positive

Table 9.1 Grain yields of rice variety Giza 178 in the best experimental treatments versus adjacent farmer's fields at different locations in Kafr El-Sheikh, Nile delta, Egypt, 2002.

Farm location	Best experimental treatment: inoculated strains + kg of nitrogen fertilizer[a]	Grain yield of best experimental treatment (kg ha^{-1})	Yield in adjacent field (no researcher supervision) (kg ha^{-1})[b]	Increase over farmer's yield (%)
Baltem	E11 + E12 + 96 N	8623	8330	3.5
Beila	E11 + E12 + 96 N	11 309	9520	18.8
Metobas	E11 + E12 + 96 N	12 400	9520	30.3
Sidi Salem	144 N	11 118	9068	22.6

Source: Dazzo and Yanni [59].
[a]The method of inoculation was direct broadcast of the inoculum (10^9 CFU g^{-1}) at the rate of 720 g peat-based inoculum per hectare, 3 days after transplanting of the rice seedlings and during a period of calm wind at sunset. Nitrogen, kg N ha^{-1}, was applied as urea (46% N) in two equal doses, 15 days after transplanting and at the mid-tillering stage.
[b]Recommended rate of nitrogen fertilizer for the tested rice varieties when used without inoculation with biofertilizers is 144 kg N ha^{-1}. This rate was used by the farmer in the adjacent field who was not supervised by the research personnel.

Table 9.2 Grain yields of wheat in the best experimental treatments versus grain yields in adjacent farmers' fields at different locations in Kafr El-Sheikh, Nile delta, Egypt, 2002–2003.

Farm location	Wheat variety	Best experimental treatment: inoculated strain + kg N fertilizer	Yield of the best experimental treatment (kg ha^{-1})	Yield in the rest of the same farmer's field (no researcher supervision) (kg ha^{-1})[a]	Increase over farmer's yield
1	Sakha 93	EW 54 + 180 N	7112	6120	16.2%
2	Sakha 61	EW 72 + 120 N	7382	5712	29.2%
3	Sakha 61	EW 72 + 180 N	8247	6936	18.9%
4	Sakha 61	EW 72 + 60 N	5802	4896	18.5%

Source: Dazzo and Yanni [59].
EW: rhizobial wheat-root endocolonizer used at the rate of 720 g peat-based inoculum (10^9 CFU g^{-1}) for inoculation of 144 kg wheat seeds for cultivation of one hectare. The method of inoculation involved mixing the seeds with the inoculum in the presence of a suitable quantity of a solution with an adhesive material such as pure Arabic gum, gelatin or sucrose. The mixed seeds were left for some time in the shade and then planted as fast as possible during sunset followed by irrigation of the field area. Nitrogen, kg N ha^{-1}, was applied as urea (46% N) in two equal doses just before sowing and 75 days later.
[a]Recommended rate of nitrogen fertilizer for the tested wheat varieties when used without inoculation with biofertilizers was 180 kg N ha^{-1}. This rate was used by the farmer in the part of the field that was not supervised by the research personnel.

increases in grain yield resulting from inoculation with those selected strains of our cereal-adapted rhizobia have occurred in 16 of 18 field tests for rice (89%) and in 18 of 19 field tests for wheat (95%) [1,7] (Yanni and Dazzo, unpublished data). Tables 9.1 and 9.2 summarize the data on grain yield from the recent scaled-up experiments on farmers' fields. They illustrate the best performance obtained with field inoculation treatments using our cereal-adapted strains on the scaled-up experimental plots versus yields of the same variety obtained simultaneously by traditional agricultural practices on adjacent fields without inoculation or supervision by research personnel.

Increases in grain yields of rice and wheat ranged between 3.5 and 30.3% and from 16.2 to 29.2%, respectively, using the researchers' package of agronomic treatments rather than farmers' practices (Tables 9.1 and 9.2). The best inoculation responses for rice occurred with an inoculum combining two strains of rice-adapted rhizobia (rather than one). The wheat and rice varieties tested in most of these experiments displayed an increase in agronomic nitrogen fertilizer use efficiency (kg grain yield/kg fertilizer-N applied) indicating that those rhizobial strains can help these crops utilize the nitrogen taken up more efficiently to produce grain with less dependence on nitrogen fertilizer inputs. The results, with a few exceptions, also suggest that even after the nitrogen requirements have been satisfied, our microbial inoculants facilitate the acquisition of other nutrients, which then become the next limiting factor(s) for rice and wheat productivity in these fields. The exceptions of no inoculation response were most likely a result of the natural widespread abundance of the same inoculant strain(s) in the same field, thus needing no inoculation. This is being examined further by studies of autecological biogeography [50].

These results reflect the potential benefits of rhizobial biofertilizer inoculants for rice and wheat production in the Nile delta. This knowledge should improve agricultural production by advancing basic scientific knowledge on beneficial plant–microbe associations and also by assisting low-income farmers to increase cereal production on marginally fertile soils using biofertilizers consistent with environmental soundness and sustainable agriculture. In this regard, it must be considered that the differences between the yields of the best experimental treatment(s) and those obtained in farmers adjacent fields cannot be solely attributed to inoculation with the effective endophytes, since they include the whole research package of recommendations extending beyond inoculation.

9.6.2
Extensions of Rhizobial Endophyte Effects

9.6.2.1 Use of Rhizobial Endophytes from Rice with Certain Maize Genotypes

It is of obvious interest to know whether superior rhizobial endophyte strains that are PGP$^+$ with rice can also promote the growth of other cereal crops. Field tests conducted in Wisconsin, USA, have provided preliminary evidence for this, at least for certain genotypes of maize [7]. In that study, inoculation with one of the rice endophyte strains isolated from the Nile delta resulted in statistically significant increase in dry weight in three of the six tested maize genotypes in the greenhouse and one of the seven maize genotypes in experimental field plots that received no nitrogen fertilizer [7]. In addition, effects of rice rhizobial endophytic strains (RRE-2, RRE-5 and RRE-6) isolated from India [30] were also tested on some maize cultivars and a significant increase in root proliferation and overall plant growth were observed (R.K. Singh, unpublished data).

A cross between the high-responding maize genotype and a different nonresponding genotype resulted in a hybrid maize genotype with an intermediate level of growth responsiveness to inoculation with the rhizobial endophyte strain from Egypt (Figure 9.5) [7]. This suggests the possibility of genetic transmissibility and inheritance in corn of the ability to respond to selected endophytic rhizobia. This result reinforces the earlier finding that induction of positive growth responses in cereals by rhizobia is genotype specific (also in maize). Such experimental results may help to identify the genes in cereals necessary for expression of these growth responses to rhizobia. Also, since recent work indicates that one of our rice endophyte strains of rhizobia can promote the growth of shoots and roots, root architecture and uptake of N and Ca^{2+} for certain cotton varieties under growth-room conditions [61], a thorough screen of their potential benefit to a wide variety of nonlegume crop plants should be made.

9.6.2.2 Rhizobia–Rice Associations in Different Rice Varieties

For practical reasons, it is important to know whether the various desirable interactions of these rhizobial endophytes can be extended to cereal varieties that are preferred by farmers in cropping systems worldwide. The number of varieties tested

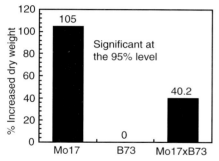

Figure 9.5 Genotype-specific inheritance and growth response of corn (*Zea mays*) to inoculation with a rice endophyte strain of rhizobia under field conditions in Wisconsin (USA) without nitrogen fertilizer application. From Yanni *et al.* [7] and reprinted with permission from CSIRO Publishing (http://www.publish. csiro.au/journals/fpb).

so far is too small to make reliable and accurate armchair predictions about genotypes not yet tested. Because many characteristics of this association exhibit high strain–variety specificity, tests of their compatibility at the laboratory bench are necessary before they are tested in the field. Studies so far have included rice genotypes commonly used in Egypt (Sakha 101, 102, 104, Giza 175, 177, 178 and Jasmine rice), United States (M202 and L204), Australia (Calrose and Pelde) and India (Pankaj, Sarjoo-52 and Pant-12).

However, what about rice varieties preferred by low-income farmers who cultivate rice on marginally fertile soils and who cannot afford to purchase fertilizers? To address this question, a study was undertaken to measure how well the rhizobial endophyte strain E11 can colonize the root environment of four rice genotypes preferred by Filipino peasant farmers because of their good yielding ability and grain characteristics (Sinandomeng, PSBRC 74, PSBRC 58 and PSBRC 18). The bacteria were inoculated on axenic seedling roots, then grown gnotobiotically in hydroponic tube culture, and the resultant populations were enumerated by viable plate counts. For comparison, seedlings of equal size received an equivalent inoculum of a local, unidentified isolate BSS 202 from *Saccharum spontaneum* used as a PGP$^+$ biofertilizer inoculant for rice in the Philippines. The results indicated significant colonization potential of the *Rhizobium* endophyte test strain on roots of all four rice genotypes [7]. In some cases, the population size achieved by rhizobia was even higher than the BSS 202 isolate (Table 9.3). The implications of this experiment are significant: the *Rhizobium* test strain exhibited no obvious difficulty in its ability to intimately colonize roots of not only the superior rice varieties that have undergone significant breeding development but require high nitrogen inputs for maximum yield, but also with other rice varieties that perform acceptably on marginally fertile soil without significant nitrogen fertilizer inputs. The latter type of rice cropping could derive significant benefit from the biofertilizer inoculants that we intend to develop.

Table 9.3 Colonization potential of *R. leguminosarum* bv. *trifolii* strain E11 and isolate BSS 202 from *S. spontaneum* on four rice varieties preferred by low-income Filipino farmers.

Inoculum strain	Sample location	Colonization potential (viable plate count/three rice roots)			
		Sinandomeng	PSBRC 74	PSBRC 58	PSBRC 18
Rlt E11	External rooting medium	1.03×10^7	1.50×10^7	7.16×10^7	8.17×10^7
BSS 202	External rooting medium	9.17×10^7	7.00×10^6	8.00×10^7	1.27×10^7
Rlt E11	Root surface	6.34×10^7	1.05×10^8	6.50×10^7	6.50×10^7
BSS 202	Root surface	1.08×10^7	6.00×10^7	5.60×10^7	5.66×10^7
Rlt E11	Root interior	1.50×10^8	2.19×10^9	1.32×10^8	1.03×10^9
BSS 202	Root interior	7.50×10^7	6.30×10^6	7.34×10^7	1.23×10^8

Source: From Yanni *et al.* [7].
Reprinted with permission from CSIRO Publishing (http://www.publish.csiro.au/journals/fpb).

9.7
Mechanisms of Plant Growth Promotion by Endophytic Rhizobia

The ability of some endophytic rhizobial strains to promote the growth of rice and wheat prompted follow-up studies to identify possible mechanisms operative in this beneficial plant–microbe interaction. These studies have focused primarily on rice, addressing the following possible mechanism of rhizobial growth promotion: (1) induction of an expansive root architecture having enhanced efficiency in plant mineral nutrient uptake; (2) production of extracellular growth-regulating phytohormones; (3) solubilization of precipitated inorganic and organic phosphate complexes, thereby increasing the bioavailability of this important plant nutrient; (4) endophytic nitrogen fixation; (5) production of Fe-chelating siderophores; and (6) induction of systemic disease resistance.

9.7.1
Stimulation of Root Growth and Nutrient Uptake Efficiency

Responsive rice varieties commonly develop expanded root architectures when inoculated with candidate biofertilizer strains of rhizobia. This suggests that these rhizobial endophytes alter root development in ways that could make them better 'miners', more capable of exploiting a larger reservoir of plant nutrients from existing resources in the soil. This possibility was suggested in early studies showing significantly increased production of root biomass in plants that had been inoculated [1,7,42,57] and by studies using greenhouse-potted soil indicating significant increases in N, P, K and Fe uptake by rice plants inoculated with selected rhizobia, including rice endophyte strains [58]. More recent studies have confirmed this result using plants grown gnotobiotically with rhizobia in nutrient-poor medium (50% Hoaglands), followed by measurements of root architecture and mineral nutrient

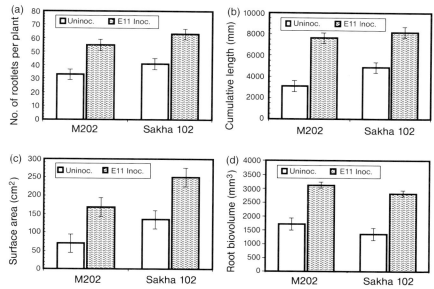

Figure 9.6 (a–d) Effect of inoculation with *R. leguminosarum* bv. *trifolii* E11 on root architecture of rice varieties. From Yanni *et al.* [7] and reprinted with permission from CSIRO Publishing (http://www.publish.csiro.au/journals/fpb).

composition using CMEIAS image analysis and atomic absorption spectrophotometry, respectively [7]. In these latter studies, inoculated plants developed more expanded root architecture, as seen in Figure 9.6a–d, and accumulated higher concentrations of N, P, K, Ca, Mg, Na, Zn and Mo than did their uninoculated counterparts (Figure 9.7).

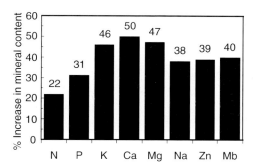

Figure 9.7 Effect of inoculation with *R. leguminosarum* bv. *trifolii* E11 on elemental composition of rice plants. From Yanni *et al.* [7] and reprinted with permission from CSIRO Publishing (http://www.publish.csiro.au/journals/fpb).

The results also showed, however, that the levels of Fe, Cu, B and Mn were not statistically different in the same inoculated and uninoculated plants grown under these microbiologically controlled experimental conditions [7]. This selectivity in terms of which plant nutrients exhibit increased accumulation as a result of inoculation argues against an across-the-board, general enhancement of mineral accumulation resulting exclusively from an expanded root architecture with increased absorptive biosurface area. The results indicate that these bacteria can modulate the rice plant's plasticity that enables it to control the adaptability of its root architecture and also physiological processes for more efficient acquisition of selected nutrient resources when they become limiting. This same mechanism is considered one of the major reasons for the beneficial growth promotion effect of *Azospirillum brasilense* on grasses [62–66].

These *Rhizobium*-induced increases in the mineral composition of rice plants raise new possibilities regarding their potential impact on the human nutritional value of this crop. A potential value-added benefit resulting from inoculation could be to increase the nutritional value of resulting grain, not only for increased nitrogen (mostly in the form of protein) but also for other macro- and micronutrients. For instance, rice is indeed an important major bioavailable source of some minerals, for example zinc in the human diet, particularly in developing countries [67]. Zinc is considered an essential micronutrient that is important for maturation of the reproductive organs in women and the developing fetus and maintenance of healthy immunity [67]. Inoculation shows a capacity to increase the zinc content of rice grain (Figure 9.7).

SDS-PAGE and RP-HPLC analyses of the protein composition in field-grown Giza 177 rice grains indicated no discernable differences in the ratios of the major nutritionally important storage proteins, particularly glutelin, albumin and globulin, as a result of inoculation with rhizobial endophyte strain E11. Field inoculation with this rhizobial endocolonizer thus did not qualitatively alter rice grain protein, as all nutritionally important proteins were present in the treated and control samples in similar ratios. However, since inoculation with rhizobia causes a significant increase in total grain nitrogen per hectare of crop (in protein form), the benefits of inoculation to small farmers will include an increase in the quantity of rice grain protein produced per unit of land used for cultivation. This increases the nutritional value of the harvested grain as a whole in comparison with uninoculated rice.

Enhanced uptake of mineral nutrient resources by inoculated rice could be a two-edged sword if accompanied by enhanced bioaccumulation of toxic metals. Therefore other rice grain sampled from the same field inoculation experiment were analyzed for their heavy metal content (Hg, Se, Pb, Al and Ag). The results indicated no significant differences in the low levels of these toxic heavy metals in the rice grain of uninoculated versus rhizobial endophyte-inoculated treatments [7]. Considered collectively, these studies indicate that rice plants inoculated with selected rhizobial biofertilizer strains produce rice grain whose human nutritional value is equal to or improved (depending on the nutrient considered) compared to that of uninoculated plants.

9.7.2
Secretion of Plant Growth Regulators

Early studies suggested that rhizobial endophyte strain E11 produced the auxin indole acetic acid (IAA) in pure culture and in gnotobiotic culture with rice [57,58]. Further studies indicated that production of IAA equivalents by this test strain was tryptophan-dependent. A simple defined medium was developed to optimize production of IAA by fast-growing rhizobia, and axenic bioassays of filter-sterilized culture supernatant from strain E11 grown in this defined medium showed an ability to stimulate rice root growth at critical concentrations [7]. These results suggest that endophytic strains of rhizobia can boost rice growth by producing extracellular bioactive metabolites that promote the development of more expansive root architecture. These results logically led to the identification of growth-regulating phytohormones produced and secreted by strain E11 in pure culture. Analysis of its culture supernatant using electrospray ionization gas chromatography/mass spectrometry (GC/MS) indicated the presence of IAA and a gibberellin (consistent with GA7) (Figure 9.8). These represent two different major classes of plant growth regulators that play key roles in plant development. Rhizobia also naturally produce other biomolecules such as lumichrome, abscisic acid, riboflavin and other vitamins that promote plant growth, and therefore their colonization and infection of cereal roots would be expected to increase plant development and grain yield [68]. This fundamentally new information has increased our understanding of the mechanisms underlying plant growth promotion in this beneficial *Rhizobium*–rice association.

9.7.3
Solubilization of Precipitated Phosphate Complexes by Rhizobial Endophytes

Phosphorous is one of the most important macro-element after nitrogen as it plays a vital role in the growth and survival of both bacteria and plants [69]. It is an important component of biomolecules such as nucleic acid, membrane lipids and protein and performs crucial roles in various enzymatic reactions responsible for normal functioning of living organisms. In plant systems, phosphate promotes root growth, grain filling and many other physiological processes and growth parameters [70]. More than 75% of applied phosphorous forms complexes and is fixed in soil in forms that are unavailable for plant use [70]. Most Nile delta soils used for rice cultivation contain about 1000 ppm phosphorus, primarily in the unavailable form of precipitated tricalcium phosphate, $Ca_3(PO_4)_2$. Although waterlogged conditions normally prevail in lowland rice fields, less than 8 ppm phosphorus (Olsen P) is available to rice. Any significant solubilization of precipitated phosphates by rhizobacteria *in situ* would enhance phosphate availability to rice in these soils, representing another possible mechanism of PGP for rice under these field conditions. The diversity of rice-adapted rhizobia was tested for phosphate-solubilizing activity on culture media impregnated with insoluble organic and inorganic phosphate complexes (Figure 9.9). An improved, double-layer plate assay indicated that some of the rhizobial endophyte strains are active in solubilizing both inorganic (calcium

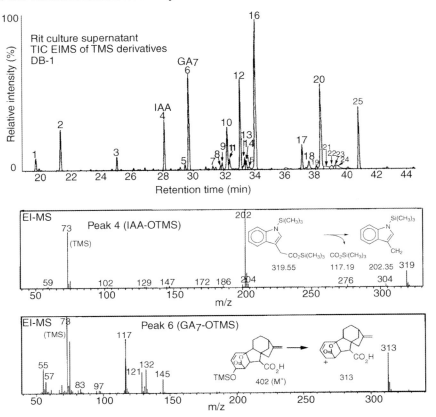

Figure 9.8 GLC/MS fractionation and identification of phyto-hormones in the culture supernatant of rice-adapted rhizobial endophyte strain E11. From Yanni *et al.* [7] and reprinted with permission from CSIRO Publishing (http://www.publish.csiro.au/journals/fpb).

phosphate) and organic (inositol hexaphosphate = phytate) insoluble phosphorus complexes [7]. These positive results indicate extracellular acidification and phos-phatase enzyme (phytase) activity, respectively. This extracellular PGP^+ activity would potentially increase the availability of phosphorus for rice in rhizosphere soil, and thereby promote rice growth when soil phosphorus is limited.

9.7.4
Endophytic Nitrogen Fixation

Rice plants accumulate more shoot and grain nitrogen when inoculated with select-ed strains of rhizobial endophytes [1,7,58]. However, this additional combined nitrogen is mainly derived from soil mineral nitrogen and not from biological nitrogen fixation (BNF). This conclusion is based on several lines of evidence:

Figure 9.9 Solubilization of insoluble phosphate complexes by rice-adapted endophytes of rhizobia. From Mishra [31].

1. Growth benefits by rhizobia are enhanced rather than suppressed when nitrogen fertilizer is provided [1,42,57,58]. Studies usually show an inverse relationship between high nitrogen fertilizer inputs and BNF activity.

2. The degree of growth benefit linked to inoculation with rhizobial endophytes does not correlate with their degree of nitrogen-fixing activity in symbiosis with their normal nodulated legume host (e.g. berseem clover), since some rice-adapted strains of rhizobia that are ineffective on clover are nevertheless PGP$^+$ on rice.

3. Acetylene reduction tests on rice plants whose growth is promoted by rhizobial endophytes indicate no associative nitrogenase activity [1,57].

4. Greenhouse studies using the ^{15}N-based isotope dilution method indicate that the increased nitrogen uptake in inoculated plants is by and large not derived from BNF [58].

5. Measurements of the natural abundance of nitrogen isotope ratios (Δ^{15}N) on field-grown plants indicate that their greater proportion of nitrogen resulting from inoculation with rhizobial endophytes is not derived from BNF [7].

Considered collectively, these results indicate that biological nitrogen fixation is not a significant factor responsible for the positive growth response of rice to inoculation with these rice-adapted rhizobia.

9.7.5
Production of Fe-Chelating Siderophores

Siderophore production potentially provides a dual mechanism of PGP: enhancing uptake of Fe for the plant and suppressing rhizosphere pathogens unable to utilize the Fe–siderophore complex. However, none of the PGP^+ genotypes of rhizobial endophyte strains isolated from the Nile delta produced siderophores detectable on CAS differential medium [7]. Nonetheless, production of rhizobial siderophores in the rhizosphere of inoculated plants remains to be examined before reaching any conclusions about the possible contribution of this mechanism to their commonly found growth-promoting benefit to rice.

9.7.6
Induction of Systemic Disease Resistance

Recent studies showed that rhizobial inoculation of rice may trigger the biochemical pathways (particularly enhanced production of phenolic acids) involved in defense reactions during pathogenic ingress [71]. An HPLC analysis of the different rice plant parts after inoculation with two *Rhizobium* sp. (*R. leguminosarum* bv. *phaseoli* RRE6 and *R. leguminosarum* bv. *trifolii* ANU 843) as well as *Rhizoctonia solani* (which causes blast disease of rice) revealed the induction of phenolic acids such as gallic, tannic, ferulic and cinnamic acids [71]. These phenolics mediate defense responses of crop plants against phytopathogens that cause various devastating diseases [72].

The exact mechanism used by *Rhizobium* endophytic strains to alter the phenolic profiles is still not very clear. However, bacterial endophytic biocontrol agents are reported to benefit crop plants via disease resistance by two possible ways: (i) by extensive colonization of internal plant tissues and suppression of invading pathogens by niche occupation, antibiosis or both; and (ii) by colonization of the root cortex, where they stimulate general plant systemic defenses/resistances [73]. It is quite possible that endophytic rhizobia employ one or more of these mechanisms to protect plants and promote their growth while colonizing their root tissues.

9.8
Summary and Conclusion

Studies completed thus far indicate that superior candidate strains of rhizobial endophytes suitable for use as biofertilizers for rice under field conditions have been widely developed and are now being used in cereal production worldwide. Information on the spatial distribution of candidate strains at scales relevant to the rice farmer is currently under examination to fully exploit their benefits for sustainable agriculture. The rationale for these spatial ecology studies is that a thorough understanding of their natural spatial distribution within rice agroecosystems should assist our biofertilization strategy program by helping to predict and interpret results of tests to evaluate their efficacy as inoculants [50].

The cumulative information derived from the studies described here indicates that rhizobia have evolved an alternate ecological niche that enables them to maintain a three-component life cycle that includes a free-living heterotrophic phase in soil, a nitrogen-fixing endosymbiont phase within the root nodules of legumes and a beneficial growth-promoting endocolonizer phase within cereal roots in the same crop rotation (Figure 9.2). The results further indicate an opportunity to exploit this newly described plant–rhizobia association by developing biofertilizer inoculants that have potential to increase cereal production, including rice and wheat, with less nitrogen fertilizer inputs, which would be supportive of both sustainable agriculture and environmental safety. The situation exemplified in the Egyptian Nile delta indicates that inoculation of the cereal crop with the appropriate cereal endophyte strain(s) of rhizobia would follow rather than replace a crop rotation with the legume. This way the cereal crop would gain maximum benefit from its indirect biological association with the nitrogen-fixing *Rhizobium*–legume root-nodule symbiosis and then directly as an alternate host that reaps all the benefits of rhizobia as an efficient plant growth promoting rhizobacterium colonizing within the cereal root's interior. Interestingly, some rhizobial endophytes can benefit multiple cereal crop species as previously mentioned. Assuming that the past results summarized here accurately reflect the potential benefits of this new agricultural biotechnology based on exploitation of a natural resource (natural rhizobial endophytes of cereals), the following outcomes can be expected:

1. Increased cereal crop yield above what is reached using synthetic fertilizers alone, with a reduction of up to 33–50% of the fertilizer input previously recommended and currently under use when there was no biofertilizer inoculation. Economically, field management programs that include this new biofertilization technology could assist farmers to increase their production by 3.5–30.3% for rice and 16.2–29.2% for wheat, with a saving of one third or more of their fertilization costs (Tables 9.1 and 9.2).

2. Decreased environmental pollution and health risks originating from excessive use of synthetic nitrogen fertilizers. However, this needs further medical economics studies to verify the reduction in costs involved in dealing with diseases associated with the excessive use of agrochemicals and the economical benefits from increasing individuals' work abilities.

3. Decreased energy needed for production, transportation and distribution of fertilizers, directing this energy to other socioeconomic and industrial uses.

4. A better understanding of how farmers can practice sustainable agriculture by utilizing biofertilizers as a safe and effective alternative to fertilizer application.

5. Promotion of cooperation between governmental and university research institutions on one side and private sectors represented by farmers and agricultural biotechnology industries on the other.

It is interesting here to note that farmers who hosted the experiments in their fields in the Nile delta were initially suspicious about the validity of using rhizobia as inoculants for cereals, but having experienced the benefits, they are very enthusiastic about inoculating their fields with the biofertilizer formulation and also ask for advice about the use of other microbial inoculant preparations. Other farmers near the experimentation areas who learned directly or indirectly about the results that the biofertilizer inoculants produce have become curious about this innovation. However, a well-designed agricultural extension program is still needed and in great demand to meet the needs of potential beneficiaries of this microbial biotechnology.

In summary, our studies indicate that certain strains of rhizobia that are natural endophytes of rice significantly enhance plant growth and development in ways that can be utilized to increase crop production in sustainable agriculture. One of the many lessons we have learned from these studies on the *Rhizobium*–cereal association was well stated centuries ago by Leonardo da Vinci: 'Look first to nature for the best design before invention.' This is particularly relevant for the design of biofertilizer inoculant strategies for sustainable agriculture.

References

1 Yanni, Y.G., Rizk, R.Y., Corich, V., Squartini, A., Ninke, K., Philip-Hollingsworth, S., Orgambide, G., de Bruijn, F.D., Stoltzfus, J., Buckley, D., Schmidt, J.M., Mateos, P.M., Ladha, J.K. and Dazzo, F.B. (1997) *Plant and Soil*, **194**, 99–114.

2 IRRI (1996) IRRI Towards 2020, Manila, The Philippines.

3 IRRI (1993) Rice Research in a Time of Change. Medium Term Plan for 1994–1998. International Rice Research Institute, Manila, The Philippines.

4 Ladha, J.K., de Bruijn, F. and Malik, K.A. (1997) *Plant and Soil*, **194**, 1–10.

5 James, E.K., Gyaneshwar, P., Barraquio, W.L., Mathan, N. and Ladha, J.K. (2000) *The Quest for Nitrogen Fixation in Rice* (eds J.K. Ladha and P.M. Reddy), IRRI, Manila, The Philippines, pp. 119–140.

6 Perrine, F.M., Prayitno, J., Weinman, J.J., Dazzo, F.B. and Rolfe, B.G. (2001) *Australian Journal of Plant Physiology*, **28**, 923–937.

7 Yanni, Y.G., Rizk, R.Y., Fatah, F.K.A., Squartini, A., Corich, V., Glacomini, A., Bruijn, F., Rademaker, R., Maya-Flores, J., Ostrom, P., Hernandez, M.V., Hollingsworth, R.I., Martinez-Molina, E., Mateos, P., Velazquez, E., Wopereis, J., Triplett, E., Umali-Garcia, M., Anarna, J.A., Rolfe, B.G., Ladha, J.K., Hill, J.H., Mujoo, R., Ng, K. and Dazzo, F.B. (2001) *Australian Journal of Plant Physiology*, **28**, 845–870.

8 Gyaneshwar, P., James, E.K., Reddy, P.M. and Ladha, J.K. (2002) *New Phytologist*, **154**, 131–146.

9 James, E.K., Gyaneshwar, P., Mathan, N., Barraquio, W.L., Reddy, P.M., Lannetta, P. P.M., Olivares, F., Ladha, L. and Mol, J.K. (2002) *Molecular Plant–Microbe Interactions*, **15**, 894–906.

10 Elbeltagy, A., Nishioka, K., Sato, T., Suzuki, H., Ye, B., Hamada, T., Isawa, T., Mitsui, H. and Minamisawa, K. (2001) *Applied and Environmental Microbiology*, **67**, 5285–5293.

11 Borlaug, N.E. (1970) The Green Revolution: Peace and Humanity (Nobel Peace Prize), CIMMYT, Mexico DF.

12 Kennedy, I.R. and Cocking, E.C. (1997) Biological nitrogen fixation: the global challenge and future needs. Position Paper, The Rockefeller Foundation Bellagio Conference Center, Italy, 8–12 April, 1-86451-364-7.

13 Shrestha, R.K. and Ladha, J.K. (1998) *Soil Science Society of America Journal*, **62**, 1610–1619.

14 Kennedy, I.R. and Tchan, Y. (1992) *Plant and Soil*, **141**, 93–118.

15 Bronson, K.F., Singh, U., Neu, H.U. and Abao, E.B. (1997) *Soil Science Society of America Journal*, **61**, 988–993.

16 Stoltzfus, J.R., So, R., Malarvith, P.P., Ladha, J.K. and de Bruijn, F. (1997) *Plant and Soil*, **194**, 25–36.

17 Khush, G.S. and Bennett, J. (1992) *Nodulation and Nitrogen Fixation in Rice: Potential and Prospects*, International Rice Research Institute Press, Manila, The Philippines.

18 Balandreau, J. and Knowles, R. (1978) The rhizosphere, in *Interactions Between Non-Pathogenic Soil Microorganisms and Plants* (eds Y.R. Dommergues and S.V. Krupa), Elsevier, Amsterdam, pp. 243–268.

19 Klein, D.A., Salzwedel, J.L. and Dazzo, F.B. (1990) *Biotechnology of Plant Microbe Interaction* (eds J.P. Nakas and C. Hagedorn), McGraw-Hill Publishing Company, New York, pp. 189–225.

20 Old, K. and Nicolson, T. (1975) *New Phytologist*, **74**, 51–58.

21 Old, K. and Nicolson, T. (1978) *Microbial Ecology* (eds M. Loutit and J. Miles), Springer-Verlag, New York, pp. 291–294.

22 Kloepper, J.W., Schippers, B. and Bakker, P.A. (1992) *Phytopathology*, **82**, 726–727.

23 McCully, M.E. (2001) *Australian Journal of Plant Physiology*, **28**, 983–990.

24 Chaintreuil, C., Giraud, E., Prin, Y., Lorquin, J., Ba, A., Gillis, M., de Lajudie, P. and Dreyfus, B. (2000) *Applied and Environmental Microbiology*, **66**, 5437–5544.

25 Alazard, D. (1985) *Applied and Environmental Microbiology*, **50**, 732–734.

26 Dreyfus, B. and Dommergues, Y.R. (1981) *FEMS Microbiology Letters*, **10**, 313–317.

27 Fleischman, D. and Kramer, D. (1998) *Biochimica et Biophysica Acta*, **1364**, 17–36.

28 Malouba, F., Lorquin, J., Willems, A., Hoste, B., Giraud, E., Dreyfus, B., Gillis, M., de Lajudie, P. and Masson-Boivin, C. (1999) *Applied and Environmental Microbiology*, **65**, 3084–3094.

29 Singh, R.K., Mishra, R.P.N. and Jaiswal, H.K. (2005) *International Rice Research Notes*, **30**, 28–29.

30 Singh, R.K., Mishra, R.P.N., Jaiswal, H.K., Kumar, V., Pandey, S.P., Rao, S.B. and Annapurna, K. (2006) *Current Microbiology*, **52**, 345–349.

31 Mishra, R.P.N. (2005) Genetic and biochemical studies on some endophytic rhizobia in rice plants, PhD thesis submitted to Banaras Hindu University, Varanasi, India.

32 Antoun, H. and Prevost, D. (2000) Proceedings of the 5th International Plant Growth-Promoting Rhizobacteria Workshop, Villa Carlos Paz, Cordoba, Argentina, 29 October–3 November (available at http://www.ag.auburn.edu/argentina/pdfmanuscripts/antoun.pdf).

33 Matiru, V., Jaffer, M.A. and Dakora, F.D. (2000) Proceedings of the 9th Congress of the African Association for Biological Nitrogen Fixation: Challenges and Imperatives for BNF Research and Application in Africa for 21st Century Nairobi, Kenya, 25–29 September. pp. 99–100.

34 Gutierrez-Zamora, M.L. and Martinez-Romero, E. (2001) *Journal of Biotechnology*, **91**, 117–126.

35 Hilali, A., Prevost, D., Broughton, W.J. and Anton, H. (2000) Proceedings of the 17th North American Conference on Symbiotic Nitrogen Fixation, University of Laval, Quebec City, Canada, p. 81.

36 Lupwayi, N.Z., Clayton, G.W., Hanson, K.G., Rice, W.A. and Biederbeck, V.O. (2004) *Canadian Journal of Plant Science*, **84**, 37–45.

37 Yanni, Y.G., Squartini, A. and Dazzo, F.B. (2007) Natural endophytic association between *Rhizobium leguminosarum* bv. *trifolii* and wheat and its potential to promote wheat plant growth and crop performance. Poster presentation: Rhizosphere International Congress-2, Montpellier, France, 26 August–1 September.

38 Fischer, H.M. (1994) *Microbiological Reviews*, **58**, 352–386.

39 Dazzo, F.B. (2004) *Plant Surface Microbiology* (eds A. Varma, L. Abbott, D. Werner and R. Hampp), Springer-Verlag, Germany, pp. 503–550.

40 Reddy, P.M., Ladha, J.K., So, R., Hernandez, R., Dazzo, F.B., Angeles,

O.R., Ramos, M.C. and de Bruijn, F. (1997) *Plant and Soil*, **194**, 81–98.

41 Chi, F., Shen, S.H., Cheng, H.P., Jing, Y.X., Yanni, Y.G. and Dazzo, F.B. (2005) *Applied and Environmental Microbiology*, **71**, 7271–7278.

42 Prayitno, J., Stefaniak, J., Melver, J., Weinman, J.J., Dazzo, F.B., Ladha, J.K., Barraquio, W., Yanni, Y.G. and Rolfe, B.G. (1999) *Australian Journal of Plant Physiology*, **26**, 521–535.

43 Cocking, E.C. (2003) *Plant and Soil*, **252**, 169–175.

44 Sabry, R.S., Sahel, S.A., Batchelor, C.A., Jones, J., Jotham, J., Webster, G., Kothari, S.L., Davey, M.R. and Cocking, E.C. (1997) *Proceedings of the Royal Society of London Series B – Biological Sciences*, **264**, 341–346.

45 Webster, G., Gough, C., Vasse, J., Batchelor, C.A., O'Callaghan, K.J., Kothari, S.L., Davey, M.R., Denarie, J. and Cocking, E.C. (1997) *Plant and Soil*, **194**, 115–122.

46 Saxena, S., Ladha, J.K., Gyaneshwar, P., Reinhold-Hurek, B., Hernandez, R.J. and Biswas, J.C. (2000) *Indian Journal of Microbiology*, **40**, 15–20.

47 Webster, G., Jain, V., Davey, M.R., Gough, C., Vasse, J., Denarie, J. and Cocking, E.C. (1998) *Plant Cell Environment*, **21**, 373–383.

48 Govindarajan, M., Balandreau, J., Kwon, S.-W., Weon, H.-Y. and Lakshminarasimhan, C. (2008) *Microbial Ecology*, **55**, 21–37.

49 Senthilkumar, M., Madhaiyan, M., Sundaram, S.P. and Kannaiyan, S. (2007) *Microbiological Research*, 10.1016/j. micres.2006.10.007, in press.

50 Dazzo, F.B., Joseph, A.R., Gomaa, A.-B., Yanni, Y.G. and Robertson, G.P. (2003) *Symbiosis*, **35**, 147–158.

51 Dazzo, F.B. and Wopereis, J. (2000) *Prokaryotic Nitrogen Fixation: A Model System for the Analysis of a Biological Process* (ed. E. Triplett), Horizon Scientific Press, Norwich, UK, pp. 295–347.

52 Liu, J., Dazzo, F.B., Yu, B., Glagoleva, O. and Jain, A. (2001) *Microbial Ecology*, **41**, 173–194.

53 McDermott, T. and Dazzo, F.B. (2002) *Manual of Environmental Microbiology* (eds C. Hurst, R.C. Crawford,G.R. Knudsen, M.J. McInerney and L.D. Stetzenbach), American Society for Microbiology Press, Washington, DC, pp. 615–626.

54 Mateos, P.F., Baker, D.L., Petersen, M., Velazquez, E., Jimenez-Zurdo, J.I., Martinez-Molina, E., Squartini, A., Orgambide, G., Hubbell, D.H. and Dazzo, F.B. (2001) *Canadian Journal of Microbiology*, **47**, 475–487.

55 Rolfe, B.G., Mathesius, U., Prayitno, J., Perrine, F., Weinman, J.J., Stefaniak, J., Djordjevic, M., Guerrero, N. and Dazzo, F.B. (2000) *The Quest for Nitrogen Fixation in Rice* (eds J.K. Ladha and P.M. Reddy), IRRI, Manila, The Philippines,291–309.

56 Sreevidya, V.S., Rao, C.S., Sullia, S.B., Ladha, J.K. and Reddy, P.M. (2006) *Journal of Experimental Botany*, **57**, 1957–1969.

57 Biswas, J.C., Ladha, J.K. and Dazzo, F.B. (2000) *Soil Science Society of America Journal*, **64**, 1644–1650.

58 Biswas, J.C., Ladha, J.K., Dazzo, F.B., Yanni, Y.G. and Rolfe, B.G. (2000) *Agronomy Journal*, **92**, 880–886.

59 Dazzo, F.B. and Yanni, Y.G. (2006) *Biological Approaches to Sustainable Soil Systems* (eds N. Uphoff, A. Ball, E. Fernandes, H. Herren, O. Husson, M. Laing, C. Palm, J. Pretty, P. Sanchez, N. Sanginga and J. Thies),CRC Taylor & Francis, Boca Raton, FL, pp. 109–127.

60 Peng, S., Biswas, J.C., Ladha, J.K., Gyaneshwar, P., Chen, Y. and Agron, J. (2002) *Agronomy Journal*, **94**, 925–929.

61 Hafeez, F.Y., Safdar, M.E., Chaudhry, A.U. and Malik, K.A. (2004) *Australian Journal of Experimental Agriculture*, **44**, 617–622.

62 Bashan, Y., Harrison, S.K. and Whitmoyer, R.E. (1990) *Applied and Environmental Microbiology*, **56**, 769–775.

63 Okon, Y. and Labandera-Gonzalesz, C.A. (1994) *Soil Biology and Biochemistry*, **26**, 1591–1601.

64 Okon, Y. and Kapulnik, Y. (1986) *Plant and Soil*, **90**, 3–16.

65 Tien, T., Gaskins, M.H. and Hubbell, D.H. (1979) *Applied and Environmental Microbiology*, **37**, 1016–1024.

66 Umali-Garcia, M., Hubbell, D.H., Gaskins, M.H. and Dazzo, F.B. (1980) *Applied and Environmental Microbiology*, **39**, 219–226.

67 IRRI (1999) More nutrition for women and children, in *Rice: Hunger or Hope?* IRRI, Manila, The Philippines, pp. 22–25.

68 Matiru, V.N. and Dakora, F.D.((2004) *African Journal of Biotechnology*, **3**, 1–7.

69 Vance, C.P. (2001) *Plant Physiology*, **127**, 390–397.

70 Tomar, N.K.J. (2000) *Indian Society of Soil Science*, **48**, 640–673.

71 Mishra, R.P.N., Singh, R.K., Jaiswal, H.K., Kumar, V. and Maurya, S. (2006) *Current Microbiology*, **52**, 383–389.

72 Nicholson, R.L. and Hammerschmidt, R. (1992) *Annual Review of Phytopathology*, **30**, 369–389.

73 Hallmann, J., Quadt-Hallmann, A., Miller, W.G., Sikora, R.A. and Lindow, S.E. (2001) *Phytopathology*, **91**, 415–422.

10
Principles, Applications and Future Aspects of Cold-Adapted PGPR

Mahejibin Khan and Reeta Goel

10.1
Introduction

The concept of optimal temperature for an organism is a fundamental principle in biology; however, organisms are unable to control temperature variations. Therefore, they either avoid stress or tolerate it. Since plants are not able to move from one place to another to handle stress, they, therefore, use different strategies to deal with temperature fluctuations within their habitat. Chilling may occur at temperature below 15 °C in the absence of ice nucleation in plant cells. The symptoms of chilling may gradually appear afterwards, especially when the plants return to optimal growth temperature, and include loss of vigor, reduction in growth rate, autolysis of cell and loss of chlorophyll [1]. Moreover, disruption of cell membranes at low temperature allows the leakage of solutes and nutrients, providing an excellent growth medium for opportunistic pathogens such as bacteria and fungi. The rates of CO_2 and ethylene production usually increase before vital symptoms such as seed germination [2] appear. If the soil temperature is very low at the time of planting of seeds, the initial uptake of water disrupts membrane integrity and increases electrolyte leakage and blocks seed germination. Nevertheless, in vegetative stages, seedlings are generally more sensitive to chilling than mature plants.

To cope with chilling injuries and to protect plants such injuries, various methods have been investigated, such as prevention of exposure of plants to chilling and use of tolerant cultivars. Genetic variability and transfer of chilling tolerance into commercially well-adapted cultivars is a complex and time consuming process; therefore, a solution for the protection of plants from chilling and for their growth enhancement involves the application of cold-adapted plant growth promoting rhizobacteria (PGPR). This term was initially used to describe strains of naturally occurring nonsymbiotic soil bacteria having the ability to colonize the plant rhizosphere and stimulate plant growth. PGPR activity has been reported in strains belonging to several genera such as *Azotobacter, Arthrobacter, Bacillus, Clostridium,*

Plant-Bacteria Interactions. Strategies and Techniques to Promote Plant Growth
Edited by Iqbal Ahmad, John Pichtel, and Shamsul Hayat
Copyright © 2008 WILEY-VCH Verlag GmbH & Co. KGaA, Weinheim
ISBN: 978-3-527-31901-5

Hydrogenophaga, Enterobacter, Serratia and *Azospirullum* [3]. The major application of PGPR strains is plant growth improvement in agriculture, horticulture, forestry and environmental restoration. The mode of action of PGPR strains can be classified into two major categories: indirect and direct. The indirect mode includes antibiotic production, reduction of iron availability to phytopathogens and synthesis of cell-wall lysing enzymes. The direct mode involves provision of bioavailable phosphorus for plant uptake, nitrogen fixation, sequestration of iron for plant uptake by siderophores, production of plant hormones such as auxin, cytokinin and gibberellins and lowering of plant ethylene production [4].

To utilize PGPR for growth promotion, it is inevitable that PGPR must colonize and survive in the rhizosphere of the host plants. Colonization of PGPR in the rhizosphere is influenced by a number of factors such as soil temperature and type, predation by protozoa, production of antimicrobial compounds by other soil microorganisms, bacterial growth rate and utilization of exudates. Of these abiotic and biotic factors, the most important is soil temperature. Kemp and coworkers [3] studied the influence of soil temperature on the leaching of inoculated rhizobacteria in soil microcosms and showed that the whole process is favored at low temperature (15 °C) than at higher temperature (35 °C).

The role of seed and root exudates as the source of nutrients for microorganisms has been demonstrated as an important factor responsible for their colonization. The biosynthesis of antagonistic compounds by PGPR strains might also play an important role in establishing bacterial populations in the rhizosphere [12]. It has also been reported that PGPR strains isolated from the native rhizosphere colonize faster and show maximum increase in germination and yield as compared to PGPR strains isolated from nonrhizospheric soil or rhizosphere of other plants [5]. Therefore, for better colonization and higher plant growth promotion at low temperature, PGPR strains must be tolerant to cold.

10.2
Cold Adaptation of PGPR Strains

Cold survival often requires organisms to exhibit a wide range of flexible behavior and physiological adjustments, including adaptive features in their membranes, protein structure and genetic responses to thermal shifts. Compared to its mesophilic counterparts, a cold-active enzyme tends to have reduced activation energy, leading to a high catalytic efficiency, which may possibly be attributed to an enhanced local or overall flexibility of the protein structure. Membranes appear to incorporate specific lipid constituents to maintain fluidity and critical ability to transport substrates and nutrients under very cold, rather rigidifying, conditions [6]. Production of antifreeze proteins and accumulation of compounds that inhibit ice recrystallization (IR) can also be part of the adaptive response in some bacteria [7,8]. Various mechanisms used for cold adaptation of PGPR are discussed below.

10.2.1
Cytoplasmic Membrane Adaptation

Membrane adaptation to different growth temperatures has been a target of a large number of PGPR strains such as *Pseudomonas fluorescens, Escherichia coli, Serratia* species, *Bacillus subtilis and so on* for a long time. Cell membranes are complex heterogeneous systems whose properties are, to a large extent, determined by their composition and spatial organization as well as by external influences, of which temperature is one of the most important [9]. The membrane of a bacterial cell defines not only the boundaries of the cell and delineates its compartments but also serves as a brain for specific functions such as regulating movement of substances into and out of the cell and its compartments, electron flow in respiration or photosynthesis and ATP synthesis [10].

A membrane is a permeable barrier consisting of hydrophobic phospholipids and proteins. Fatty acids are an integral part of the membrane structure, because their long hydrocarbon tails form an effective hydrophobic barrier to the diffusion of polar solutes. Membrane fatty acids normally contain even-numbered hydrocarbon chains, which are either fully saturated or contain varying numbers of *cis*-double bonds that make them unsaturated [11]. The heterogeneity of the fatty acid structure results in a bend in the hydrocarbon tail due to *cis*- or *trans*-formation, and this confers unique thermodynamic properties to cells, such as setting the transition temperature. The transition temperature is a critical temperature below which the membrane is rigid and above which the membrane is fluid. The fluidity of the membrane mainly depends on the length of the fatty acids present and the degree of unsaturation of their side chains (number of double bonds present). To survive at low temperature, a cell must have a cytoplasmic membrane that retains sufficient fluidity to maintain a physical state supportive of multiple functions of the membrane; this phenomenon is called homeoviscous adaptation [13]. Another adaptation is homeoproton adaptation [14], in which psychrophilic bacteria adjust the lipid composition of their membrane so that their proton permeability remains within a narrow range. Khan and coworkers [15] have studied comparative cellular morphology of PGPR strains between the wild type and the $CRPF_8$, a cold-tolerant mutant of *P. fluorescens*, using transmission electron microscopy. The results indicate thickening of the cell wall of $CRPF_8$ and reduction in the size of the cell(s). The cell wall of $CRPF_8$ becomes more osmophilic as compared to the wild type. Perhaps, the thickened cell wall helps cells to survive in cold conditions and acts as a protective coat or mantle, thus offering protection against a hostile environment such as low temperatures.

For the cell to function normally at low temperature, the membrane lipid bilayers need to be largely fluid so that the membrane proteins can continue to pump ions, take up nutrients and perform respiration [16]. Therefore, it is essential the membrane lipids are in the liquid crystalline state. When the growth temperature of a microorganism is reduced, some of the normally fluid components become gel-like, which prevents the proteins from functioning normally; therefore, for these

Table 10.1 Percent abundance of membrane fatty acids at different temperatures.

Fatty acids (IUPAC name)	No. of carbon	Percent abundance of fatty acids				
		20 °C	18 °C	15 °C	10 °C	5 °C
Didecanoic	C12 (0)	1.94	2.97	1.18	1.0	0.83
Tridecanoic	C13 (0)*	2.08	2.24	2.17	1.46	0.84
n-Tetradecanoic	C14 (0)	2.53	3.78	2.57	1.8	0.97
Pentadecanoic	C15 (0)	1.50	1.69	1.66	2.41	2.50
cis,cis-9,12-Hexadecanoic	C16 (2)	0.65	1.35	1.18	0.67	1.63
cis-9-Hexadecanoic	C16 (1)	33.8	33.3	34.7	40.5	49.8
n-Hexadecanoic	C16 (0)	27.5	25.4	24.9	20.2	13.2
cis,cis-9,12-Octadecadienoic	C18 (2)	0.92	1.08	0.99	2.00	1.00
cis-9-Octadecanoic	C18 (1)	21.2	22.5	23.9	26.6	25.9
n-Octadecanoic	C18 (0)	7.89	4.30	4.9	3.06	3.31

*Source: Russell [21].

components to remain fluid, a number of changes must occur in the pattern of fatty acids. Unsaturation of fatty acid chains is the most common change that occurs when the temperature is reduced; this increases the fluidity of the membrane because unsaturated fatty acid groups create more disturbance to the membrane than saturated chains. This process is achieved by desaturases situated in the membrane itself and thus is able to react quickly. In cyanobacteria, four desaturase genes (desA–desD) have been reported; moreover, desA, desB and desD have been demonstrated to be cold inducible in *Synechocystis* [17].

There are, however, a number of other alterations that can occur after a decline in temperature [18]. The average fatty acid chain length may be shortened, which would have the effect of increasing the fluidity of the cell membrane because there are fewer carbon–carbon interactions between the neighboring chains [19]. A psychrophilic organism, for example *Micrococcus cryophilus*, which contains high proportions of unsaturated fatty acids under all growth conditions, responds to a decrease in temperature, from 20 to 0 °C, by a reduction in the average chain length of the fatty acids [20]. All these changes, as summarized in Table 10.1, result in the membrane maintaining its fluidity by producing lipids with a lower gel-to-liquid crystalline transition temperature and by incorporating proportionally more low melting point fatty acids into membrane lipids.

10.2.2
Carbon Metabolism and Electron Flow

Sardesai and Babu [22] reported that carbon metabolism and electron flow are also affected by low temperature. Cold stress induces a change from respiratory metabolism to anaerobic lactate formation in a psychrophillic *Rhizobium* strain. Analysis of glucose-6-phosphate dehydrogenase and 6-phosphogluconate dehydrogenase of the pentose phosphate pathway showed an upward regulation of an alternative pathway of carbohydrate metabolism under cold stress that resulted in a rapidly generated energy to overcome the stress. Cold stress resulted in a decrease in the poly-β-hydroxybutyrate in a psychrotolerant *Rhizobium*, owing to PHB inhibition rather than an increase in its breakdown at low temperature. *P. fluorescens* accumulated 2-ketogluconate in medium (consisting of 0.3% $NH_4H_2PO_4$; 0.2% K_2HPO_4; 0.05% $MgSO_4 \cdot 7H_2O$; 0.5 µg ml^{-1} $FeSO_4 \cdot 7H_2O$; and 0.2% filter-sterilized glucose as carbon source) as the major oxidation product of glucose [23].

10.2.3
Expression of Antifreeze Proteins

There are a number of substances described in the published literature that inhibit ice nucleation. Certain bacterial strains, mostly found in the nonfluorescent pseudomonad species, release materials into the growth medium that reduce the nucleation temperature of water droplets to below that of distilled water [24]. These substances include sucrose, unsaturated fatty acids and phospholipids; however, in psychrophilic bacteria and some other psychrophilic organisms, specific proteins are produced that reduce freezing temperature and protect them from freeze injury. These are known as antifreeze proteins and help bacteria to survive the freezing conditions.

Antifreeze proteins (AFPs) are structurally a diverse group of proteins with the ability to modify ice crystal structure [25] and inhibit recrystallization of ice by adsorbing onto the surface of ice crystals via van der Waals interactions and/or hydrogen bonds [26–28]. During cold acclimation, many freeze-tolerant organisms accumulate antifreeze proteins [29–31].

A novel AFP assay, designed for high-throughput analysis in Antarctica, demonstrated putative activity in 187 of the cultures tested. Subsequent analysis of the putative positive isolates showed 19 isolates with significant recrystallization inhibition (RI) activity. The 19 RI-active isolates were characterized using ARDRA (amplified rDNA restriction analysis) and 16S rDNA sequencing. They belong to genera from the α-proteobacteria, with genera from the γ subdivision being predominant. The 19 AFP-active isolates were isolated from four physicochemically diverse lakes [32].

The structural and functional features of AFPs enable them to protect living organisms by suppressing the effect of freezing temperatures and modifying or suppressing ice crystal growth. The plant growth promoting rhizobacterium *Pseudomonas putida* GR12-2 was isolated from the rhizosphere of plants growing in the Canadian High Arctic. This bacterium was able to grow and promote root elongation of spring and winter canola at 5 °C, a temperature at which only a relatively small number of bacteria were able to proliferate and function. In addition, the bacterium survived exposure to freezing temperatures, ranging from −20 to −50 °C; and it was discovered that at 5°C, *P. putida* GR12-2 synthesized and secreted an antifreeze protein into the growth medium [33,34].

Katiyar and Goel [35] reported the presence of antifreeze proteins in a cold-tolerant mutant of *P. fluorescens*. It was observed that AFP was capable of protecting

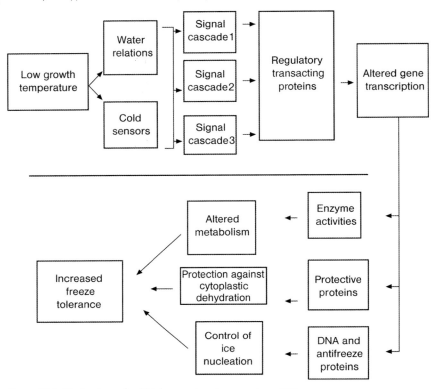

Figure 10.1 Mechanism of cold tolerance in microorganisms [98].

freeze-labile alkaline phosphatase (ALP) enzyme under freezing conditions as compared to non-AFP-producing strains of *P. fluorescens*.

Khan [36] studied the presence of AFP from $CRPF_1$, a cold-tolerant mutant of *P. fluorescens*. It was demonstrated that cold shock protein from $CRPF_1$ resulted in 75% protection of ALP against freeze inactivation. BSA was used as positive control, which was also effective in protecting ALP against freeze inactivation. However, CSP was superior to BSA; furthermore, sucrose did not protect ALP against freeze inactivation and served as a negative control.

Walker and coworkers [37] reported that, at low temperatures, different microorganisms show different types of adaptive responses, which either alter their cellular structure or affect gene expression with decrease in temperature. Those microorganisms that are exposed to frequent freeze-thaw challenges or are present in temperature ranges from psychrophilic to mesophilic are more resistant to low temperatures. It was demonstrated that when a mixture of soil isolate collected from the Chinook zone (where temperature ranges from 35 to −35 °C) along with two control isolates, *E. coli* and *Pseudomonas*, was subjected to regular freeze-thaw up to 48 cycles, control isolates were no longer viable but bacteria from soil isolates were still viable and showed inhibition of IR. IR can be mediated by polysaccharides such as xanthum gum or by AFPs that have been reported in few bacteria, for example, Antarctic *Moraxella*, [38] Arctic *Rhizobacterium* [39,40], cold-acclimated *Micrococcus* and *Rhodococcus* [41]. Therefore, it was hypothesized that IR activity can potentially contribute to overall viability of a microbial consortium under Chinook conditions.

10.3
Mechanism of Plant Growth Promotion at Low Temperature

When deciding on the type of bacterial strains to be used with a plant for a given climatic condition, understanding of the mechanism of plant growth promotion is essential; for example, the overwintering ability of PGPR is fundamental when considering them to use in colder climates. There is evidence that *Pseudomonas* sp. are able to overwinter in sufficient quantities on roots of winter wheat [42,43]. The principal mechanisms of plant growth promotion in colder regions include phytostimulation and frost injury protection.

10.3.1
Phytostimulation

Phytostimulators are chemical compounds produced by a number of bacteria that directly enhance plant growth. Different genera of bacteria such as *Proteus mirabilus*, *Pseudomonas vulgaris*, *Klebsiella pneumoniae*, *Bacillus cereus*, *Escherichia coli* and so on produce auxin cytokinins, gibberellins and abscisic acid [30]. Quantitatively, auxins are the most abundant phytohormone secreted by PGPR strains such as *Azospirullum* and is the major factor that stimulates root generation and enhances root growth. ABA and ethylene have been shown to play an essential role in plant stress signaling [43]. A more direct correlation is evident between the level of ABA and the increasing freeze tolerance [44,45]. Exogenous application of ABA can increase freeze tolerance in both woody and herbaceous plants [46,47]. Several studies have convincingly demonstrated that exogenous application of ABA increases cold tolerance. Application of ABA at room temperature increased cold resistance in callus explants of tobacco, cucumber, winter wheat and alfalfa. Therefore, the PGPR strains capable of producing ABA can be used for protecting seedlings and plants from freeze injuries and hence can contribute to growth enhancement at low temperature.

Another important factor in phytostimulation is lowering the plant ethylene level, which gets elevated during stress in plants. Higher concentrations of ethylene are inhibitive to plant growth. Any factor or stimulus that causes changes in endogenous levels of ethylene in plants leads to enhanced growth and development [43]. It has been discovered that certain microorganisms contain an enzyme, ACC deaminase, that hydrolyzes ACC into ammonia and α-ketobutyrate [48]. Hall *et al.* [49] reported that a soil isolate *P. putida* GR12-12 contains a gene for ACC deaminase, which hydrolyzes ACC, the immediate precursor of ethylene synthesized in plant tissues, and thus inhibits ethylene synthesis and eliminates the potential inhibitory effect of ethylene concentration accumulation. This mechanism is most effective in plants that are susceptible to the ethylene effect, such as dicotyledonous plants.

10.3.2
Frost Injury Protection

In most low-temperature or freezing climates, plants are extensively damaged because of not only less nutrient availability or poor hormone production but also ice crystallization within cells. Water can cool to several degrees below $0\,^{\circ}C$ without freezing. Most PGPR strains produce either antifreeze proteins or ice-nucleating protein complexes that inhibit ice recrystallization [50] or cold acclimation proteins that are believed to be responsible for cold tolerance (Figure 10.1). These proteins appear to be a membrane-bound substances that protect leaves and roots from frost injuries. Lindow [51] identified the ice-nucleating factor from *Pseudomonas syrengae* by deletion mutation.

To reduce frost damage, a biological control strategy has also been developed by Lindow [41] on the basis of the competition between populations that help to initiate ice nucleation. A strain of naturally occurring *P. fluorescens* has been registered commercially as Frostban B for the protection of pear trees [52,53]. Lindow and Panopoulous [54] carried out field experiments using *P. syringe* on strawberries and potatoes and concluded that the incidence of frost injury was significantly lower in inoculated potato plants than in uninoculated control plants in several natural field frost events.

10.4
Challenges in Selection and Characterization of PGPR

Selecting a good PGPR strain requires an understanding of the dynamics and composition of bacterial communities that colonize the rhizosphere and characterization of its PGPR-related properties. Screening of rhizobacterial isolates includes host-plant specificity and adaptation to particular soil and climatic conditions [55,56]. Several methods have been devised for the selection of PGPR; for example, an enrichment technique, that is, the spermosphere model used for the selection and isolation of promising nitrogen-fixing rhizospheric bacteria from rice [57]. The isolation and selection of rhizospheric bacteria is also done on the basis of the traits known to be associated with PGPR; for example, root colonization [1], ACC deaminase activity [58], antibiotic production [59] and siderophore production [60]. Selection of superior strains of PGPR may be facilitated by the development of high-throughput assay systems and effective bioassays [61].

10.5
Challenges in Field Application of PGPRs

PGPR show great promise in laboratory and greenhouse conditions, but considerable challenges are encountered upon their application to the field. PGPR used for controlling fungal pathogens show promise in greenhouses because of consistent environmental conditions and a high incidence of fungal diseases in greenhouses [62]. Variations in field results are because of the heterogeneity of abiotic and biotic factors and competition with indigenous microorganisms. A thorough knowledge of these factors can help in determining the optimal concentrations, timing and placement of inoculants and soil and crop management strategies to enhance survival and proliferation of the inoculants [56]. Another problem associated with PGPR is their low survival in the field or too high a concentration needed to exert the desired activity. Like chemical pesticides, the practical use of PGPR as microbial fertilizers or pesticides and their efficiency is strongly dose dependent [57]. Engineering the rhizosphere by manipulating their host plant or their substrates or altering agronomic practices is a better option today for enhancing the PGPR function and properties of [56].

Other areas that need to be focused on in the field of PGPR are the development of better formulations to ensure PGPR survival and activity in the field and compatibility with chemical and biological seed treatments [61–64]. The toxicological risk and the environmental impact associated with the introduction of PGPR into the food chain or the environment are also a matter of concern.

10.6
Applications of PGPRs

PGPR can be used in a variety of ways when plant growth enhancement is required, especially in agriculture, horticulture, forestry and environmental restoration.

10.6.1
Applications of PGPR in Agriculture

PGPR are most commonly used in agriculture. Up to 50–70% increase in crop yields has been reported by researchers upon addition of PGPR. Plant growth benefits, occurring upon PGPR addition, are as follows:

- increase in germination rates;
- increase in root growth;
- increase in yield including grain size, leaf area;
- increase in chlorophyll, magnesium, nitrogen and protein contents;
- increase in hydraulic activity, that is, fluid movement within the plant;
- tolerance to drought and low temperature;
- delayed leaf senescence; and
- disease resistance.

As the soil is an unpredictable environment, sometimes unexpected results are observed owing to a low soil pH, high mean temperature and/or low rainfall during the growing season. These undesirable conditions lead to low root colonization by PGPR [65–67]. It was reported that climatic variability also plays a role in the effectiveness of PGPR. Although field results vary, if a PGPR is found to be effective under greenhouse conditions, it is expected to show positive results in the field [67].

The main mechanism of plant growth promotion by *Azospirillum* was thought to occur by providing fixed nitrogen; however, it has been reported that plant growth promotion by *Azospirillum* only occurs under nitrogen-limited conditions [65,69]. Another way in which *Azospirillum* improves plant growth is by producing the plant hormone IAA [70,52].

The effect of PGPR on plant growth also varies with soil types [71]. De-Frietas and Germida [72] reported that the less fertile the soil, the greater is the plant growth stimulation by the PGPR. In further studies, it was found that maximum increase in plant germination and yield often occurs in crops inoculated with PGPR strains from their own rhizosphere [5]. It was mentioned by Bashan [64] that there should be threshold numbers of bacteria to be inoculated to a given plant, as excessively large numbers of bacteria could be detrimental to the germination and growth of seeds or plants. Jacoud [73], however, stated that growth promotion effects could still occur with low bacterial populations. It was also reported that some PGPR are able to counteract irrigation problems by reducing the negative effect of irrigation of crops with highly saline water. Other studies [74,75] revealed that growth promotion effects are seen early in plant development and these subsequently translate to higher yields. The majority of PGPR available commercially are biocontrol agents that indirectly promote crop growth [75]. Effects of various PGPR on different crops are summarized in Table 10.2.

The knowledge of indirect mechanisms of plant growth promotion by PGPR also aids in the cultivation of certain legumes. Evans [76] reported that hydrogen gas, produced as a by-product of nitrogen fixation by rhizobacteria within legume nodules, may be recaptured by those rhizobial strains that contain a hydrogen uptake system. A comparative analysis of crop types and different rhizobacteria species or strains is missing from the current research on PGPR in agriculture.

10.6.2
Application of PGPR in Forestry

PGPR are less commonly used in forestry than in agriculture. Although the effects of PGPR on angiosperms were the focus of the initial research through the 1980s, the effects of PGPR on gymnosperms have been the focus of research since 1990s [78]. PGPR can benefit the commercial forestry sector as well as reforestation efforts worldwide. The following considerations must be taken into account when evaluating the performance of PGPR on tree species compared to agricultural crops:

- increase in biomass due to inoculation;
- emergence of seedling;
- reduction in seedling transplant injury during transfer from nursery to field.

Some PGPR are sensitive to low-pH conditions [79]; therefore, soil type should be a major consideration when testing PGPR in a forest environment, as many forest

Table 10.2 Examples of free-living plant growth promoting rhizobacteria tested on various crop types.

Bacteria	Plant	Conditions	Results of addition of bacteria to plant	Reference
Xanthomonas maltophila	Sunflower	Laboratory and greenhouse	Increased germination rate	[5]
Enterobacter cloacae CAL3	Mungbean tomato, pepper	Greenhouse	Positive seedling growth response by all three plant species to PGPR treatment, especially tomato, where no exogenous mineral nutrients were added Early stimulation effect on seedlings	[48]
Azospirillum local isolates From Argentina	Maize, wheat	Field	In wheat cultivars over seven seasons, increases in yield from 15 to 30% and increases of 50–60% in yield when fertilized Over six seasons, maize yield increased from 15 to 25% and up to 40% with fertilization	[68]
Pseudomonas chlororaphis 2E3, O6	Spring wheat field	Laboratory	Increased seedling emergence at two different sites by 8–6% Strong inhibition of *Fusarium culmorum* No promotion effect of inoculated plants evident in soils free of *Fusarium* infection	[72]
Pseudomonas fluorescens	Mungbean, wheat	Greenhouse	Root and shoot elongation	[76]
Pseudomonas putida, Pb and Cd resistant	Mungbean	Greenhouse	Root and shoot elongation in presence of $CdCl_2$ and $(CH_3COO)_2$	[94]

(Continued)

Table 10.2 (*Continued*)

Bacteria	Plant	Conditions	Results of addition of bacteria to plant	Reference
Azospirillum brasilense	Finger millet, sorghum, pearl millet	Field	Average of up to 15% increase in yield for finger millet For sorghum, average increase is 19%	[99]
Bacillus amyliquefaciens IN937 *Bacillus subtilis* GB03 *Bacillus cereus* C4	Tomato, pepper	Field	Statistically significant increases in plant growth in two growing years in terms of stem diameter, stem area, leaf surface area, weights of roots and shoots and number of leaves Transplant vigor and fruit yield improved Pathogen numbers and disease not reduced in tomatoes or pepper with the exception of reduction of galling in pepper by root-knot nematode	[100]
Pseudomonas putida R104 *Pseudomonas cepacia* R85 *Pseudomonas fluorescens* R104, R105	Winter wheat	Potted plants chamber	Two soil types tested under simulated fall conditions 5 °C Response of wheat nutrient uptake to inoculation depends on soil composition Grain yield increased by 46–75% in more fertile soil	[101]
Pseudomonas putida R111 *Pseudomonas corrugate, Azotobacter chroococcum*	*Amaranthus paniculatus* *Eleusine coracana*	Field	Plant growth and nitrogen content increased Hypothesized that the growth promotion effect is due to the stimulation of native bacterial communities	[102]

soils are acidic, especially those of conifer forests. Zaddy [80] reported that the medium in which PGPR is prepared before inoculation might affect the root colonization pattern of the inoculated bacteria. This study revealed that malate-grown bacteria were better in promoting growth in oak (*Quercus*) than bacterial cells of the same strain grown in a fructose-based medium. This is because the malate-grown cells have a tendency to aggregate, whereas the fructose-grown cells disperse throughout the soil substrate. Therefore, fructose-grown cells were superior for growth promotion of surface-rooted plants such as maize but were inadequate for growth promotion of trees with deep tap roots such as oak. For reforestation, PGPR strains have been used as an emerging technology. It was found that seedling performance could be significantly enhanced (32–49% within 1 year) through PGPR inoculation of root systems in pine and spruce [78].

Enabak [55] reported that a PGPR is specific to a particular tree species. It was also reported that the tree ecovar also plays an important role; for example, a bacterial strain, which was effective at promoting growth in one type of pine species, was ineffective for the other. Bacterial inoculations to ecotypes, that is, trees of the same species from different regions or altitudes, also showed different response to PGPR [81]. Researchers usually tend to inoculate seedlings or older plants, which is in contrast to the agricultural use of PGPR, where there is a tendency to inoculate seeds or the substrate surrounding the seed.

10.6.3
Environmental Remediation and Heavy Metal Detoxification

Heavy metals are widespread pollutants of the surface soil, which are being added to the environment by a variety of sources including municipal, industrial and agricultural wastes [82]. Heavy metals have been a major concern because of toxic effects on diversity and activity of beneficial microorganisms and also because they impart detrimental effects on plants and humans. It has been reported that sterile soil containing (anomalous, hazardous) metals inhibit nitrogen fixation when inoculated with pure cultures of *Azotobacter*. Copper and cadmium inhibited denitrification in three environmental isolates of *Pseudomonas* [83] and similarly nickel severely inhibited aerobic growth of various microorganisms [84]; therefore, effective and affordable technologies are needed for remediation of contaminated environments.

Several rhizobacteria such as PGPR have been shown to be tolerant to heavy metals. These are believed to be an effective solution for detoxification of the environment. It is better to use indigenous microorganisms, as they have already been adapted to survival requirements in polluted soil. Bacteria have developed a variety of resistance mechanisms, which include the sequestration of heavy metals, chemical reduction of the metal to a less toxic species and direct efflux of metal out of the cell [85,86]. Gupta *et al.* [87] developed Cd-, Ni- and Cr-resistant mutants of phosphate-solubilizing *Pseudomonas* sp. NBRI 4014 and characterized them on the basis of their PGPR properties. It was reported that these mutants were able to promote root and shoot elongation in *Glycine max* to a significant level. The persistence and stability of these mutants in the rhizosphere supported their exploitation in contaminated soil. In an experiment conducted by Engqvist *et al.* [88], it was reported that successful production of plants in a Cd-polluted site requires active interaction between the microbial component and plants. There is also evidence that inoculation with rhizobia can increase nodulation, nitrogen fixation, biomass and nitrogen uptake by plants grown in soil contaminated with heavy metal [89–91]; furthermore, positive effects of PGPR on nutrient uptake by plants exposed to heavy metal stress have been described [92,93]. Tripathi and coworkers [94] isolated siderophore-producing lead- and Cd-resistant *P. putida* that significantly stimulated root and shoot growth of mungbean in the presence of $CdCl_2$.

Lucy [95] studied the degradative potential of *Azospirillum*, a plant-associated nitrogen-fixing rhizobacteria, toward oil hydrocarbons. The results indicated that oil served as carbon and energy source for *Azospirillum* and its concentration did not affect the production of phytohormone IAA. Therefore, the strain was capable of degrading the pollutant and also enhancing the growth by developing a plant root system in the oil-contaminated environment.

10.7
Prospects

In the future, new PGPR products will become available as our understanding of the complex environment of the rhizosphere, the action mechanisms of PGPR and the practical aspects of inoculants formulation and delivery increases. The success of PGPR in environmental management depends on our ability to manage the rhizosphere to enhance PGPR survival and competitiveness with indigenous microflora. To enhance the colonization and effectiveness of PGPR, genetic manipulations can be carried out, which involve addition of one or more traits associated with PGPR. Manipulation in specific genes that contribute to the colonization of PGPR with roots, such as motility, chemotaxis to seed and root exudates, production of specific cell surface components, ability to use specific components of root exudates, protein secretion and quorum-sensing signals, can be helpful in understanding the precise role of each gene and their potential for exploitation in other PGPR strains.

Strategies such as reporter transposons and *in vitro* expression technology can be used to detect genes PGPR express. The inoculated strains should also be labeled (with lux or gfp genes) so that they can be readily detected in the environment after their release. Using mixed consortia as inoculants of PGPR with known functions is of interest in increasing their consistency in the field. PGPR offer an environmentally sustainable approach to increase crop production and health, and the application of molecular tools will enhance our ability to understand and manage the rhizosphere and lead to new products with improved properties.

References

1 Sylvia, D.M., Alagely, A.K., Kane, M.E. and Philman, N.L. (2003) *Mycorrhiza*, **13**, 177–183.

2 Lyons, J.M. (1973) *Annual Review of Plant Physiology*, **24**, 445–566.

3 Kemp, J.S., Paterson, E., Gammack, S.M., Cresser, M.S. and Killham, K. (1992) *Biology and Fertility of Soils*, **13**, 218–224.

4 Glick, B.R. (1995) *Canadian Journal of Microbiology*, **41**, 109–117.

5 Fages, J. and Arsac, J.F. (1991) *Plant and Soil*, **137**, 87–90.

6 Allen, D., Huston, A.L., Wells, L.E. and Deming, J.W. (2001) Biotechnological use of psychrophiles in *Encyclopedia of Environmental Microbiology*, (ed Bitson G.), pp. John Wiley & Sons, Ltd, New York.

7 Mazur, P. (1966) Physical and chemical basis of injury in single-celled microorganisms subjected to freezing and thawing, in (ed. H.T. Merman), *Cryobiology*, Academic Press, New York, NY.

8 Gilbert, J.A., Hill, P.J., Dodd, C.E. and Laybourn-Parry, J. (2004) *Microbiology*, **150**, 171–180.

9 Alberts, B., Bray, D., Lewis, J., Raff, M., Roberts, K. and Watson, J.D. (1994) The membrane structure in *Molecular Biology of the Cell*, 3rd edn, Garland Publishing Inc., London, pp. 477–506.

10 Becker, R.P. (1996) *The World of the Cell*, 3rd edn The Benjamin Cummings Publishing Company, Menlo Park, CA.

11 Pringle, M.J. and Chapman, D. (1981) *Effects of Temperature on Biological Systems*, Academic Press, New York.

12 De Weger, L.A., Van Der, A.J., Dekkers, Bij., Simons, L.C., Wijffelman, M.C.A. and Lugtenberg, B.J.J. (1995) *FEMS Microbiology Ecology*, **17**, 221–228.

13 Van de-Vossenberg, J.L.C.M., Diessen, A.J.M., Da-Costa, M.S. and Konings, W.N. (1999) *Biochemica et Biophysica Acta*, **1419**, 97–104.

14 Sinesky, M. (1974) *National Academy of Sciences of the United States of, America* **71**, 522–525.

15 Khan, M., Bajpai, V.K., Ansari, S.A., Kumar, A. and Goel, R. (2003) *Microbiology and Immunology*, **47** (12), 895–901.

16 Berry, E.D. and Foegeding, P.M. (1997) *Journal of Food Protection*, **60** (12), 1583.

17 Tasaka, Y.Z., Gombos-Nishiyama, Y., Mohanty, P., Ohba, T., Ohki, K. and Murata, N. (1996) *EMBO Journal*, **15**, 6416–6425.

18 Russell, N. (2000) *Journal of Extremophiles*, **4**, 83–90.

19 Russell, N. (1990) *Philosophical Transactions of the Royal Society of London, Series B, Biological Sciences*, **326**, 595–608.

20 McGibbon, L. and Russell, N.J. (1983) *Current Microbiology*, **9**, 241–244.

21 Russell, N.J., Evans, R.I., Tersteeg, P.F., Hellemons, J., Verheul, A. and Abee, T. (1995) *International Journal of Food Microbiology*, **28**, 255–261.

22 Sardesai, N. and Babu, C.R. (2001) *Current Microbiology*, **42**, 53–58.

23 Lynch, W.H., MacLeod, J. and Franklin, M. (1975) *Canadian Journal of Microbiology*, **21**, 1553–1559.

24 Strom, C.S., Liu, X.Y. and Jia, Z. (2004) *Journal of Biological Chemistry*, **279** (31), 32407–32417.

25 Urrutia, M., Duman, J.G. and Knight, C.A. (1992) *Biochemica et Biophysica Acta*, **1121**, 199–206.

26 DeVries, A.L. (1986) *Methods in Enzymology*, **127**, 293–303.

27 Ewart, K.V., Lin, Q. and Hew, C.L. (1999) *Cellular and Molecular Life Sciences*, **55**, 271–283.

28 Krembs, C. and Engel, A. (2001) *Marine Biology*, **138**, 173–185.

29 Griffith, M., Ala, P., Yang, D.S., Hon, W.C. and Moffatt, B.A. (1992) *Plant Physiology*, **100**, 593–597.

30 Griffith, M. and Ewart, K.V. (1995) *Biotechnology Advances*, **13**, 375–402.

31 Duman, J.G. and Olsen, T.M. (1993) *Cryobiology*, **30**, 322–328.

32 Jack, A. and Gilberta Daviesa, P.L. (2005) *FEMS Microbiology Letters*, **245**, 67–72.

33 Sun, X., Griffith, M., Pasternak, J.J. and Glick, B.R. (1995) *Canadian Journal of Microbiology*, **41** (9), 776–784.

34 Xu, H., Griffith, M., Patten, C.L. and Glick, B.R. (1998) *Canadian Journal of Microbiology*, **44**, 64–73.

35 Katiyar, V., Mishra, D.P., Kumar, S. and Goel, R. (2004) *Indian Journal of Biotechnology*, **3**, 378–381.

36 Khan, M. (2005) Cloning and characterization of cold shock gene/protein of fluorescent pseudomonad mutant CRPF$_1$, PhD thesis. G.B.P.U.A. &T, Pantnagar, India.

37 Walker, V.K., Palmer, G.R. and Voordouw, G. (2006) *Applied and Environmental Microbiology*, **72** (3), 1784–1792.

38 Yamashita, Y., Nakamura, N., Omiya, K., Nishikawa, J., Kawahara, H. and Obata, H. (2002) *Bioscience Biotechnology, and Biochemistry*, **66**, 239–247.

39 Muryoi, N., Sato, M., Kaneko, S., Kawahara, H., Obata, H., Yaks, M.W.F., Griffith, M. and Glick, B.R. (2004) *Journal of Bacteriology*, **186**, 5661–5671.

40 Kawahara, H., Li, L., Griffith, M. and Glick, B.R. (2001) *Current Microbiology*, **43**, 365–370.

41 De-Freitas, J.R. and Germida, J.J. (1992) *Soil Biology & Biochemistry*, **24**, 1127–1131.

42 De-Freitas, J.R. and Germida, J.J. (1992) *Soil Biology & Biochemistry*, **24**, 1137–1146.

43 McCourt, P. (1999) *Annual Review of Plant Physiology*, **50**, 219–243.

44 Lang, V., Mantyla, E., Welin, B., Sundberg, B. and Palva, E.T. (1994) *Plant Physiology*, **104**, 1341–1349.

45 Chen, H.H., Li, P.H. and Brenner, M.L. (1983) *Plant Physiology*, **71**, 362–365.

46 Li, C., Puhakainen, T., Welling, A., Vihera-Aarnio, A., Ernstsen, A.J.O., Heino, P. and Palva, E.T. (2002) *Physiologia Plantarum*, **116**, 478–488.

47 Heino, P., Sandman, G., Lang, V., Nordin, K. and Palva, E.T. (1990) *Theoretical and Applied Genetics*, **79**, 801–806.

48 Mayak, S. (1999) *Journal of Plant Growth Regulation*, **18**, (2),49–53.

49 Hall, J.A., Peirson, D., Ghosh, S. and Glick, B.R. (1996) *Israel Journal of Plant Sciences*, **44**, 37–42.

50 McGrath, J., Waegner, S. and Gilichinski, D. (1994) in *Viable organisms in permafrost*, (ed. D. Gilichiniski), Pushino, Russia, pp. 48–67.

51 Lindow, S.E. (1983) *Annual Review of Phytopathology*, **21**, 363–384.

52 Lindow, S.E. (1997) in *Biotechnology of Plant–Microbe Interactions*, (eds J.P. Naks and C. Hagedron), McGraw Hill, NewYork, pp. 85–110.

53 Wilson Lindow, S.E. (1993) *Annual Review of Microbiology*, **47**, 913–944.

54 Lindow, S.E. and Panopoulos, N.J. (1988) Proceedings of First International Conference on Release of Genetically Engineered Microorganisms, Academic Press, London, pp. 121–138.

55 Enebak, S.A., We, I.G. and Kloepper, J.W. (1998) *Forest Science*, **44**, 139–144.

56 Bowen, G.D. and Rovira, A.D. (1999) *Advances in Agronomy*, **66**, 1–102.

57 Montesinos, E. and Bonaterra, A. (1996) *Phytopathology*, **86**, 464–472.

58 Cattelan, A.J., Hartel, P.G. and Fuhrmann, J.J. (1999) *Soil Science Society of America Journal*, **63**, 1670–1680.

59 Giacomodonato, M.N., Julia, P.M., Guadalupe, I.S., Beatriz, S.M. and Nancy, I.L. (2001) *World Journal of Microbiology*, **17**, 51–55.

60 Gupta, A., Rai, V., Bagadwal, N. and Goel, R. (2005) *Microbiological Research*, **160**, 385–388.

61 Mathre, D., Cook, E. and Callan, N. (1999) *Plant Dis.*, **83** (11), 972–983.

62 Paulitz, T.C. and Be?langer, R. (2001) *Annual Review of Phytopathology*, **39**, 103–133.

63 Mansouri, H., Petit, A., Oger, P. and Dessaux, Y. (2002) *Applied and*

Environmental Microbiology, **68** (5), 2562–2566.

64 Bashan, Y. (1998) *Biotechnology Advances,* **16** (4), 729–770.

65 Dobbelaere, S., Croonenborghs, A., Thys, A., Ptacek, D., Vanderleyden, J., Dutto, P., Labandera-Gonzalez, C., Caballero-Mellado Aguirre, J.F., Kapulnik, Y., Brener, S., Burdman, S., Kadouri, D., Sarig, S. and Okon, Y. (2001) *Australian Journal of Plant Physiology,* **28**, 871–879.

66 Klein, D.A., Salzwedel, J.L. and Dazzo, F. B. (1990) in *Biotechnology of Plant–Microbe Interactions* (eds J.P. Nakas and C. Hagedorn), McGraw-Hill, New York, USA, pp. 189–225.

67 Parke, J.L. (1991) in *The Rhizosphere and Plant Growth* (eds D.L. Keister and P.B. Cregan), Kluwer Academic Publishers, Dordrecht, The Netherlands, pp. 33–42.

68 Okon, Y. and Labandera-Gonzalez, C.A. (1994) *Soil Biology & Biochemistry,* **26**, 1591–1601.

69 Fallik, E. and Okon, Y. (1996) *World Journal of Microbiology & Biotechnology,* **12**, 511–515.

70 Dobbelaere, S., Croonenborghs, A., Thys, A., Vande-Broek, A. and Vanderleyden, J. J. (1999) *Plant and Soil,* **212**, 155–164.

71 Kloepper, J.W., Schoth, M.N. and Miller, T.D. (1980) *Phytopathology,* **70**, 1078–1082.

72 De-Freitas, J.R. and Germida, J.J. (1990) *Canadian Journal of Microbiology,* **36**, 265–272.

73 Jacoud, C., Faure, D., Wadoux, P. and Bally, R. (1998) *FEMS Microbiology Ecology,* **27**, 43–51.

74 Glick, B.R., Liu, C., Ghosh, S. and Dumbroff, E.B. (1997) *Soil Biology & Biochemistry,* **29**, 1233–1239.

75 Polyanskaya, L.M., Vedina, O.T., Lysak, L. V. and Zvyagintev, D.G. (2000) *Microbiology,* **71**, 109–115.

76 Katiyar, V. and Goel, R. (2004) *Plant Growth Regulation,* **42**, 239–244.

77 Evans, H.J., Harker, A.R., Papen, H., Russell, S.A., Hanus, F.J. and Zuber, M.

(1987) *Annual Review of Microbiology,* **41**, 335–361.

78 Chanway, C.P. (1997) *Forest Science,* **43**, 99–112.

79 Brown, M.E. (1974) *Annual Review of Phytopathology,* **12**, 181–197.

80 Zaady, E.A., Perevoltsky, A. and Okon, Y. (1993) *Soil Biology & Biochemistry,* **25**, 819–823.

81 Chanway, C.P. (1995) *Soil Biology & Biochemistry,* **27**, 767–775.

82 Timoney, J.F., Port, J., Giles, J. and and Spanier J. (1978) *Journal of Applied and Environmental Microbiology,* **36**465–472.

83 Bollag, J.M. and Barabasz, W. (1979) *Journal of Environmental Quality,* **8**, 196–201.

84 Babich, H. and Sotzky, G. (1983) in *Advances in Advance Microbiology,* vol. 29 (ed. Ai. Laskin), Academic Press, New York, pp. 1995–1265.

85 Nies, D.H. and Silver, S. (1995) *Journal of Industrial Microbiology,* **14**, 186–199.

86 Nucifera, G., Ghu, L., Mishra, T.K. and Silver, S. (1989) *Proceedings of the National Academy of Sciences of the United States of America,* **86** (10), 3544–3548.

87 Gupta, A., Meyer, J.E. and Goel, R. (2002) *Current Microbiology,* **45**, 323–327.

88 Engqvist, L.G., Martensson, A., Orlowska, E., Turnau, K., Belimov, A.A., Borisov, A. and Gianinazzi-Pearson, Y.V. (2006) *Acta Agriculturae Scandinavica, B,* **56**, 9–16.

89 Angle, J.S., McGrath, S.P., Chaudhri, A. M., Chaney, R.L. and Giller, K.E. (1993) *Soil Biology & Biochemistry,* **25**, 575–580.

90 Giller, K.E., McGrath, S.P. and Hirsh, P. R. (1989) *Soil Biology & Biochemistry,* **25**, 841–848.

91 Ibekwe, A.M., Angle, J.S., Chaney, R.L. and Van Berkum, P. (1995) *Journal of Environmental Quality,* **24**, 1199–1204.

92 Belimov, A.A., Safronova, V.I., Sergeyeva, T.A., Egorova, T.N., Matveyeva, V.A. and Tsyganov, V.E. (2001) *Canadian Journal of Microbiology,* **47**, 642–652.

93 Burd, G.I., Dixon, D.G. and Glick, B.R. (2000) *Canadian Journal of Microbiology,* **46**, 245–247.

94 Tripathi, M., Munot, H.P., Shouche, Y., Meyer, J.E. and Goel, R. (2005) *Current Microbiology*, **50**, 233–237.

95 Lucy, M., Reed, E. and Glick, B.R. (2004) *Antonie van Leeuwenhoek*, **86**, 1–25.

96 Lindow, S.E. (1987) *Applied and Environmental Microbiology*, **53**, 2520–2527.

97 Russell, N.J. (2002) *International Journal of Food Microbiology*, **79**, 27–34.

98 Yamanka, K. (1999) *Journal of Molecular Microbiology and Biotechnology*, **1**, 193–202.

99 Subba Rao, N.S. (1986) *Soil Microorganisms and Plant Growth*, Oxford and IBH Publishing Company, New Delhi.

100 Kokalis-Burelle, N., Vavrina, C.S., Rosskopf, E.N. and Shelby, R.A. (2002) *Plant and Soil*, **238**, 257–266.

101 Kropp, B.R., Thomas, E., Pounder, J.I. and Anderson, A.J. (1996) *Biology and Fertility of Soils*, **23**, 200–206.

102 Pandey, A. and Durgapal, A. (1999) *Microbiological Research*, **154** (3), 259–266.

11
Rhamnolipid-Producing PGPR and Their Role in Damping-Off Disease Suppression

Alok Sharma
Dedicated to Professor Bhavdish N. Johri on his 63rd birthday.

11.1
Introduction

In the course of industrialization, several problems have been created such as excessive use of synthetic products to achieve greater crop yield and protect plants from phytopathogens. This has resulted in the pollution of the environment, while pathogens have adapted themselves to such chemicals. Realizing the severity of the problem, several of these chemicals have been banned globally. Therefore, the issue of sustainable agriculture, one that is based on the use of eco-friendly agents, has gained more currency today. It is here that biological agents derived from antagonistic and plant growth promoting rhizobacteria or as supplements to chemical pesticides have been promoted in a system of integrated plant disease management, acquiring considerable acceptance in recent times [1].

Over the past few years, agricultural policies in developing countries have undergone major changes to meet the increased demand of food through diversification in general and emphasis on sustainable production systems in particular. The latter is a consequence of problems associated with nonjudicious use of fertilizers and pesticides, as well as the low purchasing power of the marginal farmer. Intervention of biotechnologies in development strategies is therefore of prime significance to crop management. Biological control, using microorganisms to suppress plant diseases, offers a powerful and inevitable alternative to the application of synthetic chemicals. With the growing importance of such control systems in plant disease management, study of the mechanisms involved is paramount [2,3]. Consequently, it is imperative to establish the molecular basis behind the success of these approaches in the context of their interaction with nature and their potential applicability. Despite this complexity, benefits of biocontrol mechanisms and their growth promotion attributes have been demonstrated in several plant systems [4–8].

The increasing use of chemical inputs causes several negative effects, development of pathogen resistance to the applied agents and nontarget environmental

Plant-Bacteria Interactions. Strategies and Techniques to Promote Plant Growth
Edited by Iqbal Ahmad, John Pichtel, and Shamsul Hayat
Copyright © 2008 WILEY-VCH Verlag GmbH & Co. KGaA, Weinheim
ISBN: 978-3-527-31901-5

impacts [9,10]. Furthermore, the growing cost of pesticides, particularly in less-affluent regions of the world, and consumer demand for pesticide-free food have led to a search for substitutes for these products. There are also a number of fastidious diseases for which chemical solutions are few, ineffective or nonexistent [10]. Biological control is thus being considered as an alternative or a supplemental means of reducing the use of chemicals in agriculture [9–12]. Despite their different ecological niches, free-living rhizobacteria and endophytic bacteria use some of the same mechanisms to promote plant growth and control phytopathogens [13–18]. The widely recognized mechanisms of biocontrol mediated by plant growth promoting bacteria (PGPB) include competition for an ecological niche or substrate, production of inhibitory allelochemicals and induction of systemic resistance (ISR) in host plants to a broad spectrum of pathogens [13,15,19–22] and/or abiotic stresses (reviewed in [23,24]). This chapter reviews the advances of plant–PGPB interaction research focusing on the principles and mechanisms of action of PGPB, both free-living and endophytic, and their use or potential use in biological control of plant diseases, especially damping-off disease in vegetable crops.

11.2
Biocontrol

Biocontrol is a broad term that refers to control of disease proliferation either by inhibition or by lysis, via biological means. It is a complex phenomenon based on the principle of negative interaction (competition) between the two inhabitants of the same niche. Several modes of action of microbial biological agents have been identified [1,15,25], none of which are mutually exclusive. These can involve interactions between the antagonist and the pathogen, directly or indirectly, either associated with roots and seeds or free in soil. Often, one antagonist may exhibit several modes of action simultaneously or sequentially. Also, in the case of natural disease suppressive soils, several antagonists exhibiting a range of actions in concert control a disease [26].

11.2.1
Antibiotic-Mediated Suppression

The production of antibiotics by microorganisms is considered a major event in the soilborne disease suppression by rhizospheric microbial strains. Microbes produce a vast range of antibiotics under different physiological conditions. 2,4-Diacetylpholoroglucinol (DAPG), phenazines, pyocyanin, pyoluteorin, pyrrolnitrin and topolone – like secondary metabolites – have been characterized from varied groups of soil bacteria including pseudomonads.

Shanahan *et al.* [27] reported that a Tn5-induced DAPG⁻ mutant of fluorescent *Pseudomonas* strain had lost the ability to protect sugarbeet roots from *Pythium* spp. infection. Mazzola *et al.* [28] reported that DAPG-producing strain Q2-87 suppressed take-all disease caused by three DAPG sensitive isolates of *Gaeamannomyces graminis* var. *tritici* but failed to suppress two other isolates and the pathogen was resistant to DAPG at 3 enzyme units per ml. DAPG-producing fluorescent *Pseudomonas* spp. have been shown to be responsible for take-all decline, a natural biological control system found to develop in soils following extended monoculture of wheat or barley [29,30]. It is possible to isolate DAPG from the rhizosphere. Therefore, the positive role of DAPG in the biological control of plant disease can be assessed by genetic approaches. Raaijmakers *et al.* [31] have demonstrated that the level of DAPG in the rhizosphere is directly related to the DAPG-producing population. Phloroglucinol (Phl) is a phenolic metabolite produced by bacteria and plants with broad-spectrum antibacterial, antifungal, antiviral, antihelmintic and phytotoxic properties [32]. This polyketide antibiotic has been identified to be largely responsible for the prevention of 'damping-off' in sugarbeet and cotton caused by *Pythium ultimum* and *Phytium* spp., respectively [27,33].

The biocontrol strain *P. fluorescens* F113 is an effective antagonist of *Pythium ultimum* under laboratory conditions [27] besides reducing the severity of damping-off in soil naturally infested with *Pythium* spp. [34]. *P. fluorescens* F113G22, a Phl-negative Tn5::lacZY mutant derivative, does not inhibit *P. ultimum* grown *in vitro* or reduce the severity of damping-off [27]. The *Phl* biosynthetic locus has been cloned in several pseudomonads [34–40]. In microcosm studies, these two *Phl* overproducing strains proved to be as effective in controlling damping-off disease as a proprietary fungicide treatment, indicating enhanced potential of genetic modification in plant disease control [41].

Pyoluteorin, an aromatic polyketide antibiotic, is produced by several *Pseudomonas* species including strains that suppress plant diseases caused by phytopathogenic fungi [33,42,43]. Howell and Stipanovic [33] reported that pyoluteorin treatment was effective in providing protection against damping-off caused by *Pythium*. Of the antibiotics known to be produced by *Pseudomonas fluorescens* Pf-5 or CHA0, pyoluteorin is most toxic to *Pythium ultimum* [42], although 2,4-diacetylphloroglucinol [44,45] and pyoverdine siderophores [46] also suppress mycelial growth [33].

Howell and Stipanovic [47] reported that pyrrolnitrin plays an important role in providing protection against *Rhizoctonia solani* infection in cotton seedlings. Several studies suggest that pyrrolnitrin production by *Burkholderia cepacia* and *Pseudomonas* spp. is closely related to biocontrol of plant diseases. Jayaswal *et al.* [48,49] generated a Tn5-induced mutant strain of *B. cepacia* deficient in pyrrolnitrin and showed that the mutant completely lost antifungal activity. Furthermore, Hill *et al.* [50] cloned a gene responsible for pyrrolnitrin production in *P. fluorescens* and demonstrated that interruption within the gene region resulted in the loss of biocontrol activity against *Rhizoctonia* damping-off in cotton. A chemically induced overproducing mutant of *P. aeruginosa* exhibited 30-fold increase in synthesis of pyrrolnitrin [51]. Replacing the native promoter with a more active promoter within Prn gene cluster also increased pyrrolnitrin production in *P. aureginosa* with enhanced biocontrol of *Rhizoctonia* damping-off [52].

1-Aminocyclopropane-1-carboxylic acid (ACC) deaminase, which is found only in microorganisms, catalyzes cleavage of ACC to α-ketobutyrate and ammonia by cyclopropane ring opening. ACC is a key intermediate in the biosynthesis of ethylene produced by almost all plants. Ethylene mediates a range of plant responses and developmental steps including seed germination [53], tissue differentiation, formation of root and shoot primordia, root elongation, lateral bud development, flowering initiation, anthocyanin synthesis, flower opening and senescence, fruit ripening and degreening, production of volatile organic compounds responsible for aroma formation in fruits, storage product hydrolysis, leaf and fruit abscission and response of plants to biotic and abiotic stresses [54–57].

11.2.2
HCN Production

HCN production by certain rhizospheric microorganisms influences root and soil-borne pathogens [58]. Generally, these are known as deleterious rhizobacteria (DRB). In their study, Voisard *et al.* [58] demonstrated that insertional inactivated HCN⁻ mutants lost the ability to suppress black root rot in tobacco, whereas the same could not be repeated in the case of take-all disease. Several DRB that reduce seed germination, seedling vigor and subsequent plant growth have been isolated from roots and rhizosphere of various weeds [59]; some cyanogenic rhizobacteria are typically host specific and associated with the roots of their host plants. Therefore, HCN produced in the rhizosphere of seedlings by selected rhizobacteria is a potential and environmentally compatible mechanism for biologically controlling weeds and minimizing deleterious effects on the growth of desired plants [60]. Successful establishment of cyanogenic DRB in the weed rhizosphere would be more economical than chemical synthesis and/or field application of growth-suppressive compounds.

11.2.3
Induced Systemic Resistance

Plants in general possess active defense mechanisms against pathogen attack in nature. Induced resistance occurs naturally as a result of limited infection by a pathogen, particularly when the plant develops a hypersensitive reaction [61]. Induced resistance can be triggered by certain chemicals, nonpathogens and also by avirulent forms of the pathogen. It can be systemic, as it increase the defensive capacity not only in primary infected plant parts but also in noninfected, spatially separated tissues; therefore, it is referred to as 'systemic acquired resistance' (SAR), which is characterized by the accumulation of salicylic acid (SA) and pathogenesis-related proteins (PRs) [62].

Resistance is also induced in plants by some strains of nonpathogenic rhizobacteria resulting in the suppression of the disease. This has been termed as induced systemic resistance [63,64]. Common procedures to accomplish induced resistance include use of a suspension of bacteria on a plant surface, mixing it with autoclaved soil, dipping the roots of seedlings in a bacterial suspension while transplanting or coating seeds with a large number of bacteria before sowing [65].

11.3
Damping-Off

Vegetables are severely attacked at nursery stages by many soilborne pathogenic fungi and oomycetes; particularly, damping-off causes great agricultural damage. Soilborne diseases caused by *Fusarium*, *Phytophthora*, *Pythium* and *Verticillium* spp. are very difficult to eliminate by any of the known methods of control. Chemical control, when available, is often too expensive to be economically practical and moreover is hazardous to the environment. Damping-off generally refers to a sudden plant death at the seedling stage because of a fungal attack. Such fungi are soilborne and are stimulated to grow by nutrients released from a germinating seed and infect the seedling.

Damping-off diseases of seedlings are widely distributed and are a problem worldwide. They occur in moist soils, temperate and tropical climates and greenhouses. The disease affects seeds and seedlings of various crops such as bean, sweet corn, tomato, pea, cucurbits (squash, cucumber, pumpkin and melon), sugarbeet, maize and cotton. The amount of damage caused to seedlings depends on the fungus, soil moisture and temperature. Normally, however, cool wet soils favor the development of a disease. Seedlings in seedbeds are often completely destroyed by damping-off or die after transplantation.

When seeds are planted in infested soils, damping-off fungi may attack at any stage: prior to germination (preemergence damping-off) or after the seed has germinated but before the seedling has emerged above the soil line (postemergence damping-off). Infected seeds usually become soft and mushy, turn brown or black and eventually disintegrate. Seeds that have germinated and become infected develop water-soaked spots that enlarge and turn brown. The infected tissue collapses, resulting in the death of the seedling. The death of seeds before they emerge is termed preemergence damping-off.

Seedlings that have emerged are usually attacked below the soil line. Since fungal pathogens can easily penetrate the young, soft stem tissue, which becomes discolored and begins to shrink, the supportive strength of the invaded portion of the stem is lost and the seedling usually topples. However, the fungi continue to invade the remaining portion of the seedling, resulting in its death. This phase of the disease is termed as postemergence damping-off. Older plants can also be attacked by damping-off fungi. Usually, the developing rootlets are infected, resulting in root rot. Infected plants show symptoms of wilting and poor growth.

11.3.1
Causal Organisms

Species of the genus *Pythium* are natural inhabitants of soil, where they occur as low-grade parasites on fibrous roots. The oospores serve as the overwintering or oversummering organs. They are not vigorous competitors and their saprophytic activities are generally restricted [66]. Soil moisture is important for saprophytic growth of *Pythium* and survival by the formation of resistant structures is more important than saprophytic persistence. Royle and Hickman [67] showed that zoospores in a water suspension are attracted toward the region of elongation behind the tips of pea roots, to wounds in the epidermis and to the exposed stele at cut ends of roots where they encyst and germinate. Factors that influence infection include inoculum density, soil moisture, temperature, pH, cation composition, light intensity and presence of other microorganisms. Soil temperature and moisture (or experimentally tested environmental factors) are known to favor the seasonal activity of *Pythium* spp. [68,69]. Generally, wet soil conditions (0 to −0.3 bar matric water potential) are necessary for development of pathogen. Among *Pythium* species, *P. irregulase*, *P. spinosum* and *P. ultimum* are more damaging at lower temperatures, whereas *P. aphanidermatum*, *P. arrhenomanes*, *P. myriotylum* and related species cause greater damage at higher temperatures. Available evidence suggests that high soil moisture per se does not necessarily favor the activity of *Pythium*. High matric water potential and accompanying poor aeration conditions indirectly favor disease development by (i) decreasing host vigor and increasing host exudation and (ii) providing a suitable environment for rapid diffusion [70,71] of host exudates necessary for germination of dormant propagules and/or vegetative growth.

11.3.2
Control

Most species of *Pythium* produce oospores and chlamydospores and persist for many years; therefore, field-level elimination of the disease is not only difficult but also expensive. On a small scale, however, *Pythium* can be eliminated from soil by steam treatment or pasteurization. Fumigation with chloropicrin or methyl bromide has been a standard practice in nursery and horticultural operations [72].

Drenching the soil with suitable fungicides, such as chestnut compound, 1% Bordeaux mixture, 0.1% ceresin, 0.3% Blitox-50 or 0.2% Esso Fungicide 406 has been useful in eliminating soilborne infection. Crop rotation is another method of reducing populations of soil pathogens; but on account of its wide host range, *Pythium* spp. are difficult to eliminate through this approach. Biological control of soilborne pathogens is yet another remedy. Microorganisms that can grow in the rhizosphere are ideal for use as biocontrol agents, since the rhizosphere provides a frontline defense for roots against attack by pathogens. Pathogens encounter antagonism from rhizosphere microorganisms before and during primary infection and also during the secondary spread on the root. In some soils described as microbiologically suppressive to pathogens [73], microbial antagonism of the pathogen is especially strong, leading to substantial disease control.

Greenhouse methods have been developed to screen antagonists of *Pythium* spp. [74] on wheat and *Phytophthora megasperma* f. sp. *glycinea* on soybean [75]. Kloepper *et al.* [76] demonstrated the importance of siderophore production as a mechanism of biological control. Siderophores have been shown to be involved in suppression of *Pythium* spp. [77,78]. Howell and Stipanovic [33,47] demonstrated that the purified antibiotics pyoluteorin and pyrrolnitrin, obtained from *P. fluorescens* Pf-5, provided the same protection of cotton against damping-off by *P. ultimum* or *R. solanii* as did the bacterium.

11.4
Rhamnolipids

Microorganisms produce a wide variety of secondary metabolites that have a diverse spectrum of activity in environmental remediation. More recently, rhamnolipids of bacterial origin have found special application in biocontrol of phytopathogenic fungi. Such bioactive molecules are likely to influence biotic and abiotic processes in the natural environment including soil and rhizosphere ecosystems.

Biosurfactants or surface-active compounds are produced by a variety of microorganisms including bacteria, yeast and fungi (Table 11.1). Their broad range of potential applications include enhanced oil recovery; surfactant-aided bioremediation of water-insoluble pollutants; facilitation of industrial processes such as emulsification, phase separation and viscosity reduction [79–81]; replacement of

Table 11.1 Major types of biosurfactants produced by microorganisms.

Biosurfactant type	Microorganism
Alasan	*Acinetobacter radioresistens*
Arthrofactin	*Arthrobacter* sp. MIS38
Biosur Pm	*Pseudomonas maltophilla* CSV 89
Glycolipid	*Serratia rubudia*
Glycolipid	*Serratia marcesens*
Glycolipid	*Alcanivorax borkumenis*
Glycolipid	*Tsukamurella* sp.
Lychenysin A	*Bacillus licheniformis* BAS50
Lychenysin B	*Bacillus licheniformis* JF-2
Mannosylerythritol lipids	*Candida antarctica*
Mannosylerythritol lipids	*Candida* sp. SY16
Mannosylerythritol lipids	*Candida antarctica* KCTC 7804
PM factor	*Pseudomonas marginalis* PD 14B
Rhamnolipid	*Pseudomonas aeruginosa* GS3
Rhamnolipid	*Pseudomonas aeruginosa* UW-1
Rhamnolipid	*Pseudomonas aeruginosa* GL1
Sophorose lipid	*Candida apicola* IMET 43 747
Sophorose lipid	*Candida bombicola*
Streptofactin	*Streptomyces tendae* TU901/8c
Surfactin	*Bacillus pumilus* A1
Surfactin	*Bacillus subtilis* C9
Surfactin	*Bacillus subtilis*
Surfactin	*Lactobacillus* sp.
Surfactin	*Bacillus subtilis* ATCC 21 332
Trehalose dimycolate	*Rhodococcus erythropolis*
Trehalose lipid	*Nocardia* SFC-D
Trehalose lipid	*Rhodococcus* sp. H13 A
Trehalose lipid	*Rhodococcus* sp. ST-5
Trehalose tetraester	*Arthrobacter* sp. EK1
Viscosin	*Pseudomonas fluorescens*

chlorinated solvents used in cleaning up oil-contaminated pipes; vessels and machinery used in the detergent industry; formulations of herbicides and pesticides; formation of stable oil-in-water emulsions for food and cosmetic industries [80,82–91].

Surfactants are substances that adsorb and alter the conditions prevailing at interfaces. The surfactants concentrate at interfaces as they are amphipathic; that is, they contain both hydrophilic and hydrophobic groups. Hydrophilic groups consist of mono-, oligo- or polysaccharides, amino acids, peptides, carboxylate or phosphate groups and hydrophobic groups are made up of saturated or unsaturated (hydroxy) fatty acids and fatty alcohols. The main classes of biosurfactants are glycolipids, lipopeptides and high-molecular-weight biopolymers such as lipoproteins, lipopolysaccharides and others. Among these, glycolipids contain various sugar moieties, for example rhamnose, sophorose and trehalose, attached to long-chain fatty acids. Lipopeptides, however, consist of a short polypeptide of S-12 amino acids attached to a lipid moiety and include lipoproteins, lipopolysaccharide–protein complexes and polysaccharide–protein–fatty acid complexes as part of high-molecular-weight biopolymers [81,92].

Rhamnolipids of various bacterial groups have been studied in detail; therefore, explained below are the developments that relate largely to pseudomonads since this group has found special application in bioremediation, biocontrol and plant growth promotion.

As early as 1946, Bergstrom *et al.* [93] grew *Pseudomonas pyocyanea* on glucose and detected glycolipids containing rhamnose and β-hydroxydecanoic acid; the researchers were, however, unable to describe the molar ratio of the two components. This was subsequently determined by Jarvis and Johnson [94] who demonstrated the presence of a glycosidic linkage of β-hydroxydecanoyl-β-hydroxydecanoate with two rhamnose molecules after cultivating *Pseudomonas aeruginosa* on 3% glycerol. Edwards and Hayashi [95] were, however, the first to elucidate the complete structure of this molecule and demonstrated the presence of a 1,2-linkage after periodate oxidation and methylation.

The first rhamnolipid identified was rhamnolipid 2 (R2), a dirhamnolipid. Hisatsuka *et al.* [96] reported that *P. aeruginosa* S$_7$B$_1$ produced only one type of rhamnolipid molecule, that is dirhamnolipid (R2) when grown on a mixture of *n*-hexadecane and *n*-paraffins (C$_4$–C$_8$).

In the same year, Itoh *et al.* [97] isolated yet another rhamnolipid, rhamnolipid 1 (R1), a monorhamnolipid from the culture supernatant of *P. aeruginosa* KY 4025 on 10% *n*-paraffins. Rhamnolipids 1 and 2 are chemically referred to as L-rhamnosyl-β-hydroxydecanoyl-β-hydroxydecanoate and L-rhamnosyl-L-rhamnosyl-β-hydroxydecanyl-β-hydroxydecanoate, respectively, and are the principal glycolipids produced in liquid cultures of *P. aeruginosa*. Yamaguchi *et al.* [98] described rhamnolipids A and B as acylated products (acylation by α-decanoic acid) of rhamnolipids 1 and 2, respectively. In 1982, Hirayama and Kato [99,100] purified methyl esters of rhamnolipids 1 and 2 from *P. aeruginosa* strain 158 grown in Difco trypticase soya medium. Syldatk *et al.* [101,102] detected rhamnolipids 3 and 4 containing only one β-hydroxydecanoyl moiety in culture supernatant of the resting cells of

Table 11.2 Glycolipids found in the crude ethyl acetate extract of *Pseudomonas* strain GRP$_3$ by positive ion mode EMI-MS [107].

Structure	Molecular mass (M+ H+)	Relative amount
Rha–Rha–C$_{10}$–C$_{10}$	673	100
Rha–C$_{10}$–C$_{10}$	527	21
Rha–C$_{10}$–C$_{12}$	701	21
Rha–Rha–C$_{12}$–C$_{10}$	701	7
Rha–Rha–C$_{10}$–C$_{12}$	699	17
Rha–Rha–C$_{12}$–C$_{10}$ (with double bond in C$_{12}$ unit)	699	1

P. aeruginosa sp. DSM 2874; however, these could have been degradation products of rhamnolipids 1 and 2 [103].

Additional types of rhamnolipids harboring alternative fatty acid chains have been purified from culture broths of a clinical isolate of *P. aeruginosa* [104]. The fatty acid homologues present in these rhamnolipids, as identified by fast atom bombardment and electron impact mass spectrometry, include β-hydroxyoctanoyl-β-hydroxyde-canoate, β-hydroxydecanolyl-β-hydroxydodecanoate and β-hydroxydecanyl-β-hydro-xydodec-5-enoate. However, it is generally believed that the latter variants represent minor components. The chain length of the carbon substrate employed has no effect on the structure of the rhamnolipids produced. The predominant types of rham-nolipids appear to be strain specific and seem to depend, to an extent, on the environmental and cultural conditions, especially the medium composition [81]. In fluorescent pseudomonad GRP$_3$, for example, a rhizoplane isolate from soybean that has been studied by our group as a plant growth promoting agent [105,106], the total rhamnolipids comprised of a number of mono- and dirhamnolipids with satu-rated and unsaturated fatty acid side chains of varying chain length (Table 11.2) [107]. The structures of these rhamnolipids are described in Figure 11.1.

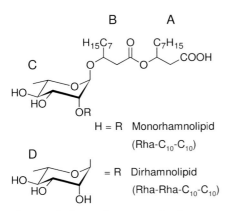

Figure 11.1 Chemical structure of major mono- and dirhamno-lipids from *Pseudomonas* sp. strain GRP$_3$.

11.4.1
Biosynthesis of Rhamnolipids

The biosynthesis of rhamnolipids in *P. aeruginosa* was initially studied *in vivo* employing radioactive precursors such as [^{14}C] acetate and [^{14}C] glycerol [108,109]. The synthesis of rhamnolipids proceeds by sequential glycosyl transfer reactions, each catalyzed by a specific rhamnosyltransferase. Monorhamnolipids are synthesized by the action of rhamnosyltransferase 1 with thymidine-diphospho-rhamnose (TDP-rhamnose) acting as the rhamnosyl donor and β-hydroxydeca-noyl-β-hydroxydecanoate or L-rhamnosyl-β-hydroxydecanoyl-β-hydroxydecanoate acting as the acceptor [110]. Dirhamnolipids are synthesized by the action of rham-nosyltransferase 2 on TDP-rhamnose and monorhamnolipid. TDP-rhamnose is pres-ent in Gram-negative bacteria since rhamnose is incorporated into lipopolysaccharide side chains. After several enzymatic steps, glucose is converted into TDP-glucose, which is finally transformed into TDP-rhamnose [111,112]. The acceptor substrate precursor, β-hydroxydecanonic acid, can be synthesized either as an intermediate of fatty acid degradation via the β-oxidation cycle or *de novo* as an intermediate during the fatty acid biosynthesis.

11.4.2
Genetics of Rhamnolipid Synthesis

Ochsner *et al.* [113] isolated and characterized *rhlAB* genes from *P. aeruginosa* PG201 and presented evidence to show that they encode rhamnosyltransferase 1. They also showed that a regulatory gene *rhlR*, encoding a transcriptional activator, affected rhamnolipid biosynthesis in *P. aeruginosa* 65E12. Ochsner and Raiser [114] reported a 21-kDa protein, RhlI, encoded by the *rhlI* gene, which was present on the same operon that worked as an autoinducer of the rhamnolipid biosynthetic pathway. The activity of the *rhlR* regulator protein was enhanced, in a mechanism that depended on a cell density-dependent system, that is quorum sensing (QS). These workers cloned and transferred the *rhlAB* gene from *P. aeruginosa* to a *P. putida* heterologous host, KT 2442 and achieved a monorhamnolipid concentration of $0.6 \, \text{g} \, \text{l}^{-1}$ with recombinant *P. putida* KT 2442. The QS system encoded by *rhlR* (*vsmR*) and *rhlI* (*vsmI*) positively regulates *rhlA* and *rhlB* and is located immediately downstream of the structural genes in one operon, transcribed in the same direction by a different promoter [114–117]. Pearson *et al.* [118] demonstrated that with an increase in cell density, the concentration of N-butyryl-homoserine lactone synthesized by *rhlI* reached a threshold level where it attached itself to the transcriptional activator RhlR, which was bound to a 'lux box'-like consensus sequence located upstream of *rhlAB* and enhanced their transcription [117]. Campos-Gracia *et al.* [119] reported that insertional mutation in *rhlG*, homologous to the *fabG* gene, produced no effect on growth rate and total lipid content of *P. aeruginosa* W51D and *P. aeruginosa* PAO1 cells, but the production of rhamnolipids was completely checked. Recently, rham-nosyltransferase 2 gene, *rhlC* has been identified and characterized by Rahim *et al.* [120] and found to be responsible for rhamnolipid biosynthesis. In addition, these

Table 11.3 Major genes involved in the biosynthesis of rhamnolipids in *Pseudomonas aeruginosa* [121].

Gene	G + C %	promoter	Size (nt)	Peptide length	Gene product (kDa)	pI	Function
RhlA	65.8	54	887	296	32.5	7.4	Rhamnosyl- transferase 1
RhlB	67.9	54	1280	427	47	8.4	Rhamnosyl- transferase 1
RhlC	70.7	54	975	325	35.9	nd	Rhamnosyl- transferase 2
RhlG	Nd	54	768	256	26.8	nd	NADPH-dependent ketoacyl reductase
RhlR	61.7	70	726	242	26.5	7.0	Transcriptional regulator
RhlI	64.8	64.8	NA	NA	NA	NA	NA

researchers showed that *rhlC* is coordinately regulated with a *rhlAB* quorum-sensing system. Major genes involved in this system are listed in Table 11.3.

The *rhl* system also regulates the stationary phase sigma factor encoded by *rpoS*, which is involved in the regulation of numerous genes important for survival under adverse conditions [116]. A second quorum-sensing system, located at a different region in the *P. aeruginosa* chromosome, also influences the expression of rhamnolipid biosynthesis. This second system is encoded by *lasR* (31% homology to *rhlR*) and *lasI* (28% homology to *rhlI*) wherein *lasI* encodes the autoinducer N-(3-oxodo-decanoyl)-L-homo-serine lactone [122]. The interaction between both systems was described as a hierarchical quorum-sensing cascade, with *lasR* and *lasI* as the master regulators [123].

Olvera *et al.* [124] proved that the *algC* gene, involved in alginate production through its phosphomannomutase activity and in LPS synthesis through phospho-glucomutase activity, participates in rhamnolipid biosynthesis in *P. aeruginosa*. The phosphoglucomutase activity of AlgC is responsible for the production of glucose-1-phosphate, the precursor of dTDP-glucose and ultimately of dTDP-l-rhamnose, whereas products of other *alg* genes are involved neither in rhamnolipid production nor LPS synthesis. Pearson *et al.* [117] have shown that both quorum-sensing systems can induce expression of rhamnolipids and other virulence genes but with different efficiencies due to specificity of the transcriptional activator with its cognate autoinducer.

11.4.3
Regulation

The synthesis of rhamnolipids in *P. aeruginosa* under nitrogen depletion, when the cells shift into stationary growth phase, has been reported [125,126]. Ochsner *et al.* [115] observed the expression of *P. aeruginosa* genes for rhamnolipid synthesis in *P. fluorescens* and *P. putida* only under nitrogen-limited conditions. Addition of nitrogen inhibits the production of rhamnolipids by resting cells in *Pseudomonas* sp. DSM 2874 [102]. Guerra-Santos *et al.* [127] reported that higher levels of rhamnolipids could be achieved in *P. aeruginosa* DSM 2659 under limiting conditions of magnesium, calcium, potassium, sodium, iron and trace elements. Glycerol, glucose, *n*-alkanes, triglycerides are suitable carbon sources for overproduction of rhamnolipids. An essential precondition for overproduction of the above glycolipids is growth limitation, induced by appropriately limiting the concentration of nitrogen sources or multivalent ions, and an excess of carbon sources. Environmental factors and growth conditions such as temperature, agitation and oxygen availability also affect biosurfactant production through their effect on cellular growth or activity. The pH of the medium plays an important role in rhamnolipid production in *P. aeruginosa*; maximum production is achieved in the pH range 6.0–6.5, with a sharp decrease above 7.0 [125]. While the optimum temperature is in the 31–34 °C range, lower and higher temperatures resulted in significant reduction. Mulligan and Gibbs [128] observed a direct correlation between rhamnolipid synthesis and glutamine synthetase activity in *P. aeruginosa*. This enzyme showed maximum activity at the end of the exponential phase of growth, that is at the start of rhamnolipid production. Mulligan *et al.* [129] reported that a shift in phosphate metabolism coincided with biosurfactant production.

11.4.4
Rhamnolipid-Mediated Biocontrol

The efficacy of synthetic surfactants to control the zoosporic plant pathogen *Olpidium brassicae* was first demonstrated in 1980 during a study of big-vein disease of hydroponically grown lettuce by Tomlinson and Faithful [131]. Further studies demonstrated that a nonionic surfactant, Agral 90 (ICI) was responsible for inhibition of disease in commercial lettuce production facilities [132]. In these studies, it was shown that surfactants were active against the zoosporic stage of the pathogen, which is devoid of cell wall. Similar results were reported by Stanghellini *et al.* [133] for the major root infecting zoosporic fungus, *Pythium aphanidermatum* in a hydroponic system. They have subsequently reported that rhamnolipids were effective against three genera of zoosporic plant pathogens: *Pythium aphanidermatum*, *Phytophthora capsici* and *Plasmopara lactuae-radicis*. Purified mono- and dirhamnolipids from *P. aeruginosa* at 5–30 mg l^{-1} caused cessation of motility and lysis of the entire zoospore population in less than 1 min. When *Pseudomonas* strains were used *in situ* along with olive oil as substrate in a hydroponic recirculating cultural system, rhamnolipids were produced but the performance was variable [134].

The idea to use biosurfactants against zoosporic fungi was pioneered by the work of Tomlinson and Faithful [131,132] who used the fungicide benzimidazole against zoospores of *Olpidium brassicae*; this worked as a vector for big-vein disease virus. Later on it was found that inert ingredients of a fungicide formulation, which contained synthetic surfactant, exhibited lytic activity against zoospores of *Olpidium*. As a sequel to this observation, anionic (manoxol O/T, Marasperse CB, sodium lauryl sulfate), nonionic (Agral, Triton X-100, Spredite, Ethylan CPX) and cationic (cetrimide, Deciquam 222, Hyamine 1622) synthetic surfactants were used, which showed differential effects. The concentration of biosurfactant for zoospore lysis was a function of both the sensitivity of zoospore and the type of rhamnolipid. Stanghellini and Miller [134] demonstrated that dirhamnolipid was equal to or better than monorhamnolipid in causing lysis of zoospores of *Pythium aphenidermatum*, *Phytophthora capsici* and *Plasmopara lactucae-radicis*. These workers proposed that the rhamnolipids interacted and disrupted the plasma membrane of zoospores. Addition of rhamnolipid to zoospore suspension (concentration $600\,\mu g\,ml^{-1}$) resulted in cessation of zoospore motility, which declined (lysed) in less than 1 min. By contrast, in the absence of rhamnolipid, zoospores of each tested species swam for approximately 20 h. On the basis of this and subsequent work, Stanghellini and coworkers [133,134] obtained patents on the control of zoospore fungi employing rhamnolipids.

Currently, our research effort is directed toward helping poorer farming communities, particularly those in the state of Uttarakhand in the Central Himalayan region, who depend exclusively on vegetable cultivation for their livelihood. In one study, we focused on the role of rhamnolipid-producing bacterial agents against zoosporic phytopathogenic oomycetes [7]. A pool of promising rhizobacteria was screened through *in vitro* antagonism against prevalent phytopathogenic oomycetes and their plant growth promoting properties to evaluate the influence of the most promising bacterial strains on plant growth. Bacterized seeds were planted in an artificially created nursery and in a natural field nursery, with and without a history of fungicide use. The potential of such bacteria has been analyzed to provide empirical field data of their effects on growth parameters and on pre- and postemergence damping-off disease in chili and tomato.

The experimental conditions were chosen to compare the effect on two pertinent crops, namely chili and tomato, of a relatively large number of promising rhamnolipid-producing isolates under severe environmental conditions prevalent in the Central Himalayan region. In general, vegetable nurseries are established twice per year (in winter and wet seasons) in this region. Therefore, field trials were performed in both the seasons to evaluate seasonal variations in terms of efficacy of bacterial strains to reduce disease occurrence. We demonstrated that rhamnolipid-producing plant growth promoting *Pseudomonas* sp. strains effectively checked pre- and post-emergence disease in chili and tomato plants [7].

Pseudomonas sp. GRP_3 showed significant reduction in the postemergence damping-off index in chili and tomato under natural field conditions (without fungicide), closely followed by FQP PB-3 and FQA PB-3. However, *Pythium* spp. have poor competitive ability among the rhizospheric populations and often act only as primary colonizers [135]. Hence, low competitive ability and early pathogenesis offer possibilities of developing efficient biocontrol agents employing the type of isolates used here. Our data clearly showed that rhamnolipid-producing bacterial isolates cause a significant degree of suppression of both preemergence and postemergence damping-off [7]. Taken together, our data have identified at least *Pseudomonas* sp. strains FQA PB-3, FQP PB-3 and GRP_3 to hold considerable promise for the treatment of damping-off in chili and tomato in both natural and artificial nurseries [7,146].

11.4.5
Other Agricultural Applications

Surface-active agents are needed for hydrophilization of heavy soils to obtain good wettability and also to achieve equal distribution of fertilizers and pesticides. These compounds expose field sites to microbial degradation and help in lowering soil toxicity.

11.5
Quorum Sensing in the Rhizosphere

In this context, various species of *Pseudomonas* are of interest as they are dominant in soil and are known to secrete antimicrobials and siderophores. In addition, they are also important microbial producers of biosurfactants such as rhamnolipids [7]. Here, we discuss QS operations in rhizosphere, keeping *Pseudomonas* as the model system. *Pseudomonas aeurginosa* is an opportunistic human pathogen that infects immuno-compromised individuals and people with cystic fibrosis. Pathogenicity of *P. aeruginosa* depends on its ability to secrete virulent compounds and degradative enzymes, including toxins, proteases and hemolysins. These factors are not expressed until the late logarithmic phase of growth, when cell density is high; this occurs through QS. There are two known QS systems in *P. aeruginosa*; the *las* and the *rhl* system. Each system has a transcriptional activator and an autoinducer synthetase. The *P. aeruginosa* autoinducers (PAI-1 and PAI-2) bind to specific target proteins, the transcriptional activators, and these complexes activate a large number of virulence factors.

11.5.1
The Dominant System (*las*)

The two QS systems of *P. aeruginosa* are linked by the *las* system dominant over the *rhl* system. The *las* system regulates the expression of *lasB* elastase. This system is composed of *las*, the autoinducer synthetase gene responsible for synthesis of 3-oxo-C_{12}-HSL-(*N*-[3-oxododecanoyl]-L-homoserine lactone, previously named PAI-1 or OdDHL) and the *lasR* gene that codes for a transcriptional activator protein [122,136]. The *las* cell-to-cell signaling system regulates *lasB* expression and is required for optimal production of other extracellular virulence factors such as *LasA* protease and exotoxin A [137]. The *las* cell-to-cell signaling system is positively controlled by GacA [138], as well as by vfr, which is required for the transcription of *lasR* [139]. An inhibitor, RsaL, repressing the transcription of *lasI*, has also been described [140].

11.5.2
The *rhl* System

It is so named because of its ability to control the production of rhamnolipids. This system is consists of *rhlI*, the 4-HSL, *N*-butyrylhomoserine lactone, previously named PAI-2 or BHL, autoinducer synthase gene and the *rhlR* gene encoding a transcriptional activator protein [110,115]. This system regulates the expression of the *rhlAB* operon that encodes a rhamnosyltransferase required for rhamnolipid biosurfactant production [113]. The presence of these compounds reduces surface tension and thereby allows *P. aeruginosa* cells to swarm over semisolid surfaces [141]. The *rhl* system is also necessary for optimal production of LasB elastase, Las A protease, pyocyanin, cyanide and alkaline protease [116,117,138]. Significantly, transcription of *rhlI* is enhanced in presence of RhlR–BHL, creating a further autoregulatory loop within *LasRI/RhlRI* regulons. Latifi *et al.* [116] have reported that the *rhl* system also regulates the expression of *rpoS*, which encodes a stationary phase sigma factor (σ^3) involved in the regulation of various stress-response genes. However, QS regulation of σ^5 in *P. aeruginosa* has recently been questioned [142]. According to this group, the sigma factor negatively regulates *rhlI* transcription. Like the *las* cell-to-cell signaling system, the *rhl* system, also referred to as VSm (virulence secondary metabolites), regulates the expression of various extracellular virulence factors of *P. aeruginosa*. Studies conducted to determine how the *LasRI* and *RhlRI* systems interact with each other demonstrated the subordinate nature of the *RhlRI* system in the hierarchy of regulatory command that exists between two QS regulons. Two independent studies have shown that the *RhlRI* system functionally depends on the *LasRI* system, as transcriptional activation of rhlR is dependent on *LasR–OdDHL* [116,123]. Thus, activation of the Las system leads to subsequent activation of the *Rhl* system and together the two LuxR homologues regulate the transcription of genes within their respective regulons.

Molecules involved in QS have gained special attention in nitrogen fixation and associated symbiotic processes. Several gene products required in symbiosis are encoded by the Sym plasmid, which also carries many important AHL synthase genes. Among these, *rhlABC* genes, which are implicated in rhizosphere establishment, are also controlled by the QS system [143]. Also, QS genes, *bisR* and *triR*, are responsible for the transfer of plasmid in *Agrobacterium tumefaciens* [144], an organism endowed with excellent properties to serve as a model in gene transfer.

The operation of the QS system in the rhizosphere appears to hold great promise in controlling the damping-off in vegetables caused by zoosporic fungi in nurseries and elsewhere. Rhamnolipid production controlled by the *rhlRI* system plays a crucial role in controlling the spread of zoospore in the rhizosphere, which causes rapid and severe seedling loss [110,114,145].

Indeed, in terms of AHL-mediated communication, the intricacies of a language that once seemed alien, appears now to enter a new era, with innovative technologies presenting even more opportunities to rapidly enhance our understanding of QS systems. Among these innovations, high-throughput analysis of bacterial genes and proteins that fall under the regulatory umbrella of proteins such as LuxRI homologues is of particular interest. Further, it is likely that many more physiological processes, regulated by bacterial QS systems, will be characterized in future.

11.6
Conclusions and Future Directions

Since rhamnolipids are involved in zoospore lysis of soilborne pathogens such as *Pythium*, *Phytophthora* and *Plasmopara* spp., application of such rhamnolipid-producing rhizobacterial strains should facilitate control of damping-off especially at vegetable cultivation nursery sites. The PGP rhizobacterial isolates are significantly effective in protecting plants against soilborne pathogens by enhancing peroxidase and PAL activities in plant tissues [146]. For vegetable nurseries, strain such as *Pseudomonas* sp. GRP$_3$ should now be tested for developing an effective management strategy to control damping-off diseases affecting vegetable nurseries. Using rhamnolipid-producing plant growth promoting rhizobacteria would open a new way to combat damping-off disease in vegetables during nursery stage. Furthermore, other plant growth promoting properties such as siderophore production, phosphate solubilization and IAA production would be beneficial for plant health and growth. It would be advantageous to isolate and characterize indigenous rhamnolipid-producing PGPR to maximize climate and natural adaptation. Such bacterial strains can also be exploited in hydroponics and recirculating water systems. This strategy could play an immensely important role in protecting vegetable nursery crops against attacks by damping-off disease, by using native rhizobacteria having plant growth promoting activity and rhamnolipid-producing capabilities in such highly humid geographical regions as the Central Himalayas.

Although current efforts are directed toward laboratory-based assays of molecules involved in QS systems, their *in situ* operation in the rhizosphere appears imminent. Such information will permit not only the delivery of more appropriate and effective bioinoculants for plant and soil health but also the cell density-dependent control of *in situ* biological equilibrium, a feature of consequence in minimizing competition with indigenous microorganisms for the limited resources available in this unique ecosystem.

References

1 Johri, B.N., Sharma, A. and Virdi, J.S. (2003) *Advances in Biochemical Engineering/Biotechnology*, **84**, 49–89.

2 Martin, F.N. (2003) *Annual Review of Phytopathology*, **41**, 325–350.

3 Moenne-Loccoz, Y. and Defago, G. (2004) in *Pseudomonas*, vol.1 (ed. J.L. Ramos), Kluwer/Plenum, New York, pp. 457–476.

4 Ciccillo, F., Fiore, A., Bevivino, A., Dalmastri, C., Tabacchioni, C. and Chiarini, L. (2002) *Environmental Microbiology*, **4**, 238–245.

5 Compant, S., Duffy, B., Nowak, J., Clément, C. and Barka, E.A. (2005) *Applied and Environmental Microbiology*, **71**, 4951–4959.

6 Sessitsch, A., Reiter, B. and Berg, G. (2004) *Canadian Journal of Microbiology*, **50**, 239–249.

7 Sharma, A., Wray, V. and Johri, B.N. (2007) *Archives of Microbiology*, **187**, 321–335.

8 Kloepper, J.W. (1991) *Phytopathology*, **81**, 1006–1013.

9 De Weger, L.A., van der Bij, A.J., Dekkers, L.C., Simons, M., Wijffelman, C.A. and Lugtenberg, B.J.J. (1995) *FEMS Microbiology Ecology*, **17**, 221–228.

10 Gerhardson, B. (2002) *Trends in Biotechnology*, **20**, 338–343.

11 Postma, J., Montanari, M. and van den Boogert, P.H.J.F. (2003) *European Journal of Soil Biology*, **39**, 157–163.

12 Welbaum, G., Sturz, A.V., Dong, Z. and Nowak, J. (2004) *Critical Reviews in Plant Sciences*, **23**, 175–193.

13 Bloemberg, G.V. and Lugtenberg, B.J.J. (2001) *Current Opinion in Plant Biology*, **4**, 343–350.

14 Dobbelaere, S., Vanderleyden, J. and Okon, Y. (2003) *Critical Reviews in Plant Sciences*, **22**, 107–149.

15 Glick, B. (1995) *Canadian Journal of Microbiology*, **41**, 109–117.

16 Hallman, J., Quadt-Hallman, A., Mahafee, W.F. and Kloepper, J.W. (1997) *Canadian Journal of Microbiology*, **43**, 895–914.

17 Lodewyckx, C., Vangronsveld, J., Porteous, F., Moore, E.R.B., Taghavi, S., Mezgeay, M. and van der Lelie, D. (2002) *Critical Reviews in Plant Sciences*, **21**, 583–606.

18 Sturz, A.V., Christie, B.R. and Nowak, J. (2000) *Critical Reviews in Plant Sciences*, **19**, 1–30.

19 Haas, D., Blumer, C. and Keel, C. (2000) *Current Opinion in Biotechnology*, **11**, 290–297.

20 Haas, D., Keel, C. and Reimmann, C. (2002) *Antonie Leeuwenhoek*, **81**, 385–395.

21 Lugtenberg, B.J.J., Dekkers, L. and Bloemberg, G.V. (2001) *Annual Review of Phytopathology*, **39**, 461–490.

22 Ryu, C.M., Murphy, J.F., Mysore, K.S. and Kloepper, J.W. (2004) *The Plant Journal*, **39**, 381–392.

23 Mayak, S., Tirosh, T. and Glick, B.R. (2004) *Plant Physiology and Biochemistry*, **42**, 565–572.

24 Nowak, J. and Shulaev, V. (2003) *In Vitro Cellular & Developmental Biology – Plant*, **39**, 107–124.

25 O'Sullivan, D.J. and O'Gara, F. (1992) *Microbiological Reviews*, **56**, 662–676.

26 Alabouvette, C., Lemanceau, P. and Steinberg, C. (1993) *Pesticide Science*, **37**, 365–373.

27 Shanahan, P., O'Sullivan, D.J., Simpson, P., Glennon, J.D. and O'Gara, F. (1992) *Applied and Environmental Microbiology*, **58**, 353–358.

28 Mazzola, M., Cook, R.J., Thomashow, L. S., Weller, D.M. and Pierson, L.S. III (1992) *Applied and Environmental Microbiology*, **58**, 2616–2624.

29 Raaijmakers, J.M., Weller, D.M. and Thomashow, L.S. (1997) *Applied and Environmental Microbiology*, **63**, 881–887.

30 Raaijmakers, J. and Weller, D.M. (1998) *Molecular Plant–Microbe Interactions*, **11**, 142–152.

31 Raaijmakers, J.M., Bonsall, R.F. and Weller, D.M. (1999) *Phytopathology*, **89**, 470–475.

32 Thomashow, L.S. and Weller, D.M. (1996) in *Strategies for Managing Soilborne Plant Pathogens*, (ed. R. Hall), American Phytopathological Society.

33 Howell, C.R. and Stipanovic, R.D. (1980) *Phytopathology*, **70**, 712–715.

34 Fenton, A.M., Stephens, P.M., Crowley, J., O'Callaghan, M. and O'Gara, F. (1992) *Applied and Environmental Microbiology*, **58**, 3873–3878.

35 Vincent, M.M., Harrison, L.A., Bracin, J.M., Kovacevich, P.A., Mukerji, P., Weller, D.M. and Pierson, E.A. (1991) *Applied and Environmental Microbiology*, **57**, 2928–2937.

36 Cook, R.J., Thomashow, L.S., Weller, D.M., Fujimoto, D., Mazzola, M., Bangera, G. and Kim, D. (1995) *Proceedings of the National Academy of Sciences of the United States of America*, **72**, 4197–4201.

37 Bangera, M.G. and Thomashow, L.S. (1996) *Molecular Plant–Microbe Interactions*, **9**, 83–90.

38 Thomashow, L.S., Bangera, M.G., Bonsall, R.F., Kim, R.F., Raaijmakers, J. and Weller, D.M. (1997) Proceeding of the International Symposium on Molecular Plant–Microbe Interactions Knoxville, TN, USA.

39 Delany, I.R., Sheman, M.M., Fenton, A., Bardin, S., Aarons, S. and O'Gara, F. (2000) *Microbiology*, **146**, 537–546.

40 Schnider-Keel, U., Seematter, A.,
Maurhofer, M., Blumer, C., Duffy, B.,
Gigot-Bonnefoy, C., Reimann, C., Notz,
R., Defago, G., Haas, D. and Keel, C.
(2000) *Journal of Bacteriology*, **182**, 1215–
1225.

41 Delany, I.R., Walsh, U.F., Fenton, A.M.,
Corkery, D.M. and O'Gara, F. (2001)
Plant and Soil, **232**, 195–205.

42 Maurhofer, M., Keel, C., Schnider, U.,
Haas, D. and Defago, G. (1992)
Phytopathology, **82**, 190–195.

43 Maurhofer, M., Keel, C., Haas, D. and
Defago, G. (1994) *Plant Pathology*, **44**,
44–50.

44 Keel, C., Schnider, U., Maurhofer, M.,
Voisard, C., Laville, J., Burger, U.,
Wirther, P., Haas, D. and Defago, G.
(1992) *Molecular Plant–Microbe
Interactions*, **5**, 4–13.

45 Nowak-Thompson, B. and Gould, S.J.
(1994) *Biometals*, **7**, 20–24.

46 Kraus, J. and Lopa, J.E. (1995) *Applied and
Environmental Microbiology*, **61**, 849–
854.

47 Howell, C.R. and Stipanovic, R.D. (1979)
Phytopathology, **69**, 480–482.

48 Jayaswal, R.K., Fernandez, M.A.,
Visintin, L. and Upadhyay, R.S. (1992)
Canadian Journal of Microbiology, **38**, 309–
312.

49 Jayaswal, R.K., Fernandez, M., Upadhyay,
R.S., Visintin, L., Kurz, M., Webb, J. and
Rinehart, K. (1993) *Current Microbiology*,
26, 17–22.

50 Hill, D.S., Stein, J.I., Torkewitz, N.R.,
Morse, A.M., Howell, C.R., Pachlatko, J.
P., Beker, J.O. and Ligon, J.M. (1994)
Applied and Environmental Microbiology,
60, 78–85.

51 Salcher, O. (1984) *Journal of General
Microbiology*, **118**, 509–514.

52 Ligon, J.M., Hill, D.S., Hammer, P.E.,
Torkewitz, N.R., Hofmann, D., Kampf,
H. and Van Pee, K. (2000) *Pest
Management Science*, **56**, 688–695.

53 Abeles, F.B., Morgan, P.W. and Saltveit,
M.E. Jr (1992) *Ethylene in Plant Biology*,
2nd edn, Academic Press, New York.

54 Mattoo, A.K. and Suttle, C.S. (1991) *The
Plant Hormone Ethylene*, CRC Press, Boca
Raton, FL, p. 337.

55 Glick, B.R., Penrose, D.M. and Li, J.
(1998) *Journal of Theoretical Biology*, **190**,
63–68.

56 Ma, J.H., Yao, J.L., Cohen, D. and Morris,
B. (1998) *Plant Cell Reports*, **17**, 211–214.

57 Li, J., Ovakim, D.H., Charles, T.C. and
Glick, B.R. (2000) *Current Microbiology*,
41, 101–105.

58 Voisard, C., Keel, C., Haas, D. and
Defago, G. (1989) *EMBO Journal*, **8**, 351–
358.

59 Kremer, R.J. and Kennedy, A.C. (1996)
Weed Technology, **10**, 601–609.

60 Kremer, R.J. and Souissi, T. (2001)
Current Microbiology, **43**, 182–186.

61 Hammerschmidt, R. and Kuc, J. (1995)
Induced Resistance to Disease in Plants,
Kluwer Academic Publishers, Dordrecht,
p. 182.

62 Sticher, L., Mauch-Mani, B. and Metraux,
J.P. (1997) *Annual Review of
Phytopathology*, **35**, 235–270.

63 Kloepper, J.W., Tuzun, S. and Kuc, J.A.
(1992) *Biocontrol Science and Technology*, **2**,
349–351.

64 Pieterse, C.M.J., van Wees, S.C.M.,
Hoffland, E., van Pelt, J.A. and van Loon,
L.C. (1996) *Plant Cell*, **8**, 1225–1237.

65 Kloepper, J.W. (1996) *Bioscience*, **46**, 406–
409.

66 Barton, R. (1961) *Transactions of the
British Mycological Society*, **44**, 105–118.

67 Royle, D.J. and Hickman, C.J. (1964)
Canadian Journal of Microbiology, **10**, 151–
162.

68 Watson, A.G. (1966) *New Zealand Journal
of Agricultural Research*, **9**, 956–963.

69 Robertson, G.I. (1973) *New Zealand
Journal of Agricultural Research*, **16**, 357–
365.

70 Stanghellini, M.E. and Burr, T.J. (1973)
Phytopathology, **63**, 1493–1496.

71 Stanghellini, M.E. and Burr, T.J. (1973)
Phytopathology, **63**, 1496–1498.

72 Whipps, J.M. (1997) *Advances in Botanical
Research*, **26**, 1–134.

73 Schneider, R.W. (1982) *American Phytopathological Society*, 88.

74 Weller, D.M. and Cook, R.J. (1986) *Canadian Journal of Plant Pathology*, **8**, 328–334.

75 Lifshitz, R., Simonson, C., Scher, F.M., Kloepper, J.W., Rodrick-Semple, C. and Zaleska, I. (1986) *Canadian Journal of Plant Pathology*, **8**, 102–106.

76 Kloepper, J.W., Schroth, M.N. and Miller, T.D. (1980) *Phytopathology*, **70**, 1078–1082.

77 Becker, O. and Cook, R.J. (1988) *Phytopathology*, **78**, 4212–4217.

78 Loper, J.E. (1988) *Phytopathology*, **78**, 166–172.

79 Banat, I.M., Makkar, R.S. and Cameotra, S.S. (2000) *Applied Microbiology and Biotechnology*, **53**, 495–508.

80 Lin, S.C. (1996) *Journal of Chemical Technology and Biotechnology*, **66**, 109–120.

81 Desai, J.D. and Banat, I.M. (1997) *Microbiology and Molecular Biology Reviews*, **61**, 47–64.

82 Greek, B.F. (1990) *Chemical & Engineering News*, **68**, 37–38.

83 Hommel, R.K. and Ratledge, C. (1993) *Biosurfactants: Surfactant Science Series*, vol. 48 (ed. N. Kosaric) Dekker, New York, Basel, Hong Kong, pp. 3–63.

84 Lang, S. and Wagner, F.E. (1993) *Biosurfactants: Surfactant Science Series*, vol. 48 (ed. N. Kosaric), Dekker, New York, Basel, Hong Kong, pp. 205–207.

85 Lang, S. and Wagner, F.E. (1993) in *Biosurfactants: Surfactant Science Series* vol. 48 (ed. N. Kosaric), Dekker, New York, Basel, Hong Kong, pp. 251–268.

86 Banat, I.M. (1995) *Bioresource Technology*, **51**, 1–12.

87 Banat, I.M. (1995) *Acta Biotechnologica*, **15**, 251–267.

88 Wagner, F.E. and Lang, S. (1996) Proceedings of the 4th World Surfactants Congress Barcelona, vol. 1, AEPSAT, Barcelona, pp. 125–137.

89 Rosenberg, E. and Ron, E.Z. (1997) *Current Opinions in Biotechnology*, **8**, 313–316.

90 Rosenberg, E. and Ron, E.Z. (1998) in *Biopolymers from Renewable Resources* (ed. D.L. Kaplan), Springer, Berlin, Heidelberg, New York, pp. 281–291.

91 Rosenberg, E. and Ron, E.Z. (1999) *Applied Microbiology and Biotechnology*, **52**, 154–162.

92 Neu, T. (1996) *Microbial Review*, **60**, 151–166.

93 Bergstrom, S., Theorell, H. and Davide, H. (1946) *Arkiv Kem Mineral und Geology*, **23**, 1–12.

94 Jarvis, F.G. and Johnson, M.J. (1949) *Journal American Chemistry Society*, **71**, 4124–4126.

95 Edwards, J.R. and Hayashi, J.A. (1965) *Archives of Biochemistry and Biophysics*, **111**, 415–421.

96 Hisatsuka, K., Nakahara, T., Sano, N. and Yamada, K. (1971) *Agricultural and Biological Chemistry*, **35**, 686–692.

97 Itoh, S., Honda, H., Tomita, F. and Suzuki, T. (1971) *Journal of Antibiotics*, **24**, 855–859.

98 Yamaguchi, M., Sato, A. and Yakuyama, A. (1976) *Chemistry & Industry*, **17**, 741–742.

99 Hirayama, T. and Kato, A. (1982) *FEBS Letters*, **139**, 81–85.

100 Hirayama, T. and Kato, I. (1982) *Agricultural and Biological Chemistry*, **35**, 686–692.

101 Syldatk, C., Lang, S. and Wagner, F. (1985) *Zeitschrift für Naturforschung*, **40**, 51–60.

102 Syldatk, C., Lang, S., Matulovic, U. and Wagner, F. (1985) *Zeitschrift für Naturforschung*, **40**, 61–67.

103 Lang, S. and Wullbrandt, D. (1999) *Applied Microbiology and Biotechnology*, **51**, 22–32.

104 Rendell, N.B., Taylor, G.W., Somerville, M., Todd, H., Wilson, R. and Cole, J. (1045) *Biochemica et Biophysica Acta*, **1990**, 189–193.

105 Rao, Ch.V.S. and Johri, B.N. (1999) *Indian Journal of Microbiology*, **39**, 29–36.

106 Rao, Ch.V.S., Sachan, I.P. and Johri, B.N. (1999) *Indian Journal of Microbiology*, **39**, 23–28.

107 Sharma, A., Jansen, R., Johri, B.N. and Wray, V. (2007) *Journal of Natural Product,* **70**, 941–947.

108 Hauser, G. and Karnovsky, M.L. (1957) *Journal of Biological Chemistry,* **224**, 91–105.

109 Hauser, G. and Karnovsky, M.L. (1958) *Journal of Biological Chemistry,* **233**, 287–291.

110 Ochsner, U.A. and Reiser, J. (1995) *Proceedings of the National Academy of Sciences of the United States of America,* **92**, 6424–6428.

111 Burger, M.M., Glaser, L. and Burton, R.M. (1966) *Methods in Enzymology,* **8**, 441–445.

112 Ochsner, U.A., Hembach, T. and Fiechter, A. (1996) *Advances in Biochemical Engineering/Biotechnology,* **53**, 89–118.

113 Ochsner, U.A., Fiechter, A. and Reiser, J. (1994) *Journal of Biological Chemistry,* **1994**, 269, 19787–19795.

114 Ochsner, U.A., Reiser, J., Fiechter, A. and Witholt, B. (1995) *Applied and Environmental Microbiology,* **61**, 3503–3506.

115 Ochsner, U.A., Fiechter, A. and Reiser, J. (1994) *Journal of Bacteriology,* **176**, 2044–2054.

116 Latifi, A., Foglino, M., Tanaka, K., Williams, P. and Lazdunski, A. (1996) *Molecular Microbiology,* **21**, 1137–1146.

117 Pearson, J.P., Pesci, E.C. and Iglewski, B.H. (1997) *Journal of Bacteriology,* **179**, 5756–5767.

118 Pearson, J.P., Passador, L., Iglewski, B.H. and Greenberg, P. (1995) *Proceedings of the National Academy of Sciences of the United States of America,* **92**, 1490–1494.

119 Campos-Garcia, J., Caro, A.D., Najera, R., Miller-Maier, R.M., Al-Tahhan, R.A. and Soberon-Chavez, G. (1998) *Journal of Bacteriology,* **180**, 4442–4451.

120 Rahim, R., Olvera, C., Graninger, M., Messner, P., Ochsnerm, U.A., Lam, J.S. and Soberon-Chavez, G. (2001) *Molecular Microbiology,* **40**, 708–718.

121 Sharma, A. and Johri, B.N. (2002) Proceedings of National Symposium on Developments in Microbial Biochemistry and Its Impact on Biotechnology (eds B.S. Rao, P.M. Mohan and C. Subramanyam), Osmania University, Hyderabad, India, pp. 166–184.

122 Pearson, J.P., Gray, K.M., Passador, L., Tucker, K.D., Eberhard, A., Iglewski, B.H. and Greenberg, E.P. (1994) *Proceedings of the National Academy of Sciences of the United States of America,* **91**, 187–201.

123 Pesci, E.C. and Iglewski, B.H. (1997) *Trends in Microbiology,* **5**, 132–134.

124 Olvera, C., Goldberg, J.B., Sanchez, R. and Soberon-Chavez, G. (1999) *FEMS Microbiology Letters,* **71**, 85–90.

125 Guerra-Santos, L.H., Kappeli, O. and Fiechter, A. (1984) *Applied and Environmental Microbiology,* **48**, 301–305.

126 Ramana, K.V. and Karanth, N.G. (1989) *Biotechnology Letters,* **11**, 437–442.

127 Guerra-Santos, L.H., Kappeli, O. and Fiechter, A. (1986) *Applied Microbiology and Biotechnology,* **24**, 443–448.

128 Mulligan, C.N. and Gibbs, B.F. (1989) *Applied and Environmental Microbiology,* **55**, 3016–3019.

129 Mulligan, C.N., Mahmourides, G. and Gibbs, B.F. (1989) *Journal of Biotechnology,* **12**, 199–210.

130 Sullivan, E.R. (1998) *Current Opinion in Biotechnology,* **9**, 263–269.

131 Tomlinson, J.A. and Faithfull, E.M. (1980) *Annual Applied Biology,* **93**, 13–19.

132 Tomlinson, J.A. and Faithfull, E.M. (1980) *Acta Horticulturae,* **98**, 325–331.

133 Stanghellini, M.E., Rasmussen, S.L., Kim, D.M. and Rorabaugh, P.A. (1996) *Plant Disease,* **80**, 422–428.

134 Stanghellini, M.E. and Miller, R.M. (1997) *Plant Disease,* **81**, 4–12.

135 Tedla, T. and Stanghellini, M.E. (1992) *Phytopathology,* **82**, 652–656.

136 Gambello, M.J. and Iglewski, B.H. (1991) *Journal of Bacteriology,* **173**, 3000–3009.

137 Gambello, M.J., Kaye, S. and Iglewski, B.H. (1993) *Infection and Immunity,* **61**, 1180–1184.

138 Reimman, C., Maurhofer, M., Schmidli, P., Gaille, C. and Haas, D. (1997) *Journal of the Facility of Agriculture of Hokkaido University,* **72**, 248–250.

139 Albus, A.M., Pesci, E.C., Runyen-Janecky, L.J., West, S.E.A. and Iglewski, B.H. (1997) *Journal of Bacteriology*, **179**, 3928–3935.

140 De Kievit, T.R., Gillis, R., Marx, S., Brown, C. and Iglewski, B.H. (2001) *Applied and Environmental Microbiology*, **67**, 1865–1873.

141 Ochea-Loza, F.J., Artiola, J.F. and Maier, R.M. (2001) *Journal of Environmental Quality*, **30**, 479–485.

142 Whiteley, M., Lee, K.M. and Greenberg, E.P. (2000) *Proceedings of the National Academy of Sciences of the United States of America*, **96**, 13904–13909.

143 Lithgow, J.K., Wilkinson, A., Hardman, A., Rodelas, B., Wisniewski-Dye, F., Williams, P. and Downie, J.A. (2000) *Molecular Microbiology*, **37**, 81–97.

144 Wilkinson, A., Danino, V., Wisniewski-Dye, F., Lithgow, J.K. and Downie, J.A. (2002) *Journal of Bacteriology*, **184**, 4510–4519.

145 Sharma, A., Sahgal, M. and Johri, B.N. (2003) *Current Science*, **85**, 1164–1172.

146 Sharma, A., Pathak, A., Sahgal, M., Meyer, J.-M., Wray, V. and Johri, B.N. (2007) *Archives of Microbiology*, **188**, 483–494.

12

Practical Applications of Rhizospheric Bacteria in Biodegradation of Polymers from Plastic Wastes

Ravindra Soni, Sarita Kumari, Mohd G.H. Zaidi, Yogesh S. Shouche, and Reeta Goel

12.1
Introduction

Recalcitrant plastics accumulate in the environment at the rate of about 25 million tons per year and continue to do so in the soil without much change in their structure over long periods of time [8]. In spite of their presence in the soil, microbes cannot utilize plastics as their nutrient source for two reasons: the absence of free functional groups and the complex nature of the polymeric chains in plastics. However, the long-term persistence of plastic waste in soil affects adversely not only the normal microflora of the soil but also the structure of the soil, thus rendering it unsuitable for agricultural use.

Microorganisms are components of natural ecosystems and have evolved with them over millions of years. The diversity of microorganisms is, thus, a representation of their diverse nutrient requirements and of varying capabilities to metabolize a compound. However, biodegradation of environmental pollutants appears to be an eco-friendly alternative to inefficient and costly physiochemical decontamination strategies. Low-density polyethylene (LDPE) is used to make films and packaging materials. However, polyesters and polyamides are commonly used as plasticizers for improving the flexibility and toughness of plastics [4,5].

Therefore, the vast diversity of microorganisms needs to be explored to identify potential strains for LDPE biodegradation. In this study, a total of 12 bacterial strains, adapted under the natural environment for polymer degradation, were selected. These isolates were then checked for their inherent tolerance to LDPE, LDPE-g-polymethyl methacrylate (LDPE-g-PMMA) and LDPE-g-polymethacrylic hydrazide (LDPE-g-PMH). Three isolates showing high tolerance to the polymers were used for biodegradation under *in vitro* conditions. Change in λ_{max} of the polymers within the medium was observed during 11 days of degradation studies, after which degraded product was recovered. Furthermore, comparative FTIR (Fourier transform infrared) spectra and thermal gravimetric analyses of the degraded and un-degraded products have affirmed these results.

Plant-Bacteria Interactions. Strategies and Techniques to Promote Plant Growth
Edited by Iqbal Ahmad, John Pichtel, and Shamsul Hayat
Copyright © 2008 WILEY-VCH Verlag GmbH & Co. KGaA, Weinheim
ISBN: 978-3-527-31901-5

12.2
Materials and Methods

12.2.1
Chemicals and Media

LDPE beads were purchased from Aldrich Chemical Company, USA. Polyester and polyamide graft copolymers were prepared by the Department of Chemistry, G.B. Pant University of Agriculture and Technology, Pantnagar, India. All other chemicals and solvents used in the study were of analytical grade.

12.2.2
LDPE-g-PMMA

MMA was purified through repeated washing with sodium hydroxide (10%) followed by distillation under reduced pressure (bp 98 °C/10 mm). Further, a mixture of LDPE (1.0 g), MMA (1.10 g) and benzoyl peroxide (0.05 g) was refluxed in toluene (20 ml) over 2 h. The contents were cooled and purified via extraction against benzene for a further 1.0 h. The LDPE-g-PMMA was recovered by filtration.

12.2.3
LDPE-g-PMH

In each case, 20 g LDPE-g-PMMA was taken in 70 ml of hydrazine hydrate (98%) and refluxing was done for 5 h at 70 °C, after which LDPE-g-PMH was recovered by distilling the excess hydrazine hydrate at room temperature.

12.2.4
Isolation of Bacteria

Two soil beds were prepared in which small pieces of polyethylene were added and left for 90 days. Water was intermittently sprinkled to provide moisture. In one of the soil beds, glucose (0.5%) and maleic anhydride (0.3%) were added. One gram of soil sample was collected from these four locations and bacteria were isolated on Davis minimal medium (dextrose $1 \, g \, l^{-1}$, dipotassium phosphate $7 \, g \, l^{-1}$, monopotassium phosphate $2 \, g \, l^{-1}$, sodium citrate $0.5 \, g \, l^{-1}$, $MgSO_4$ $0.1 \, g \, l^{-1}$, ammonium sulfate $1 \, g \, l^{-1}$, pH 7.0 ± 0.2) and pseudomonas agar medium (pancreatic digest of gelatin $16 \, g \, l^{-1}$, casein enzymatic hydrolysate $10 \, g \, l^{-1}$, K_2SO_4 $10 \, g \, l^{-1}$, $Mgcl_2 \cdot 6H_2O$ $1.4 \, g \, l^{-1}$, glycerol 10 ml, agar powder $20 \, g \, l^{-1}$, pH 7.0 ± 0.2) following the serial dilution method. Scientists at the National Chemical Laboratory, Pune, have reported the accelerated biodegradation of plastics by as much as 30% when sugar in the form of glucose and maleic anhydride [2] is added. Adaptation of microorganisms can play a major role in determining biodegradation rates [7]. Thus, the bacteria isolated from such soil beds are expected to possess degradative activity for polymers present in the soil bed. Moreover, the polymers were left for 3 months for photooxidation. Light is known to impact biodegradation as it causes photooxidation and thus exposes free functional groups for bacterial degradation.

12.2.5
Screening of Bacterial Isolates to Grow in the Presence of Polymer

Bacterial isolates were screened for their ability to grow in the presence of polymers, namely LDPE-g-PMMA, LDPE-g-PMH and LDPE. An aliquot of $20\,\mu l$ from overnight-grown active culture (OD = 0.40) was inoculated into a 96-well cell culture plate (Tarson, India) containing $200\,\mu l$ Davis minimal broth per well. The polymer was added in minimal broth at increasing concentrations from 0 to $10\,mg\,ml^{-1}$. The cell culture plate was then incubated at $37\,°C$ at 120 rpm. Absorbance was recorded for all the treatments at 600 nm. The experiment was performed in triplicate. Based on tolerance-level studies, three cultures, namely PN15, PN13 and S2, were sent to NCCS, Pune, India, for characterization. Optimum temperature and pH of these cultures were characterized before further use.

12.2.6
Optimization of Growth Conditions

A number of factors such as temperature, pH, microbial biomass and preexposure can affect the degradation rate; thus, it is important that rate-determining factors be understood before initiating biodegradation studies [9]. Therefore, growth conditions of all selected cultures were optimized. For the three cultures, that is, *Bacillus pumilus*, *Bacillus cereus* and other *Bacillus* species, the optimum temperature is $37 \pm 0.2\,°C$. Moreover, these cultures were neutrophils with an optimum pH of 7.0 ± 0.2.

12.2.7
Biodegradation Studies

Active cultures ($100\,\mu l$) of PN15, PN13 and S2 isolates were inoculated in 100 ml minimal broth pairs (1/10 diluted) containing $5\,mg\,ml^{-1}$ polymer. Samples were withdrawn on days 0, 2, 3, 4, 7 and 11 to determine λ_{max} (baseline was collected with MBD1/10th strength). In the case of LDPE, a consortium ($30\,\mu l$ active culture each of PN13, PN15 and S2) was inoculated in 100 ml Davis minimal broth. After 11 days when the cultures reached stationary phase, the broth was filtered and the degraded product was kept in a hot air oven at $65\,°C$ overnight. Compounds recovered upon biodegradation were sent for FTIR spectroscopy to SAIF, CDRI, Lucknow.

12.3
Results and Discussion

Plastics contain additives called plasticizers that impart characteristic properties to them, namely flexibility, thermostability, resistance to corrosion and so on [6]. These plasticizers form the bulk of the plastic material; therefore, more attention is being given to their degradation. Once the plasticizers have modified bonds/linkages (ester and/or amide), the microorganism possessing respective enzymes can easily act upon them [1]. In view of the above, this study was conducted, wherein graft copolymers of LDPE are used for *in vitro* degradation studies.

12.3.1
Growth in the Presence of Polymer

Among the 12 isolates screened for tolerance to LDPE, LDPE-g-PMMA and LDPE-g-PMH, three (namely S2, PN15 and PN13) had shown maximum growth and were thus selected for further studies. Moreover, based on 16S rRNA sequencing, these isolates have been characterized as *B. pumilus*, *B. cereus* and other *Bacillus* species, respectively.

Another important observation was that the bacteria isolated from the same site had varying levels of tolerance to the polymers mentioned above. This observation strongly supports the exploration of bacterial diversity to identify novel microbes that can help biodegrade the plastic waste accumulated in the environment. It is evident from these data that PN15 has shown maximum tolerance to LDPE-g-PMMA, S2 for LDPE-g-PMH and PN13 for both.

12.3.2
Biodegradation Studies

12.3.2.1 *B. cereus*
B. cereus had maximum tolerance ($10\,\text{mg\,ml}^{-1}$) to the polymers studied. The UV–visible spectrum of 1/10th-diluted Davis minimal broth in the presence of *B. cereus* and LDPE-g-PMMA was compared with that of the medium in the absence of the compound. After 2 days, the λ_{max} shifted from 289 to 298 nm in the presence of the bacteria, while it remained unaffected when bacteria were grown in its absence (Figure 12.1). Furthermore, in the presence of the bacteria, λ_{max} changed to 215 nm on day 4 and finally fell below 200 nm on day 11 of the experiment. As Figure 12.1 makes it evident, the cultures had reached the stationary growth phase between days 7 and 11 and hence no further biological activity was expected in the culture.

12.3.2.2 *Bacillus* sp.
When this strain was used for biodegradation of LDPE-g-PMMA, it was observed that for the first 2 days virtually no change occurred in λ_{max} and it remained constant at 289 nm. However, on day 3 λ_{max} shifted to 257 nm and remained constant until day 11 except for a minor change on day 4 – 263 nm (Figure 12.2).

On days 1 and 2 of *Bacillus* sp. growth on LDPE-g-PMH, there was no change in λ_{max} of the compound, whereas on day 3 it changed to 221 nm and further to 224 nm on day 4 and finally stabilized at 227 nm from days 7 to 11. This culture entered stationary growth phase between days 7 and 11, which corresponds to the duration when no shift in λ_{max} took place (Figure 12.3).

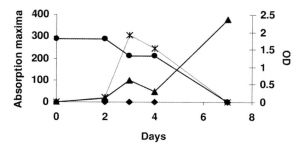

—●— Davis minimal broth + *B. cereus* + LDPE-g -
(PMMA)(λ_{max})

—▲— Davis minimal broth + *B. cereus* (OD)

—✳— Davis minimal broth + *B. cereus* + LDPE-g -
(OD)

Figure 12.1 Growth and biodegradation studies: *Bacillus* species grown in Davis minimal broth in the presence of LDPE-g-PMMA and LDPE-g-PMH separately.

12.3.2.3 *B. pumilus*

The UV–visible spectrum of 1/10th-diluted Davis minimal broth in the presence of *B. pumilus* was compared with that of 1/10th Davis minimal broth in the presence of LDPE-g-PMH and *B. pumilus*. It is evident by comparative analysis that on day 2 λ_{max} of polymer in the presence of *B. pumilus* had changed from 292 to 298 nm, which then decreased sharply to 206 nm on day 3. Further, changes took place in λ_{max} of LDPE-g-PMH and finally on day 11 of the experiment when the culture attained stationary phase, λ_{max} had shifted to 209 nm. However, λ_{max} remained unchanged

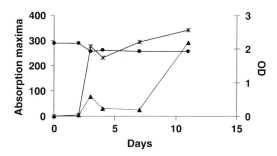

—●— Davis minimal broth + *Bacillus* sp. + LDPE-g-PMMA (λ_{max})

—▲— Davis minimal broth + *Bacillus* sp. (OD)

—✳— Davis minimal broth + *Bacillus* sp. + LDPE-g-PMMA (OD)

Figure 12.2 Biodegradation and growth studies: *B. species* grown in Davis minimal broth in the presence of LDPE-g-PMMA.

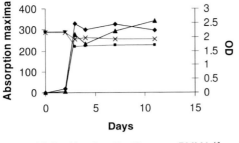

—■— Minimal broth + *Bacillus* sp. + PMMA (λ_{max})

—✕— Minimal broth + *Bacillus* sp. + PMH (λ_{max})

—◆— Minimal broth + *Bacillus* sp. + PMMA (OD)

—▲— Minimal broth + *Bacillus* sp. + PMH (OD)

Figure 12.3 Biodegradation and growth studies: *B. cereus* grown in Davis minimal broth in the presence of LDPE-g-PMMA.

for both positive and negative controls on day 2. Although λ_{max} had stabilized at 203 nm for the negative control from day 3 onward, it decreased to 206 nm and reached 203 nm on day 7 and then rose to 212 nm on day 11 in the case of the positive control (Figure 12.4).

12.3.2.4 Bacterial Consortium and LDPE

Degradation of LDPE caused by the consortium of *B. pumilus*, *B. cereus* and other *Bacillus* species was studied in 100 times diluted Davis minimal broth. Here, a

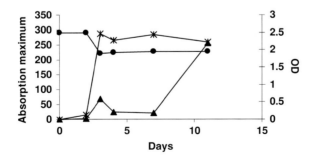

—●— Davis minimal broth + *B. pumilus* + LDPE-g-PMH (λ_{max})

—▲— Davis minimal broth + *B. pumilus* (OD)

—✳— Davis minimal broth + *B. pumilus* + LDPE-g-PMH (OD)

Figure 12.4 Biodegradation studies: *B. pumilus* grown in Davis minimal broth in the presence of LDPE-g-PMH.

greater dilution of the medium was prepared to allow very slow growth of bacteria for better adaptation. On day 1, λ_{max} of the compound was below 200 nm and thus could not be recorded. However, the consortium was allowed to grow for a period of 11 days in the presence of LDPE as observed OD suggested a stationary phase 7th day onward. For further confirmation of biodegradation of the polymers, FTIR spectra and TGA of the degraded and undegraded compounds were carried out.

12.3.2.5 FTIR Spectroscopy

The FTIR spectrum of undegraded LDPE-g-PMMA was compared with that of the polymer degraded by *B. cereus*. Formation of a graft copolymer of polymethyl methacrylate with LDPE was confirmed as an additional absorption appeared corresponding to C=O stretching (1733.1 cm^{-1}) in the undegraded sample [3]. Comparison of FTIR spectrum of the undegraded polymer with that of its degraded counterpart (acted upon by *B. cereus*) indicates a strong shift in the fingerprint region. Furthermore, an overall decrease in percent transmittance of the degraded compound is also observed from about 85% in the undegraded product to about 70% in the degraded product (Figure 12.5).

The FTIR spectrum of undegraded LDPE-g-PMH was compared with that of the polyamide degraded by *B. pumilus*. The formation of an amide linkage in the graft was confirmed due to the absorption corresponding to CO–NH stretching combined with NH bending (1595 cm^{-1}). Comparison of the FTIR spectrum of the undegraded polymer with that of the *B. pumilus* treated sample indicates a shift in the fingerprint region of *B. pumilus* degraded polymer at higher wavenumber without the appearance of any additional absorption band. On the basis of these results, a consortium of *B. pumilus*, *B. cereus* and other *Bacillus* species was used for further studies, wherein one (PN15) was found to degrade LDPE-g-PMMA, another (S2) LDPE-g-PMH and the third (PN13) degraded both (Figure 12.6).

The FTIR spectrum of LDPE showed characteristic absorption bands corresponding to CH$_2$ rocking (720.2 cm^{-1}), CH$_3$ bending (1362.6 cm^{1}), CH$_2$ bending (1465.2 cm^{-1}), CH$_3$ stretching (symmetrical, 2850.6 cm^{-1}), CH$_3$ stretching (asymmetrical, 2919.9 cm^{-1}) and CH stretching (3426.6 cm^{-1}). Comparison of the FTIR

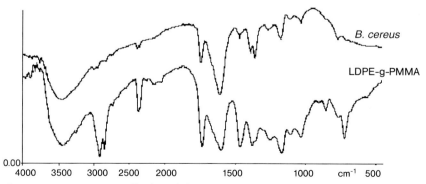

Figure 12.5 FTIR spectrum of biodegraded LDPE-g-PMMA using *B. cereus* against reference undegraded LDPE-g-PMMA.

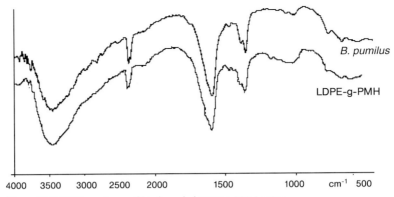

Figure 12.6 FTIR spectrum of biodegraded LDPE-g-PMH using *B. pumilus* against reference undegraded LDPE-g-PMH.

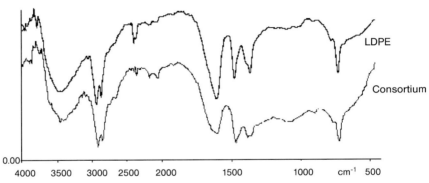

Figure 12.7 FTIR spectrum of biodegraded LDPE using *B. species* consortium against reference undegraded LDPE.

spectrum of undegraded LDPE with that of LDPE degraded by the consortium indicates a strong shift in the fingerprint region in the spectrum of LDPE acted upon by the consortium with a new additional band corresponding to C—O stretching (1056.0 cm^{-1}) (Figure 12.7). Absorption bands corresponding to CH$_3$ bending and CH$_3$ stretching (symmetrical) were shifted to a higher wavenumber, whereas others were shifted to lower wavenumbers (Figure 12.8).

12.4
Conclusions

The selection of a strain or a group of strains (consortium) with a potential to degrade plastics and their polymers is the need of the hour. Therefore, sustained efforts hold promise in identifying practical solutions for plastic biodegradation and pollution management.

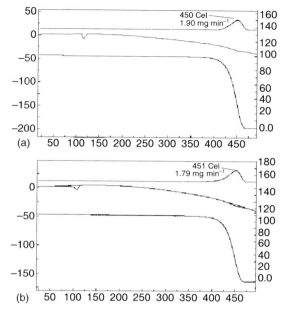

Figure 12.8 (a) DTA–DTG–DG analysis of degraded LDPE.
(b) DTA–DTG–DG analysis of undegraded LDPE.

Acknowledgments

This work was supported by a DBT grant to RG. One author (SK) also acknowledges ICAR for Junior Research Fellowship during the course of this study. We also thank CDRI (SAIF), Lucknow and NCCS, Pune, for FTIR and 16S rRNA sequencing, respectively.

References

1 Diamond, M.J., Freedman, B. and Faribaldi, J. (1975) *International Biodeterioration Bulletin*, **11**, 127–132.

2 Dua, N. (2004) Honey who shrunk the plastic, in *Down to Earth*, CSE Publishers, India, p. 22.

3 Joshi, S.K., Kapil, J.C., Rai, A.K. and Zaidi, M.G.H. (2003) *Physica Status Solidi*, **199**, 321–328.

4 Kim, D.Y. and Rhee, Y.H. (2003) *Applied Microbiology and Biotechnology*, **61** (4), 300–308.

5 Nakamura, E.M., Cordi, L., Almedia, G.S.G., Duran, N. and Mei, L.H.I. (2005)

Journal of Materials Processing Technology, **27**, 489–492.

6 Pathiana, R.A. and Seal, K.J. (1984) *International Biodeterioration Bulletin*, **20**, 229–235.

7 Piskonen, R., Nyyssonen, M., Rajmaki, T. and Itavaara, M. (2005) *Biodegradation*, **16** (2), 127–134.

8 Shivram, S. (2001) International Symposium on Biodegradable Polymers, 17 November, IICT, Hyderabad.

9 Spain, J.C. and van Veld, P.A. (1983) *Applied and Environmental Microbiology*, **45**, 428–435.

13

Microbial Dynamics in the Mycorrhizosphere with Special Reference to Arbuscular Mycorrhizae

Abdul G. Khan

13.1
The Soil and the Rhizosphere

Many soil microbes have their origin in the soil or are closely associated with the soil environment and have a substantial impact on humans [1]. We know very little about the enormous diversity of soil microbes, their properties and behavior in the soil environment. Soil microorganisms inhabiting the rhizosphere interact with plant roots and mediate nutrient availability, forming useful symbiotic associations with roots and contributing to plant nutrition. Implications of plants and their symbionts such as mycorrhizal fungi, nitrogen-fixing rhizobia and free-living rhizospheric bacterial populations that promote plant growth need to be fully exploited and encouraged by inoculating nutrient-poor agricultural soils with appropriate microbes [2].

13.2
Rhizosphere and Microorganisms

13.2.1
Glomalian Fungi

Glomales are one of the oldest groups of fungi, older than land plants. The first land plants, bryophytes, appeared in the Mid-Silurian era (476–430 million years B.P.). The oldest fossil evidence of bryophyte-like land plants, 100 million years ago in the Early Devonian, had AM-like infections even before their roots evolved [3]. The first land plants most likely evolved from algae but no fossil records are available to show if the rootless fresh water Charophycean algae, the probable ancestors of land plants, were mycorrhizal. Mosses, liverworts and hornworts often contain structures such as hyphae, vesicles and arbuscules – all characteristics of AMF [4]. Sphenophytes, lycopodophytes and pteridophytes are among the first land plants with roots, which

Plant-Bacteria Interactions. Strategies and Techniques to Promote Plant Growth
Edited by Iqbal Ahmad, John Pichtel, and Shamsul Hayat
Copyright © 2008 WILEY-VCH Verlag GmbH & Co. KGaA, Weinheim
ISBN: 978-3-527-31901-5

originated in the mid-Devonian Era, and AM associations are reported in these plants [5]. Both living and Triassic fossil Cycades contained AMF in their roots. AM associations are ubiquitous in living angiosperms, which probably arose in the early Cretaceous era [6]. The phylogenetic relationship between origin and diversification of AMF and coincidence with vascular land plants was investigated by Simon *et al.* [7] by sequencing ribosomal DNA genes (SS sequences) as a molecular clock to infer dates, from 12 Glomalean fungal species. The authors estimated that AM-like fungi originated some 354–462 million years ago, which is consistent with the hypothesis that AMF were instrumental in the colonization of land by ancient plants. This hypothesis is also supported by the observation that AM can now be found world-wide in the angiosperms and gymnosperms as well as ferns, suggesting an ancestral nature of the association.

Universal and ubiquitous fungi belonging to Glomales form symbiotic relation-ships with roots of 80–90% of land plants in natural and agricultural ecosystems [5] including halophytes, hydrophytes and xerophytes [8–10] – and, owing to greater exploration of soil for nutrients, are known to benefit plant nutrition, growth and survival [11]. These associations represent a key factor in the below-ground networks that influence diversity and plant community structure [12–14], but we know very little about the enormous AM fungal diversity in soils and their properties and behavior in the soil [2]. However, not all plants have mycorrhizal associations and not all AMF benefit host plants under all growth conditions [15]. The degree of benefit to each partner in any AMF–plant host interaction depends not only on the particular plant and AMF species involved but also on the rhizobacteria and soil abiotic factors.

Availability of modern techniques such as root organ based methodologies and rDNA techniques has made it possible to study AMF–host associations in some detail to understand the comprehensive interaction scenario of plant, rhizosphere and soil components. Much recent research showed that these symbiotic associa-tions have extreme genetic complexity and functioning diversity [16]. Based on SSU rRNA molecular phylogenetic studies, AMF have been elevated to the phylum Glomeromycota [17], a conclusion supported by the recent study [18].

Arbuscular mycorrhizal associations are important in natural and managed eco-systems because of their nutritional and nonnutritional benefits to their symbiotic partners. They can alter plant productivity, because AMF can act as biofertilizers, bioprotectants or biodegraders [19]. AMF are known to improve plant growth and health by improving their mineral intake or increasing resistance or tolerance to biotic and abiotic stresses [20,21]. Their potential role in phytoremediation of heavy metal contaminated soils and water is also becoming evident [22–26].

AMF modify the quality and abundance of rhizosphere microflora and alter overall rhizosphere microbial activity. Following host root colonization, the AMF induce changes in the host root exudation pattern, which alters the microbial equi-librium in the mycorrhizosphere [27]. These interactions can be beneficial or harm-ful to the partner microbes involved and to the plant, and sometimes may enhance plant growth, health and productivity [28,29]. Recently, Giovannetti and Avio [30] have reviewed and analyzed important data on the main parameters affecting AM fungal infectivity, efficiency and ability to survive, multiply and spread, which may help in utilizing obligate biotrophic AMF in biotechnological exploitation and sustainable agriculture. There is a need to understand and better exploit AM symbionts in different global ecosystems.

Although AMF are ubiquitous, it is probable that natural AM associations are not efficient in increasing plant growth [31]. Cropping sequences as well as fertilization and plant-pathogen management practices also dramatically affect the AMF propagules in the soil and their affects on plants [32]. The propagation system used for horticultural fruit and micropropagated plants can benefit most from AM biotechnology. Micropropagated plants can withstand transplant stress from *in vitro* to *in vivo* systems, if they are inoculated with appropriate AMF [33,34]. To use AMF in sustainable agriculture, knowledge of factors such as fertlizer inputs, pesticide use, soil management practices and so on influencing AMF communities is essential [32,35,36]. This area deserves further research because a sound scientific knowledge is necessary for improving the AM biotechnology aimed at selecting infective and efficient inoculants to be used as biofertilizers, bioprotectants and biostimulants to make agriculture, horticulture and forestry sustainable.

Although the potential of AMF in enhancing plant growth is well recognized, it is not fully exploited. AMF are rarely found in nurseries due to the use of composted soil-less media, high levels of fertilizer and regular application of fungicide drenches. The potential advantages of inoculation of plants with AMF in horticulture, agriculture and forestry are not perceived as significant by these industries, partly due to inadequate methods for large-scale inoculum production. Monoxenic root organ *in vitro* culture methods for AMF inocula production have also been attempted by various researchers under field conditions [37,38]; however, these techniques, although useful in studying various physiological, biochemical and genetic relationships, have limitations in producing inocula of AMF for commercial use. Pot culture in pasteurized soil has been the most widely used method for producing AMF inocula, but this method is time consuming and often not pathogen free. To overcome these problems, soil-free methods such as soil-less growth media, aeroponics, hydroponics and axenic cultures of AMF have been used successfully to produce AMF-colonized root inocula [37,39–41]. Substrate-free colonized roots produced by these methods can be sheared and used for large-scale inoculation purposes. Mohammad *et al.* [42] compared growth responses of wheat to sheared root and pot culture inocula of AMF at different phosphorus levels under field conditions and concluded that phosphorus fertilization can be substituted by AMF inoculum produced aeroponically to an extent of $5 \, kg \, ha^{-1}$.

13.2.2
Arbuscular Mycorrhiza–Rhizobacteria Interactions

Increased microbial activity in rhizosphere soil affects plant health and growth. A range of stimulated rhizosphere microorganisms such as saprophytes, pathogens, parasites, symbionts and so on carry out many activities important to plant health and growth. Some of these microbes affect root morphology and physiology by producing plant growth regulating hormones and enzymes. Others alter the plant nutrient availability and biochemical reactions undertaken by the plant.

AMF impart differential effects on the bacterial community structure in the mycorrhizosphere [29,43]. AMF improve phosphorus nutrition by scavenging available phosphorus through the large surface area of their hyphae. Plant growth promoting rhizobacteria (PGPR) may also improve plant phosphorus acquisition by solubilizing organic and inorganic phosphorus sources through phosphatase synthesis or by lowering soil pH [44]. Garbaye [45] defined mycorrhizal helper bacteria (MHB) as 'bacteria associated with mycorrhizal roots and mycorrhizal fungi which collectively promote the establishment of mycorrhizal symbioses'.

There is growing evidence that diverse microbial populations in the rhizosphere play a significant role in sustainability issues [46,47] and that the manipulation of AMF and certain rhizobacteria such as PGPR and MHB is important. Vivas *et al.* [48] used a dual AM fungus–bacterium inoculum to study the effect of the drought stress induced in lettuce grown in controlled-environment chambers. Their results showed that there was a specific microbe–microbe interaction that modulates the effectivity of AMF on plant physiology. The authors concluded that plants must have mycorrhizal associations in nutrient-poor soils and that mycorrhizal effects can be improved by coinoculation with MHB such as *Bacillus* spp. Results of the study by Vivas *et al.* show that coinoculation of selected free-living bacteria isolated from adverse environments and AMF can improve the formation and function of AM symbiosis, particularly when plant growth conditions are also adverse. Both AMF and PGPR complement each other in their role in nitrogen fixation, phytohormone production, phosphorus solubilization and increasing surface absorption. Behl *et al.* [49] studied the effects of wheat genotype and *Azobacter* survival on AMF and found that the genotype tolerant to abiotic stresses had higher AMF infection and noticed a cumulative effect of plant–AMF–PGPR interaction. Similar observations were made by Chaudhry and Khan [50,51] who studied the role of symbiotic AMF and PGPR nitrogen-fixing bacterial symbionts in sustainable plant growth on nutrient-poor heavy metal contaminated industrial sites and found that the plants surviving on such sites were associated with nitrogen-fixing rhizobacteria and had a higher arbuscular mycorrhizal infection, that is, a cumulative and synergistic effect.

The MHB cannot be ignored when studying mycorrhizal symbioses in their natural ecosystems. They are quite common and, as Garbaye [45] said, they are found every time they are sought and seem to be closely associated with the mycorrhizal fungi in the symbiotic organs. They are adapted to live in the vicinity of AMF as high frequencies of MHB populations have been isolated from the mycorrhizae. Some MHB isolates also promoted ectomycorrhizae formation in four conifers [52], indicating that the MHB effect is not plant specific. But various researchers have shown that MHBs are fungus selective [45]. Mosse [53] showed that cell wall degrading enzyme producing *Pseudomonas* sp. enhanced the germination of AM fungal spores of *Glomus mosseae* and promoted the establishment of AM on clover roots under aseptic conditions. These observations were later supported by other workers [54,55].

The enriched soil microbial communities in the mycorrhizosphere are often organized in the form of biofilms and probably horizontal gene transfer (HGT) among cohabiting microbial species and between plant and microbe occurs [56]. Many plant-associated *Pseudomonas* rhizobacteria produce signal molecules for quorum-sensing regulation, which were absent from soilborne strains [57]. This indicates that quorum-sensing systems exist and are required in the mycorrhizosphere. Microbial colonies on root surfaces consist of many populations or strains, and positive and negative interpopulation signaling on the plant root occur [58], which may play an important role in the efficiency of the use of biofertilizers.

In addition to the above-described interactions between AMF and rhizobacteria, certain bacterialike organisms (BLOs) reside in the AM fungal cytoplasm, first described by Mosse [53]. Khan [59] illustrated AMF spores, collected from semi-arid areas of Pakistan, containing 1–10 small spherical 'endospores' without any subtending hyphae of their own. Ultrastructural observations clearly revealed their presence in many field-collected AM fungal isolates. Because of their unculturable nature, further investigation of BLOs was hampered but current advanced electron and confocal microscopic and molecular analysis techniques have allowed us to learn more about their endosymbiotic nature. Minerdi *et al.* [60] reported the presence of intracellular and endosymbiotic bacteria belonging to the genus *Burkholderia* (a nitrogen-fixing genus) in fungal hyphae of many species of Gigasporaceae. The authors used genetic approaches to investigate the presence of nitrogen-fixing genes and their expression in this endosymbiont and found *nif*HDK genes in the endosymbiont *Burkholderia* and their RNA messengers demonstrating that they possess a molecular basis for nitrogen fixation. This discovery, as stated by Minerdi *et al.* [60], indicates that a fungus that improves phosphorus uptake might also fix nitrogen through specialized endobacteria. Endosymbiont *Burkholderia* may have an impact on AMF–PGPR–MHB–plant associations and metabolism. This finding suggests a new application scenario worth pursuing.

Recent methodological developments in molecular and microscopical techniques together with those in genomes, bioinformatics, remote-sensing, proteomics and so on will assist in understanding the complexity of interactions existing between diverse plants, microbes, climates and soil.

13.2.3
Plant Growth Promoting Rhizobacteria

Rhizobacteria include mycorrhization helper bacteria (MHB) and PGPR, which assist AMF in colonizing the plant root [61,62], phosphorus solubilizers, free-living and symbiotic nitrogen fixers, antibiotic-producing rhizobacteria, plant pathogens, predators and parasites [63]. The most common bacteria in the mycorrhizosphere are *Pseudomonas* [64], while different bacterial species exist in the hyphosphere.

Like AMF, rhizobacteria such as pseudomonads are also ubiquitous members of the soil microbial community and have received special attention as they also exert beneficial effects on plants by suppressing soilborne pathogens, synthesizing phytohormones and promoting plant growth [65–68]. Many fluorescent *Pseudomonas* strains have been reported as plant growth enhancing beneficial rhizobacteria. They are studied extensively in agriculture for their role in crop improvement as they stimulate plant growth either by producing plant growth promoting hormones, fixing atmospheric N_2 or suppressing plant pathogens.

The rhizospheric component of PGPR bacteria adheres to the root surface, uses root exudates for growth, synthesizes amino acids and vitamins and establishes effective and enduring root colonization [69]. However, quantitative and qualitative variations in root exudates during plant growth could affect the rhizospheric competency of introduced PGPRs. The development of AM symbiosis also influences PGPR dynamics. *Pseudomonas fluorescens* 92rk increased mycorrhizal colonization of tomato roots by *Glomus mosseae* BEG12, suggesting that strain 92rk behaves as MHB [70,71]. Many researchers have reported additive effects on plant growth by AMF and PGPR [72–74], but the mechanisms by which MHB and PGPR stimulate AM colonization are still poorly understood [75]. Various hypotheses have been suggested, which include physical, chemical, physiological and even direct stimulatory or antagonistic relations between AMF and other mycorrhizosphere microbes [76].

AM and PGPR symbioses not only induce physiological changes in the host plant but also modify morphological architecture of the roots such as total root length and root tip numbers [77]. Gamalero *et al.* [74] found the highest values of architecture parameters in tomato roots inoculated with two strains of *P. fluorescens* (92rk and P190r) and AMF *Glomus mosseae* BEG12. They ascribed these findings both to modification of root architecture because of PGPR and AMF and to a greater absorption surface area owing to extrametrical mycelium of AMF. PGPR have also been shown to induce systemic resistance (ISR) to fungal, bacterial and viral pathogens in various crops such as bean, tomato, radish and tobacco [78].

13.2.4
Co-occurrence of AMF and PGPR/MHB

Certain fungi and rhizobacteria have been known to coexist since 1896, when Janse reported bacteria and fungi in the same sections of legume roots. These fungi were also described as arbuscular mycorrhizae by Jones [79], and it was Asai [80] who first stated that root nodulation by rhizobacteria depends on the formation of mycorrhizae by AMF. These earlier observations have been confirmed by subsequent researchers [81]. Multifaceted interactions of AMF with various microorganisms and microfauna in the mycorrhizosphere may be positive or negative [82]. The positive/synergistic interactions between mycorrhizosphere AMF and various nitrogen-fixing and phosphorus-solubilizing bacteria form the basis of application of these microbes as biofertilizer and bioprotectant agents [76]. These microbes are regulated by AMF for their own benefit, which in turn benefit the host plant. Meyer and Linderman [83] reported enhanced mycorrhization of clover in the presence of PGPR rhizobacterium *Pseudomonas putida*. Similar observations were later made by other researchers [84]. All these studies suggest that colonization of plant roots by AMF significantly influences mycorrhizosphere microorganisms, including PGPR.

Duc *et al.* [85] were the first to report that some pea mutants defective in nodulation also did not support AM symbiosis. Since the first report of legume nodulation mutants to be nonmycorrhizal, a number of nodulation-defective mutant legumes and mutated nonlegume crops have been tested [86]. Many recent reports have confirmed the similarities between nodulation and mycorrhiza-formation processes [87,88]. There is enough evidence available now that indicates a positive interaction between AMF and PGPR [89].

Several reports address the interactions between AMF and *Rhizobium* species [90], suggesting that the interaction is synergistic, that is, AMF improve nodulation due to enhanced phosphorus uptake by the plant. In addition to this principal effect of AMF on phosphorus-mediated nodulation, other secondary effects include supply of trace elements and plant hormones, which play an important role in nodulation and N_2 fixation. Current research is directed toward understanding the role AMF play in the expression of *Nod* genes in Rhizobia [91]. Barea and associates [34,46,47,92] have made many significant findings in this regard. Synergistic interactions between AMF and asymbiotic N_2-fixing bacteria such as *Azobacter chroococum*, *Azospirillum* spp. and *Acetobacter diazotrophicus* have also been reported by many researchers [84]. Synergistic interactions between phosphorus-solubilizing bacteria and AMF and their effect on plant growth have been studied by many researchers during the last three decades. Duponnois and Plenchette [93] studied the effects of MHB *Pseudomonas monteillii* strain HR13 on the frequency of AM colonization of Australian *Acacia* species and reported a stimulatory effect. They recommend dual inoculation to facilitate controlled mycorrhization in nurseries where *Acacia* species are grown for forestation. Modern research has provided evidence that the genetic pathway of AM symbiosis is shared in part by other root–microbe symbioses such as nitrogen-fixing rhizobia [87].

Many researchers reported unsuccessful attempts to select an appropriate *Rhizobium* strain for inoculating legumes owing to the failure of the selected strain(s) to survive and compete for nodule occupancy with indigenous native strains under low phosphorus and moisture contents [94]. In this context, the role of AMF as phosphorus suppliers to legume root nodules appears to be of great relevance. Requena *et al.* [95] found a specific AM fungus–*Rhizobacterium* sp. combination for effective nodulation and N_2 fixation in a mycotrophic legume *Anthyllis cytisoides* in Mediterranean semiarid ecosystems in Spain. They reported that *Glomus intraradices* was more effective with *Rhizobium* sp. NR4, whereas *G. coronatum* was more effective when coinoculated with strain NR9. Such specificity in interactions between AMF, *Rhizobium* and PGPR have been described by various researchers, indicating that it is important to consider the specific functional compatibility relationships between AMF, PGPR and MHB and their management when using these symbiotic microbes as biofertilizers.

New techniques applied in molecular ecology have resulted in the identification of members of nonculturable Archea in the mycorrhizosphere, but their role in the hyphosphere is not known [96]. This means, as pointed out by Sen [97], that analyses of PGPR distribution and activities must now be extended to accommodate Crearchaeotal microbes as well. Areas such as host–microbe specificity and microbial-linked control of plant diversity and productivity still need to be elucidated at gene, organismal and ecosystem levels [97].

13.3
Conclusion

AM are ubiquitous and most crop plants are colonized by AMF in nature, that is, mycorrhizosphere is the rule, not the exception. Thus, if we are to understand the rhizosphere reactions and interactions, we must understand the mycorrhizosphere. MHB might be exploited to improve mycorrhization and AMF to improve nodulation and stimulate PGPR. It is anticipated that future commercial biofertilizers would contain PGPR, MHB and AMF. Requena *et al.* [95] found that AM fungus *Glomus coronatum*, native to the desertified semiarid ecosystems in the southeast of Spain, was more effective than the exotic *G. intraradices* in AM/PGPR coinoculum treatments. The indigenous isolates must be involved. This area merits greater attention. More extensive field investigations into this multiagent biofertilizer will make this a popular technology among field workers in agriculture, forestry and horticulture. Manipulation of microorganisms in the mycorrhizosphere for the benefit of plant growth requires research at the field level [98,99]. To exploit microbes as biofertilizers, biostimulants and bioprotectants against pathogens and heavy metals, the ecological complexity of microbes in the mycorrhizosphere needs to be taken into consideration and optimization of rhizosphere/mycorrhizosphere systems needs to be tailored. Smith [100] stressed the need to better integrate information on root and soil microbe distribution dynamics and activities with known spatial and physicochemical properties of soil. This, as pointed out by Smith [100], should be achieved through greater collaborative efforts between biologists, soil chemists and physicists.

References

1 Doyle, R.J. and Lee, N.C. (1986) *Canadian Journal of Microbiology*, **32**, 193–200.

2 Khan, A.G. (2002) *The Restoration and Management of Derelict Land: Modern Approaches*, World Scientific Publishing, Singapore, Chapter 8, pp. 80–92.

3 Phipps, C.J. and Taylor, T.N. (1996) *Mycologia*, **88**, 707–714.

4 SchuBler, A. (2000) *Mycorrhiza*, **10**, 15–21.

5 Brundrett, M.C. (2002) *The New Phytologist*, **154**, 275–304.

6 Taylor, T.N. and Taylor, E.L. (1993) *The Biology and Evolution of Fossil Plants*, Prentice Hall, Englewood Cliff, NJ, USA.

7 Simon, L.K., Bousquet, J., Levesque, R.C. and Lalonde, M. (1993) *Nature*, **363**, 67–69.

8 Khan, A.G. (1974) *Journal of General Microbiology*, **81**, 7–14.

9 Khan, A.G. and Belik, M. (1995) *Mycorrhiza: Structure Function, Molecular Biology and Biotechnology* (eds A. Verma and B. Hock), Springer-Verlag, Heidelberg, pp. 627–666.

10 Khan, A.G. (2004) *Developments in Ecosystems* (ed. M.H. Wong), Elsevier, Northhampton, UK, pp. 97–114.

11 Smith, S.S. and Read, D.J. (1997) *Mycorrhizal Symbiosis*, 2nd edn, Academic Press, London.

12 van der Heijden, M.G.A., Klironomos, J.N., Ursic, M., Moutoglis, P., Streitwolf-Engel, R., Boller, T., Weismken, A. and Sanders, I.R. (1998) *Nature*, **396**, 69–72.

13 Burrows, R.L. and Pfleger, F.L. (2002) *Canadian Journal of Botany*, **80** (2), 120–130.

14 O'Connor, P.J., Smith, S.E. and Smith, F. (2002) *The New Phytologist*, **154**, 209–218.

15 Francis, R. and Read, D.J. (1995) *Canadian Journal of Botany*, **73**, 1301–1309.

16 van der Heijden, M.G.A. and Sanders, I.R. (2002) Springer-Verlag Berlin, Heidelberg, p. 469.

17 Schubler, A., Shwarzott, D. and Walker, C. (2001) *Mycological Research*, **105**, 1413–1421.

18 Helgason, T., Watson, I.J., Peter, J. and Young, W. (2003) *FEMS Microbiology Letters*, **229**, 127–132.

19 Xavier, I.J. and Boyetchko, S.M. (2002) *Applied Mycology and Biotechnology vol. 2: Agriculture and Food Production* (eds G.G. Khachatourians and D.K. Arora), Elsevier, Amsterdam, pp. 311–330.

20 Clark, R.B. and Zeto, S.K.J. (2000) *Plant Nutrition*, **23**, 867–902.

21 Turnau K. and Haselwandter K. (2002) *Arbuscular Mycorrhizal Fungi: An Essential Component of Soil Microflora in Ecosystem Restoration* (eds S. Gianinazzi and H. Schuepp), Birkhauser, Basel, pp. 137–149.

22 Chaudhry, T.M., Hayes, W.J., Khan, A.G. and Khoo, C.S. (1998) *Australasian Journal of Ecotoxicology*, **4**, 37–51.

23 Khan, A.G., Kuek, C., Chaudhry, T.M., Khoo, C.S. and Hayes, W.J. (2000) *Chemosphere*, **41**, 197–207.

24 Khan, A.G. (2001) *Environment International*, **26**, 417–423.

25 Jamal, A., Ayub, N., Usman, M. and Khan, A.G. (2002) *International Journal of Phytoremediation*, **4**, 205–221.

26 Hayes, W.J., Chaudhry, T.M., Buckney, R.T. and Khan, A.G. (2003) *Australasian Journal of Toxicology*, **9**, 69–82.

27 Pfleger, F.L. and Linderman, R.G. (1994) *Mycorrhizae and Plant Health*, The American Phytopathological Society (Symposium Series), St Paul, Minnesota, USA, p. 344.

28 Paulitz, T.C. and Linderman, R.G. (1989) *The New Phytologist*, **113**, 37–45.

29 Lynch, J.M. (1990) *The Rhizosphere*, John Wiley & Sons, Ltd, West Sussex, UK.

30 Giovannetti, M. and Avio, L. (2002) Biotechnology and arbuscular mychorrizas in *Applied Mycology and Biotechnology, vol. 2: Agriculture and Food Production* (eds G.G. Khachatourians and D.K. Arora), Elsevier, Amsterdam, pp. 275–310.

31 Fitter, A.H. (1985) *The New Phytologist*, **89**, 599–608.

32 Bethlenfalvay, G.J. and Linderman, R.G. (1992) Mycorrhizae in Sustainable Agriculture Special Publication No. 54, The American Phytopathological Society, St Paul, Minnesota, USA, p. 124.

33 Lovato, P.E., Gianinazzi-Pearson, V., Trouvelot, A. and Gianinazzi, S. (1996) *Advances in Agricultural Science*, **10**, 46–52.

34 Azcon-Aguillar, C. Jaizme-Vega, M.C. and Calvet, C. (2002) The contribution of arbuscular mycorrhizal fungi to the control of soil-borne plant pathogens, in *Mycorrhizal Technology: From Genes to Bioproducts – Achievements and Hurdles in Arbuscular Mycorrhiza Research* (eds S. Gianinazzi and H. Schuepp), Birkhauser, Basel, pp. 187–198.

35 Allen, M.F. (1991) The Ecology of Mycorrhiza, Cambridge University Press, Cambridge.

36 Allen, M.F. (1992) *Mycorrhizal Functioning: An Integrative Plant-Fungal Process*, Chapman & Hall Inc., Routledge, NY, p. 534.

37 Mohammad, A. and Khan, A.G. (2002) *Indian Journal of Experimental Biology*, **40**, 1087–1091.

38 Fortin, J.A., Becard, G., Declerck, S., Dalpe, Y., St-Arnaud, M., Coughlan, A.P. and Piche, Y. (2002) *Canadian Journal of Botany*, **80**, 1–20.

39 Sylvia, D.M. and Jarstfer, A.G. (1994) *Applied Environmental Microbiology*, **58**, 229–232.

40 Sylvia, D.M. and Jarstfer, A.G. (1994) *Management of Mycorrhizas in Agriculture Horticulture and Forestry* (eds A.D. Robson, L.K. Abbott and N. Malajczuk), Kluwer Academic Publishers, Dordrecht, pp. 231–238.

41 Mohammad, A., Khan, A.G. and Kuek, C. (2000) *Mycorrhiza*, **9**, 337–339.

42 Mohammad, A., Mitra, B. and Khan, AG. (2004) *Agricultural Ecosystems & Environment*, **103** (1), 245–249.

43 Marschner, P. and Baumann, K. (2003) *Plant and Soil*, **251**, 279–289.

44 Rodriguez, H. and Fraga, R. (1999) *Biotechnological Advances*, **17**, 319–339.

45 Garbaye, J. (1994) *The New Phytologist*, **128**, 197–210.

46 Barea, J.M., Toutant, P., Balazs, E., Galante, E., Lynch, J.M., Shepers, J.S., Werner, D. and Werry, P.A. (2000) *Rhizosphere and Mycorrhiza of Field Crops*, INRA and Springer, Berlin, Heidelberg, New York.

47 Barea, J.M., Gryndler, M., Lemananceau, P., Schuepp, H. and Azcon, R. (2002) *The Rhizosphere of Mycorrhizal Plants* (eds S. Gianinazzi, H. Schuepp, J.M. Barea and K. Haselwandter), Birkhauser, Basel.

48 Vivas, A., Marulanda, A., Ruiz-Lozana, J.M., Barea, J.M. and Azcon, R. (2003) *Mycorrhiza*, **13** (5), 249–256.

49 Behl, R.K., Sharma, H., Kumar, V. and Narula, N. (2003) *Journal of Agronomy and Crop Science*, **198**, 151–155.

50 Chaudhry, T.M. and Khan, A.G. (2002) *Biotechnology of Microbes and Sustainable Utilization*, (ed. R.C. Rajak), Scientific Publishers, Jodhpur, India, pp. 270–279.

51 Chaudhry, T.M. and Khan, A.G. (2003) (eds G.R. Gorban and N. Lepp), Proceedings of the 7th International Conference on the Biogeochemistry of Trace Elements, Swedish University of Agricultural Sciences, Uppsala, Sweden, June 15–19,134–135.

52 Garbaye, J., Churin, J.L. and Duponnois, R. (1992) *Biology and Fertility of Soils*, **13**, 550–567.

53 Mosse, B. (1962) *Journal of General Microbiology*, **27**, 509–520.

54 Mayo, K., Davis, R. and Motta, J. (1986) *Mycologia*, **78** (3), 426–431.

55 Linderman R.G. and Paulitz T.C. (1990) *Biological Control of Soil-Borne Plant Pathogens* (ed. D. Hornb), CAB International, Wallingford, pp. 261–283.

56 van Elsas, J.D., Turner, S. and Bailey, M. (2003) *New Phytologist*, **157**, 525–537.

57 Elassi, M., Delorme, S., Lemanceau, P., Stewart, G., Laue, B., Glickmann, E., Oger, P.M. and Dessaux, Y. (2001) *Applied Environmental Microbiology*, **67**, 1198–1209.

58 Pierson, L.S., Pierson, E.A. and Morello, J.E. (2002) *Biology of Plant–Microbe Interactions* (eds S.A. Leong, C. Allen and E.W. Triplett), International Society for Molecular Plant–Microbe Interactions, St Paul, MN, USA, pp. 256–262.

59 Khan, A.G. (1971) *Transactions of the British Mycological Society*, **56** (2), 217–224.

60 Minerdi, D., Bianciotto, V. and Bonfante, P. (2002) *Plant and Soil*, **244**, 211–219.

61 Klyuchnikov, A.A. and Kozherin, P.A. (1990) *Microbiology*, **59**, 449–452.

62 Andrade, G., Mihara, K.L., Linderman, R.G. and Bethlenfalvay, G.J. (1997) *Plant and Soil*, **192**, 71–79.

63 Sun, Y.P., Unestam, T., Lucase, S.D., Johanson, K.J., Kenne, L. and Finlay, R. (1999) *Mycorrhiza*, **9**, 137–144.

64 Vosatka, M. and Gryndler, M. (1999) *Applied Soil Ecology*, **1**, 245–251.

65 Weller, D.M. (1988) *Plant Roots: The Hidden Half* (eds Y. Waisel, A. Eschel and U. Kafkafi), Marcel Dekker Inc., New York, pp. 769–781.

66 Glick, B.R. (1995) *Canadian Journal of Microbiology*, **41**, 109–117.

67 Kapulnik, Y. (1996) *Plant Roots: The Hidden Half* (eds Y. Waisel, A. Eshel and U. Kafkafi), Marcel Dekker Inc., New York, pp. 769–781.

68 Chin-a-woeng, T.F.C., Bloemberg, G.V. and Lugtenberg, B.J.J. (2003) *The New Phytologist*, **157**, 503–523.

69 Lugtenberg, B.J.J. and Dekkers, L.C. (1999) *Environmental Microbiology*, **1**, 9–13.

70 Toro, M., Azcon, R. and Barea, J.M. (1997) *Applied Environmental Microbiology*, **63**, 4408–4412.

71 Singh, S. and Kapoor, K.K. (1998) *Mycorrhiza*, **7**, 249–253.

72 Edwards, S.G., Young, J.P.W. and Fitter, A.H. (1998) *FEMS Microbiology Letters*, **166**, 297–303.

73 Galleguillos, C., Aguirre, C., Barea, J.M. and Azcon, R. (2000) *Plant Science*, **159**, 57–63.

74 Gamalero, E., Trotta, A., Massa, N., Copetta, A., Martinotti, M.G. and Breta, G. (2003) *Mycorrhiza*, **14** (3), 185–192.

75 Barea, J.M., Azon-Aguilar, C. and Azcon, R. (1997) *Multitrophic Interactions in Terrestrial Systems* (eds A.C. Gange and V.K. Brown), Blackwell Science, Oxford, pp. 65–77.

76 Bansal, M., Chamola, B.P., Sarwar, N. and Mukerji, K.G. (2002) *Mycorrhizal Biology*, (eds K.G. Mukerji,B.P. Chamola and J. Singh), Kluwer Academic/Plenum Publishers, New York, pp. 143–152.

77 Atkinson, D., Berta, G. and Hooker, J.E. (1994) *Impact of Arbuscular Mycorrhizas on Sustainable Agriculture and Natural Ecosystems* (eds S. Gininazzi and H. Scheupp), Birkhauser Verlag, Basal, pp. 89–99.

78 Zhang, S., Reddy, M.S. and Kloepper, J.W. (2002) *Biological Control*, **23**, 79–86.

79 Jones, F.R. (1924) *Journal of Agricultural Research*, **29**, 459–470.

80 Asai, T. (1944) *Japanese Journal of Botany*, **12**, 359–408.

81 Filion, M., St-Arnaud, M. and Fortin, J.A. (1999) *The New Phytologist*, **141**, 525–533.

82 Facelli, E., Facelli, J.M., Smith, S.E. and Mclaughlin, M. (1999) *New Phytologist*, **141**, 535–547.

83 Meyer, R.J. and Linderman, R.G. (1986) *Soil Biology & Biochemistry*, **18**, 185–190.

84 Suresh, C.K. and Bagyaraj, D.J. (2002) *Arbuscular Mycorrhizae* (eds A.K. Sharma and B.N. Johri), Scientific Publishers, Enfield, New Hampshire, USA, pp. 7–28.

85 Duc, G., Trouvelot, A., Gianinazzi-Pearson, V. and Gianinazzi, S. (1989) *Plant Science*, **60**, 215–222.

86 Barker, S.J., Duplessis, S. and Denis-Tagu, D. (2002) *Plant and Soil*, **244**, 85–95.

87 Peterson, R.L. and Guinel, F.C. (2000) *Arbuscular Mycorrhizas: Physiology and Function* (eds Y. Kapulnik and D.D. Douds), Kluwer Academic Publishers, Dordrecht, The Netherlands, pp. 147–171.

88 Resendes, C.M., Geil, R.D. and Guinel, F.C. (2001) *The New Phytologist*, **150**, 563–572.

89 Chanway, C.P., Turkington, R. and Holl, F.B. (1991) *Advances in Ecological Research*, **21**, 122–170.

90 Albrecht, C., Geurtz, R. and Bisseling, T. (1999) *EMBO Journal*, **18**, 371–374.

91 Harrison, M. (1997) *Journal of Trends in Plant Science*, **2**, 54–60.

92 Azcon-Aguilar, C. and Barea, J.M. (1994) *Mycorrhizal Functioning: An Integrative Plant-Fungus Process* (ed. M.A. Allen), Chapman & Hall, London, pp. 163–198.

93 Duponnois, R. and Plenchette, C. (2003) *Mycorrhiza*, **13**, 85–91.

94 Bottomley, P.G. (1992) *Biological Nitrogen Fixation* (eds G. Stacey, R.H. Burris and H.J. Evans), Chapman & Hall, London, pp. 293–348.

95 Requena, N., Jimenez, I., Toto, M. and Barea, J.M. (1997) *The New Phytologist*, **136**, 667–677.

96 Bomberg, M., Jurgens, G., Saano, A., Sen, R. and Timonen, S. (2003) *FEMS Microbiology Ecology*, **43** (2), 163–171.

97 Sen, R. (2003) *The New Phytologist*, **157**, 391–398.

98 Khan, A.G. (1975) *Endomycorrhizas* (eds F.E. Sanders, B. Mosse and P.B. Tinker), Academic Press, New York, pp. 419–439.

99 Khan, A.G. (2002) *The Restoration and Management of Derelict Land: Modern Approaches* (eds M.H. Wong and A.D. Bradshaw), World Scientific Publishing, Singapore, Chapter 13, pp. 149–160.

100 Smith, S.E. (2002) *The New Phytologist*, **156**, 142–144.

14

Salt-Tolerant Rhizobacteria: Plant Growth Promoting Traits and Physiological Characterization Within Ecologically Stressed Environments

Dilfuza Egamberdiyeva and Khandakar R. Islam

14.1
Introduction

Worldwide, about 380 Mha of lands that are potentially usable for agriculture are severely affected by salinity [1]. Salinity and drought are the main causes of desertification that strongly influence many properties and processes of living organisms. Salinity upsets plant–microbe interactions, constituting a critical ecological factor that helps sustain and enhance plant growth in degraded ecosystems [2]. It has been reported that salinity and drought exert negative effects on plant growth and affect the biological stability of ecosystems [3,4]. Plant productivity in saline soils is considerably reduced owing to limited biological activity in response to salt and drought stresses. Under such circumstances, it requires suitable biotechnology to improve not only crop productivity but also soil health through interactions of plant roots and soil microorganisms. Development of such a stress-tolerant microbial strain associated with roots of agronomic crops can lead to improved fertility of salt-affected soils [5]. The use of beneficial microbes in agricultural production systems started about 60 years ago and there is now increasing evidence that the use of beneficial microbes can enhance plant resistance to adverse environmental stresses; for example, drought, salts, nutrient deficiency and heavy metal contamination [6].

Microorganisms associated with soil and plant root, forming the rhizosphere, influence plant development, growth and environmental adaptation, both beneficially and detrimentally [7]. Various soil microorganisms capable of exerting beneficial effects on plants or antagonistic effects on soilborne pests and diseases can be used in agriculture to make crop production sustainable [8].

Microorganisms are capable of adapting themselves to adverse conditions, making them suitable to use in a wide range of environments including agriculture [9]. However, several environmental factors often limit the growth and activity of rhizosphere microorganisms [10]. Plant growth promoting microbes found in the rhizosphere under environmental stress, including extremely saline soils, can provide a wide range of benefits to plants [11]. Some may be able to improve plant growth by increasing the rate of seed germination and seedling emergence, minimizing the adverse effects of external stress factors, and protecting plants from soilborne pests and diseases [12]. Understanding the highly complex nature of the microbial adaptation and response to alterations in the biological, chemical and physical environment of the rhizosphere remains a significant challenge for plant biologists and microbiologists [13,14].

Interest in bacterial fertilizers has increased, as their use would substantially reduce the use of chemical fertilizers and pesticides, which often contribute to pollution of soil–water ecosystems. Presently, about 20 biocontrol products based on *Pseudomonas*, *Bacillus*, *Streptomyces* and *Agrobacterium* strains have been commercialized, but there still is a need to improve the efficacy of these biocontrol products [12]. Soil salinity, high temperatures and soil contamination often affect phytoefficiency of plant growth promoting bacterial inoculants in nature [15]. Damage to soil and plants in arid and semiarid areas is not easily repairable, because these areas are fragile and sensitive ecosystems [16]. It is important to study soil microbial activity in stressed environments to evaluate soil quality and plant productivity as affected by natural calamities and anthropogenic activities [17–19].

The challenge for the future includes understanding the behavior of microbes in their natural and often complex habitats, such as the rhizosphere [20]. Microbial processes and properties in the rhizosphere are crucial to support functional agriculture. Root-associated bacteria have a great influence on organic matter decomposition, which, in turn, is reflected in soil nutrient availability for plant growth [21]. The phosphorus- and potassium-solubilizing bacteria may enhance plant nutrient availability by dissolving insoluble phosphorus and releasing potassium from silicate minerals [22]. Plant growth promoting bacteria often help increase root surface area to increase nutrient uptake and, in turn, enhance plant production [23].

The mechanisms and interactions among these microbes are still not well understood, especially in field applications under different environments. Therefore, this requires studying plant–microbe interactions, the natural resident microorganisms and their physiological adaptation in ecologically stressed environments. Understanding the physiology, adaptation and functions of salt-tolerant bacteria in stressed environments (e.g. arid regions) may provide valuable information on plant–microbe interactions to develop such new agricultural technologies as would improve soil ecology and plant development. At present, however, the interest lies in the development and application of salt-tolerant plant growth stimulating bacterial inoculants to improve plant growth and yield; their interactions with host plants; and biological control of fungal diseases in saline environments. There have been few reports on microbial diversity and function in saline environments in different regions of the world [24–26]. However, only a limited number of studies on rhizobacteria and their physiological characterization in saline arid soils have been undertaken. This chapter intends to discuss recent developments and advances in our understanding of the high salt- and temperature-resistant rhizobacteria and their characteristics, physiology, adaptation and production of metabolites that play a synergistic role in plant growth and development under fragile and stressed environments.

14.2
Diversity of Salt-Tolerant Rhizobacteria

The microbial composition in the rhizosphere, as a result of diverse plant–microbe interactions, often differs greatly from that of the surrounding soil and from one plant species to another. Analyzing the genotypic and phenotypic characteristics of indigenous rhizobacteria can help understand better the interaction mechanisms between them and plant roots [25]. Understanding the diversity of rhizobacteria under stressed conditions and tapping the indigenous population that has managed to adjust to adverse environments can greatly help in harnessing their synergistic properties [9]. The physiological and biochemical mechanisms of adaptation to saline environments by a few plant growth promoting bacteria, for example, *Rhizobium*, *Azospirillum* and *Pseudomonas*, have been reported [27].

Salinity and desertification cause a greater disturbance to plant–microbe symbioses in degraded ecosystems. It is reported that nodule formation in legume–*Rhizobium* symbioses is more sensitive to salt stress than the rhizobia themselves [28]. Research on salt-affected soils in Egypt showed the presence of 11 species of *Bacillus* plus rhizobia and *Actinomycetes* [29]. In highly saline soils, Gram-positive spore-forming bacteria of the genus *Bacillus* and unidentified Gram-negative rods were found among nitrogen-fixing bacterial isolates [30]. The *Pseudomonas* spp. often associated with rice is a common member of the plant growth promoting bacteria present in the rhizosphere [9]. These strains were identified as *P. aureginosa*, *P. pseudoalcaligenes*, *P. alcaligenes*, *P. fluorescens*, *P. putida*, *P. stutzeri*, *P. mendocina*, *P. mallei* and *P. diminuta*. Fluorescent pseudomonads were often found in large numbers in nonsaline soils, whereas *P. alcaligenes* and *P. pseudoalcaligenes* were common in saline soils. On the contrary, *Swaminathania salitolerans* was isolated from the rhizosphere, roots and stems of salt-tolerant, mangrove-associated wild rice [33]. The isolates were found capable of fixing atmospheric nitrogen and enhancing solubilization of phosphates in the presence of salts (e.g. NaCl). However, appropriate information on taxonomic diversity of soil microbes in saline environments is lacking [31,32].

Some bacteria, such as *Pseudomonas* and *Flavobacterium* spp., tend to be more dominant in the rhizosphere than *Arthrobacter* and *Bacillus* spp. [34]. While *Azotobacter* spp. was isolated from salt-affected soils [35], the *Bacillus* was readily distributed in a wide range of natural habitats [36], suggesting an inherently remarkable degree of physiological and genetic adaptability of *Bacillus* in nature. The majority of thermophilic bacteria isolated so far belong to the genus *Bacillus*, which is well suited to arid soils because of its ability to produce endospores that are resistant to environmental stresses [37,38]. To persist and reproduce, the introduced bacterial strains are expected to be capable of rapid adaptation to a wide range of soil conditions. One such bacterial genus, *Arthrobacter*, comprised about 50% of a wheat rhizospheric microbial population [39]. Several salt-tolerant *Rhizobium* species have also been reportedly isolated that adapted to saline environments by intracellular accumulation of compatible solutes (e.g. glycine betaine, choline, low-molecular-weight carbohydrates, polyols, amino acids and amines) [40]. An exogenous supply of glycine betaine and choline was found to enhance the growth of various rhizobia such as R. tropici, R. galegae, Mesorhizobium loti and M. haukkii under salt stress [41]. Another study reported that Rhizobia can even survive in the presence of extremely high levels of salts [42]. Some strains of R. meliloti and R. fredii were able to grow at salt concentration of more than 300 mM [43]. Although the bacterial colonization and plant root hair curling were reduced in the presence of 100 mM NaCl, the proportion of root hair containing infection threads was reduced only by 30% [44]. Other studies confirmed the findings that an accumulation of compatible solutes helps to maintain osmotic regulation in Azospirillum species [45]. Azospirillum halopraeferens and A. irakense, which are known to tolerate 3% NaCl, were isolated from the rice rhizosphere in saline environments [46]. Several other researchers have isolated a number of microbial strains from saline environments, such as genera Salinivibrio, Halomonas, Chromohalobacter, Bacillus, Salinicoccus, Candida tropicalis and bacterium Alcaligenes faecalis [47,48].

A number of other salt-tolerant rhizobacteria, such as Serratia marcescens, Pseudomonas aeruginosa, Alcaligenes xylosoxidans and Ochrobactrum anthropi, were also isolated from rice roots [25]. The P. aeruginosa was found to be the most dominant member of the bacterial community associated with rice roots [25]. This species has been described as a quintessential opportunist and is generally found in soil and water ecosystems. The osmotolerance behavior of these bacteria is one of the mechanisms of rapid adaptation for their survival and colonization in the human intestine [49]. They are members of the potentially pathogenic bacterial species that survive and become enriched in the rhizosphere over time. There are reports on the presence of P. aureginosa and some other human pathogenic bacteria, such as S. marcescens and A. xylosoxidans, associated with rice roots in saline environments [25]. The pseudomorphic characteristics of the bacteria most probably accounted for their tolerance to salt stress [50]. The rhizosphere, enriched with organic substrates, stimulates microbial growth and may contain up to 10^{11} cells per plant roots. However, the growth of human pathogens as part of this microbial population is a major concern owing to the potential effects on human health [51]. It is reported that some human-associated potential pathogens, such as P. aeruginosa, S. aureus and S. pyogenes were able to colonize wheat rhizosphere in saline soils [52]. Furthermore, a greater number of P. aeruginosa cells adhered to the wheat roots than did pathogens [52]. Members of the potentially pathogenic bacterial species survived and became enriched in the rhizosphere owing to the greater availability of labile carbon as food and energy sources [53]. In another study, Acinetobacter and S. saprophytius were isolated from the rhizosphere of Western Australian orchids [54]. Similar bacterial species, such as Pseudomonas, Bacillus, Acinetobacter and Staphylococcus were also isolated in the soil and rhizosphere of different plants [55,56]. Species as diverse as Pseudomonas, Bacillus, Mycoplana, Mycobacterium, Acinetobacter, Microbacterium and Arthrobacter, were also isolated in saline soils of Uzbekistan (Table 14.1). These microorganisms are presently used in highly saline and degraded soils for supporting plant growth and development in Uzbekistan [15].

Salt-tolerant Mycobacterium phlei strains were found in association with corn planted in saline soils of Uzbekistan. The pathogenic members of this family are

Table 14.1 Diversity of salt-tolerant and temperature-resistant rhizosphere bacteria isolated from various crop roots in saline arid soils of Uzbekistan.

Bacterial species	Crops	Salt tolerant (>4% salts)	Temperature resistant (40 °C)
P. alcaligenes	Melon	+[a]	+
P. aureginosa	Wheat	+	+
P. aurantiaca	Maize	+	−
P. aureofaciens		+	+
P. denitrificans		−	+
P. fluoro-violaceus		−	+
P. mendocina		+	+
P. rathonis		+	−
P. stutzeri		−	+
B. amyloliquefaciens		+	+
B. cereus		+	+
B. circulans		+	+
B. laevolacticus		−	+
B. latvianus		+	+
B. licheniformis		+	+
B. maroccanus		+	+
B. megaterium		+	+
B. polymyxa		+	+
B. subtilis		+	+
M. phlei		+	+
Mycoplana bullata		+	−
A. globiformis		+	+
A. simplex		+	−
A. tumescens		+	+

[a]+ and − indicate the presence or absence of a particular microorganism.

responsible for human infections, including leprosy and tuberculosis. However, the Mycobacteriacea family contains a wide range of nutritional types, including saprophytic species that are present in the soil. The atypical *Mycobacteria* do not cause tuberculosis in humans or animals. They are ubiquitous and have been found practically in every part of the world, particularly in soil [57,58]. One of the important species, for example, *M. phlei* is a rapidly growing saprophyte by nature [58,59].

14.3
Colonization and Survival of Salt-Tolerant Rhizobacteria

To function as phytostimulators, a microbe must be present at the right location and the right time at the site of action [60,61]. Bacterial inoculants are often applied in seed coatings and after sowing, the bacteria must be able to establish themselves in the rhizosphere in sufficient numbers to deliver beneficial effects [62]. Therefore, efficient bacterial inoculants should survive in the rhizosphere, make use of nutrients exuded by plant roots, proliferate and efficiently colonize the root system and be able to compete with endogenous microorganisms [63].

Colonization can be considered as the delivery system of the microbes' beneficial factors [8]. It is reported that plant growth promoting rhizobacteria are able to colonize the rhizosphere, the root surface or even the superficial intercellular spaces of plants [64]. Colonization of plant roots by bacteria depends on biotic and abiotic factors such as dynamics of microbial population, plant characteristics and soil types. To persist and reproduce, bacteria introduced into a soil should be able to rapidly adapt to soil conditions.

Survival strategies of bacteria such as *Bacillus* species depend on the physiological adaptation of the introduced cells –adaptation to nutrient-limited conditions and/or other physicochemical limiting conditions, efficient utilization of root-released compounds or specific interactions with plants [65]. In general, *Bacillus* endospores have a greater survival capacity in dry soil than *Pseudomonas* cells [66]. Although soil type can influence the composition of fluorescent *Pseudomonades* in the rhizosphere [67], the action of aggressive colonizers appears to be independent of soil type [68]. Results from a number of studies show that the density of a bacterial population associated with pea roots was negatively correlated to an increase in temperature [68,69]. It is worth mentioning that [15,70] the rifampicin-resistant mutants of *P. denitrificans* PsD6 and *Bacillus amyloliquefaciens* BcA12 bacteria isolated from arid saline soils were able to establish more easily in both wheat (80 and 99.9%, respectively) and pea (93 and 99.9%, respectively) rhizosphere than in bulk soil and phyllosphere. Effects were more pronounced in peas than in wheat (Table 14.2).

A study on wheat root colonization and survival of an *Arthrobacter* spp. illustrated that the coryneforms are well adapted to long-term survival in both rhizosphere and bulk soil [71]. Efficient colonization and/or physiological adaptation to soil conditions is necessary for soil bacterial inoculants to survive under adverse environments [65]. However, microbial colonization of plant roots is often affected by biotic and abiotic factors such as root exudates, competition, nutrients, pH, electrical conductivity and temperature [72,73]. Information on the effects of different factors on rhizosphere microflora may help to understand the rhizosphere microbial dynamics in soil. It is suggested that the persistent nature of introduced bacterial inoculants in nutrient-limited habitats is closely related to their ability to resist starvation [74]. A direct relationship between starvation resistance and the microbe's ability to survive

Table 14.2 Bacterial colonization in soil, rhizosphere (RS) and phyllosphere (PS) of wheat and peas.

Bacterial strain	Wheat			Peas		
	Soil	RS	PS	Soil	RS	PS
P. denitrificans PsD6	762b	6524a	1.4c	812.5b	11909a	7.3c
B. amyloliquefaciens BcA12	152.8b	912a	31.3c	60.6b	308a	4.8c

Mean values followed by different letters in the same row for each crop were significantly different at $P \leq 0.05$.

in soil was reported in earlier studies [75]. Salt-tolerant and temperature-resistant characteristics make bacteria capable of adapting to extreme environments [25,76].

14.4
Salt and Temperature Tolerance

The natural distribution of bacteria is governed by a number of limiting factors, and each species exists where levels of these factors such as temperature, moisture, nutrient availability and pH fall within its specific range of tolerance [77]. Tolerance to high salt (e.g. NaCl) concentration and temperature is important in the survival of bacteria in arid saline and alkaline soils [76]. Extreme conditions may suppress the activities of some microorganisms but may stimulate those of others and as a result, in many extreme environments, microorganisms are the only contributors to nutrient cycling [76,77]. Stress-tolerant bacteria exhibit many mechanisms of exploiting their environments and differ greatly in their biochemical capabilities and in the capabilities to utilize various organic carbon pools as energy and food sources [72,73]. For example, free-living rhizobia have shown variations in their response to salt stress [79]. Although the growth of a number of rhizobia was found to be inhibited by 100 mM NaCl, *R. meliloti* was tolerant to 300–700 mM NaCl. However, changing the surface antigenic polysaccharides and lippolysaccharides concentration by toxic salt stress often impaired the *Rhizobium*–legume interaction in nature [10].

In nature, bacteria that are well adapted to a range of soil environments may possess an efficient response to stressful soil conditions by activating molecular mechanisms necessary for adaptation and survival. In this regard, the salt-tolerant rhizobacteria may have developed mechanisms to survive and proliferate under increased salinity [25]. So far, two categories of halophiles have been identified in nature, the moderate and extreme halophiles [80]. Among the halophiles, 18 recognized *Bacillus* species and related genera are categorized into moderate halophiles or halotolerants. As a result, rhizobacteria (e.g. *Bacillus cereus* 80 strain) associated with saline soil crops are able to grow in various levels of salinity ranges between 0 and 5% NaCl (Figure 14.1). It is reported that the majority of the salt-tolerant bacteria so far isolated can osmoregulate by synthesizing specific compatible organic osmolytes, such as glutamine, proline and glycine betaine, and a few of them accumulate inorganic solutes, such as Na^+, K^+ and Mg^{2+} to a higher intracellular concentration over time for maintaining cell turgidity pressure [10,82].

Likewise, temperature is a good parameter for illustrating both diversity and adaptation of microbes in natural environments [83]. In arid and semiarid regions, high soil temperature affects both free-living and symbiotic rhizobia [40]. While the optimum temperature for growth of rhizobia is $30 + 2\,°C$, high soil temperatures ($35–40\,°C$) in arid regions may result in the formation of ineffective nodules [84,85]. Furthermore, most of the bacterial species appear to be restricted to a temperature range of $30–40\,°C$. However, a few bacterial species (e.g. *Bacillus coagulans* and *Bacillus subtilis*) are known to survive temperatures between 30 and $60\,°C$ [83,86].

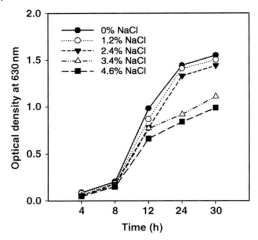

Figure 14.1 Growth of salt-tolerant *B. cereus* 80 strain at different concentration (%) of NaCl.

Many species are able to survive at temperatures higher than those that allow growth, by adopting metabolic shutdown either as vegetative cells or heat-resistant spores and cysts [83]. Salt-tolerant rhizobacteria (e.g. *Arthrobacter, Bacillus, Mycobacterium, Pseudomonas*, etc.) isolated from arid soils were able to survive at 45 °C. In earlier studies, two species of *M. phlei* and *M. thermoresistibile* were found to be able to grow at 52 °C [87]. However, bacteria isolated from alkaline soils in India during summer months are subjected to high salt concentration, pH and temperature stress. In alkaline soils, salt concentrations and pH may be as high as 2 and 10.2%, respectively, and temperatures may range between 35 and 45 °C [88]. Thus, bacterial strains that are capable of tolerating high temperatures and salt concentrations have competitive advantages over others to survive and proliferate in arid and saline regions of the world.

14.5
Physiological Characterization of Rhizobacteria

Soil conditions stressful to bacteria include nutrient starvation, salinity, temperature and low water activity. The availability of nutrients from organic matter decomposition largely depends on the sites occupied by bacterial cells or microcolonies. Oligotrophs are probably locally dispersed through most soils including arid soils. Therefore, soils in general can be regarded as a grossly oligotrophic environment [89]. Oligotrophs have been described as bacteria that are able to multiply in habitats of low nutrient flux [90]. They show relatively high growth rates in nitrogen-free medium culture and have the ability to survive in nitrogen-poor soils. In general, both starvation and low-water activity represent typical stress conditions in most soils. Other stress factors such as extreme temperatures, very high or low pH and presence of toxic metal concentrations may be more specifically related to climate, geographic location and soil management or contamination [65]. In such environments, most microorganisms have adapted through their physiological abilities to resist the stress conditions. For example, *Bacillus*, as one of the oligotrophic groups of soil bacteria, have shown relatively high growth rates in culture even at low concentrations of energy-yielding substrates (e.g. carbon and nitrogen) and have the ability to survive in nutrient-deficient soils. *Arthrobacter* are also capable of surviving in carbon- and nitrogen-deficient soils by following energy-efficient metabolic pathways [91]. Therefore, *Arthrobacter* are referred to as oligotrophs [92]. The oligotrophs can reproduce in the presence of minimum concentrations of organic matter, making them suitable candidates for plant root colonization in arid saline soils [93].

Nutrients, in general, are limited in soil and bacteria, in response, reproduce slowly and may even remain dormant for long periods. Some of these microorganisms display considerable resistance to starvation [91]. A strong relationship between cellular adaptation of *Arthrobacter globiformis* to the conditions of chemostat growth and survival of starved suspensions for up to 56 days was reported [94]. A greater starvation resistance of *Arthrobacter* than *Pseudomonas* and *Bacillus* in a carbon-free medium was also reported in a number of studies [91,94,95].

Endogenous metabolism is another important pathway for microbial survival in nature, and *Arthrobacter* may possibly use this mechanism to reduce metabolic rates to enhance their survival under starvation [96]. Arthrobacters were also considered to be the prominent members of the autochthonous microflora of soil that are able to maintain population levels over long periods in nutrient-limiting environments [97]. By exhibiting a wide range of metabolic activities, they are able to utilize various low-molecular-weight organic compounds and some more complex compounds as carbon and energy sources. In culture, they display a wide nutritional diversity and can even utilize aromatic compounds to survive under nutrient deficiencies. In our previous studies, some of the salt-tolerant bacterial strains such as *A. globiformis* ArG1 and *A. tumescens* ArT16 isolated from nutrient-poor saline and arid soils were able to grow in nitrogen-free medium and were considered as oligonitrophiles with the ability to fix organic nitrogen (Figure 14.2).

A wide variety of free-living heterotrophic bacteria are known to utilize ammonium nitrogen as the preferred nitrogen source. Under conditions of limited ammonium nitrogen, many such bacteria are able to increase their affinity for ammonium nitrogen through depression of the assimilatory system in which the task of ammonium-nitrogen assimilation is performed by glutamine synthetase [98]. Among the bacteria capable of nitrogen assimilation, glutamine synthetase is generally the catalyst for assimilating ammonium nitrogen into organic compounds resulting from fixation of atmospheric nitrogen [99]. The troposphere serves as a vast reservoir of nitrogen (78%) that can be utilized when the preferred form of nitrogen in soil becomes limited, so that oligonitrophiles, once adapted, can easily incorporate nitrogen through biological nitrogen fixation. Adding carbon-enriched materials in excess could impose nitrogen scarcity without removing the nitrogen source. Lactate, malate and succinate are all labile carbon sources for microorganisms to enhance the nitrogenase activity.

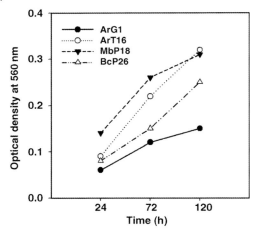

Figure 14.2 Growth of *A. globiformis* ArG1, *A. tumescens* ArT16, *M. phlei* MbP18 and *B. polymyxa* BcP26 strains in nitrogen-free medium.

Depression could allow nutrient-deficient microbial cells to use some of their limited resources for detecting and moving toward the utilizable source of nutrients, a behavior that, like nitrogen fixation, is not so useful when there is an adequate nutrient supply in the soil [99].

To adapt to saline environments, microorganisms have developed various biochemical strategies over time to maintain structural and functional stability of the cells [24]. As a result, many bacteria are able to synthesize secondary metabolites, such as extracellular enzymes and bioactive compounds. Accordingly, about 86% of the bacteria isolated from the rhizosphere of various plants were found to have the ability to produce phytohormones, enzymes, vitamins and amino acids. However, the species *Arthrobacter* isolated from arid saline soils do not produce lecithinase, cellulase, tryptophanase and arginindehydrolase activities. Instead, they produce lipase, nitrate-reductase and collagenase activities (Table 14.3). A hydrolytic reaction of *Arthrobacter* species suggests that *A. globiformis* ArG1 and *A. tumescens* ArT16 are able to hydrolyze starch (Table 14.3). Urea, an organic nitrogen compound, is hydrolyzed by *P. rathonis* PsR47 and *M. phlei* [100]. Positive gelatin liquefaction and citrate utilization patterns were also found for *Arthrobacter* (Table 14.3). In *Mycobacteria*, nitrate-reductase is used for nitrogen assimilation in which nitrate is first reduced to nitrite and then, probably via hydroxylamine, to ammonia [57]. *Mycobacterium* can utilize a wide range of carbon compounds for growth and utilize many nitrogenous compounds as sources of nitrogen [87].

Since carbohydrate fermentation patterns of salt-tolerant bacteria are different [100], the *A. globiformis*, *A. tumescens* and *P. rathonis* do not utilize carbohydrates as compared to *A. globiformis* ArG1 (Table 14.4). Other studies have suggested that the main reason for the absence of bacterial growth in most of the tested sugars could have been the oligotrophic nature of the bacteria (Table 14.4).

Table 14.3 Biochemical characteristics of some salt-tolerant rhizobacteria.

Biochemical (enzyme) characteristics	A. globiformis ArG1	A. tumescens ArT16	M. phlei MbP18	B. amyloliquefaciens BcA12	P. rathonis PsR47
Casein hydrolysis	+[a]	+	−	−	−
Pectin decomposition	−	+	+	+	−
Gelatin liquefaction	+	+	−	−	−
Citrate utilization	+	+	+	+	+
Arginine dihydrolase	−	−	−	−	+
Acetylmethylcarbinol	−	−	+	+	−
Oxidase	+	+	+	+	+
Catalase	−	−	+	+	−
Lypase	+	+	+	+	+
Tryptophanase	−	−	−	+	−
Nitrate reductase	+	+	+	+	+
Amylase	+	+	+	+	+
Collagenase	+	+	−	−	+
Protease	+	+	−	−	−
Pectinase	−	+	+	+	−
Lecithinase	−	−	−	−	−
Urease	−	+	+	+	+
Nitrogenase	+	+	+	−	−

[a]+ and − indicate the presence or absence of activity of a particular microorganism.

Bacteria that grow in nitrogen-free media are most probably unable to grow in higher concentrations of carbohydrates as carbon and energy sources. But carbohydrate utilization for M. phlei MbP18 was found to be different as they produce acids from glucose, arabinose, xylose, glycose, saccharose, maltose, glycerol and mannitol. The results of our study showed that many salt-tolerant rhizobacterial strains were capable of producing B vitamins and amino acids (Table 14.5). B vitamins and

Table 14.4 Carbohydrate fermentation patterns of salt-tolerant rhizobacteria.

Carbon source	A. globiformis ArG1	A. tumescens ArT16	M. phlei MbP18	P. rathonis PsR47
Arabinose	−[a]	−	+	−
Xylose	−	−	+	−
Glucose	−	−	+	−
Galactose	−	−	−	−
Saccharose	−	−	+	−
Lactose	−	−	−	−
Maltose	−	−	−	−
Glycerol	−	−	+	−
Mannitol	−	−	+	−

[a]+ and − indicate the presence or absence of activity of a particular microorganism.

Table 14.5 Production of B vitamins and amino acids by salt-tolerant bacterial strains.

Bacterial isolates	B$_1$	B$_3$	B$_6$	PP	Biotin	Norvaline	Gistidine	Leicin	Valine	Glutamine
P. alcaligenes PsA15	+[a]	−	−	+	−	+	+	−	+	+
P. denitrificans PsD6	+	+	−	−	+	−	−	−	+	−
P. mendocina PsM13	+	−	+	−	−	−	−	−	+	−
B. laevolacticus BcL28	+	+	−	−	−	−	+	−	−	−
B. megaterium BcM48	−	−	+	+	+	−	+	−	−	+
B. polymyxa BcP26	−	+	−	−	+	−	−	−	+	−
M. phlei MbP18	+	+	+	−	+	−	+	+	−	+
A. globiformis ArG1	+	−	−	−	+	+	−	−	+	−
A. simplex ArS50	+	+	−	+	+	+	+	−	−	−
A. tumescens ArT16	+	+	−	+	+	+	+	−	−	−

[a]+ and − indicate the presence or absence of activity by a particular microorganism.

proteins produced by microbes are regarded as important sources of catalysts for plant growth in arid saline soils.

14.6
Plant Growth Stimulation in Arid Soils

The magnitude of plant response to microbial inoculation is greatly affected by nutrient availability in soil [102]. In our previous studies, 233 bacterial strains were isolated from weakly saline arid soils in Tashkent province and 480 strains from strongly saline soils near the Aral Sea basin to evaluate their beneficial properties on plants. About 30% of the strains were isolated from non inoculated plants, 15% from weakly saline soils where they stimulated both shoot and root growth in plants and 24% were found to inhibit plant growth. From these isolates, about 21% of the bacteria were responsible for causing a number of plant root diseases. Numbers of growth stimulators were lower (9%) compared to inhibitors (21%) and disease-causing strains (16%) among the bacteria isolated from saline soils in Uzbekistan (Table 14.6).

Table 14.6 Effects of rhizosphere bacterial strains on wheat plant growth in saline soils.

Effects	Weakly saline soil (%)	Strongly saline soil (%)
Stimulators	15.2	9.3
Inhibitors	23.9	21.2
Disease-causing strains	21.0	16.2
Antagonist against *F. culmorum* 556	2.57	1.53

Microbial groups that influence plant growth by supplying combined nitrogen include the symbiotic nitrogen-fixing rhizobia associated with legumes [8]. *Rhizobium* and *Bradyrhizobium* bacteria are responsible for most of the biological nitrogen fixation in soil. It is suggested that biological nitrogen fixation is the major method for nitrogen input into desert ecosystems [10]. Particularly, the *Rhizobium*–legume symbiosis represents the major mechanism of biological nitrogen fixation in arid lands [103]. It is known that many countries are using bacterial fertilizers such as *Rhizobium* inoculants in agriculture [104,105]. Results from a number of experiments have shown positive effects on soybean growth when seeds were inoculated with certain strains of *Bradyrhizobium* [106,107]. However, factors affecting inoculums' success include temperature, soil type, nitrogen content, salt concentration and moisture content. Inoculation of legumes with salt-tolerant strains of rhizobia may enhance their nodulation and nitrogen fixation under salt stress that ultimately improves soil health and increases plant production under arid climates [10]. Generally, effective symbiotic relationships develop between rhizobia and legumes at low level of available soil nitrogen. It is reported that in arid saline soils, *B. japanicum* S2492 has significantly increased dry weight of biomass, plant height and yield (>35%) of soybeans [109]. However, protein and fat content did not vary significantly (Table 14.7).

Similar results were obtained when inoculation of legumes with *B. japanicum* increased nodule numbers, plant height and seed yield under greenhouse and field conditions [110]. Studies conducted in China have shown that salt-tolerant nitrogen-fixing bacteria promoted rice growth and increased yields under salt stress [111,112]. The nitrogen-fixing bacteria, *Azotobacter chrococcum* A2, increased cotton, wheat, rice and potato yields in saline arid soils (Table 14.8). The yield of cotton increased up to 3.5 t ha^{-1} when treated with bacterial inoculants compared to 2.9 t ha^{-1} in the control treatment. Wheat yield increased to 3.5 t ha^{-1} compared to 2.8 t ha^{-1} in the control (Table 14.8).

Salt-tolerant free-living plant growth promoting rhizobacteria are also present, which, coupled with their ability to survive in hot, saline and nutrient-poor conditions, increased growth of pea and wheat in saline arid soils. Nontreated control plants, by comparison, performed poorly under such conditions. For example, the salt-tolerant bacterial strains *P. alcaligenes* PsA15, *B. amyloliquefaciens* BcA12, *B. polymyxa* BcP26, *A. simplex* ArS7, *A. globiformis* ArG1 and *M. phlei* MbP18 have

Table 14.7 Effect of *B. japanicum* S2492 on soybean yield, protein and fat content.

Treatment	Grain yield (g plant^{-1})	Protein content (g 100 g^{-1})	Fat content (g 100 g^{-1})
Control	8.4b	35.9a	17.9a
Inoculated with B. japanicum S2492	13.4a	39.2a	20.4a

Mean values followed by the different lower case letters in the same column were significantly different at $P \leq 0.05$.

Table 14.8 Effects of *A. chrococcum* A2 on crop yields in saline soils of Uzbekistan.

Treatments	Cotton		Wheat yield (t ha^{-1})	Rice yield (t ha^{-1})	Potato yield (t ha^{-1})
	Yield (t ha^{-1})	Fiber (%)			
Control	2.95b	32a	2.8b	4b	16b
A. chrococcum A2	3.52a	35a	3.5a	4.6a	20a

Mean values followed by the different lower case letters in the same column were significantly different at $P \leq 0.05$.

significantly increased shoot growth and root length (10–43%) of wheat and peas in saline soils compared to the control treatment [15,113] (Table 14.9).

It is postulated that salt-tolerant bacteria that are adapted to saline and drought condition will develop a beneficial association with plants and are expected to produce growth-promoting substances as evidenced by the increase in plant dry matter production, yield and nutrient uptake. It is reported that *Pseudomonas* strains significantly enhance early plant growth in low-fertility soil [62]. Although bacterial inoculation marginally increased crop yields when planted under ideal climatic and soil conditions, greatest benefits in terms of improved plant growth and yield were obtained when crops encountered periods of environmental stress [114]. The beneficial effects of *Pseudomonas* strains (e.g. *P. fluorescens* ANP15 and *P. aeruginosa* 7NSK2) on plant growth were found more pronounced when plants were subjected to suboptimal conditions, such as an unfavorable climate or the presence of plant pathogens [115].

A number of bacterial strains isolated from degraded soils were found to be stimulants for wheat and rice growth in warm climates [116]. To obtain most competitive and effective bacterial strains, bacteria need be isolated and screened from the pool of indigenous microbes that supposedly are versatile to adapt to a wide range of climatic conditions [117]. In particular, the rhizobia isolated from nodules

Table 14.9 Effects of growth-promoting bacteria on relative growth (%) of shoot and root of wheat and peas.

Bacterial strains	Wheat		Peas	
	Shoot	Root	Shoot	Root
Control	100c	100d	100c	100e
P. alcaligenes PsA15	116b	125b	125a	135a
B. amyloliquefaciens BcA12	143a	137a	127a	124b
B. polymyxa BcP26	115b	119b	111b	118b
A. simplex ArS7	138a	135a	121a	134a
A. globiformis ArG1	110b	106d	108c	107d
M. phlei MbP18	113b	116c	116b	112c

Mean values followed by the different lower case letters in the same column were significantly different at $P \leq 0.05$.

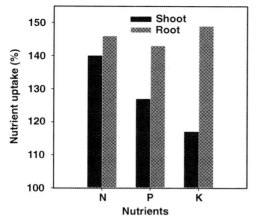

Figure 14.3 Nutrient uptake of wheat after inoculation with *B. polymyxa* BcP26.

of desert woody legumes (e.g. *Prosopis glandulosa*) grew better at 36 °C than at 26 °C [118]. These rhizobial strains are physiologically distinct, suggesting that the bacteria are highly adaptive to their respective environmental conditions. Moreover, with greater adaptation and increasing plant growth, they also enhance nutrient uptake by plants in arid saline soils. Our recent studies have shown that *Bacillus polymyxa* BcP26 and *M. phlei* MbP18 significantly increased shoot growth and root length of wheat and enhanced nitrogen uptake by 54%, phosphorus uptake by 42%, and potassium uptake by 48% over control (Figures 14.3 and 14.4).

Increased nutrient uptake by plants inoculated with effective bacteria has been attributed to the production of plant growth regulators by the bacteria at the root interface, which stimulated root growth and facilitated greater absorption of water

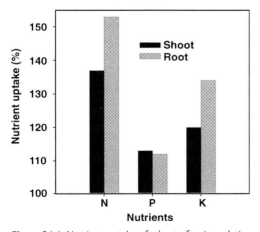

Figure 14.4 Nutrient uptake of wheat after inoculation with *M. phlei* MbP18.

and nutrients from the subsoil [119,120]. Therefore, if the increase in the amount of nutrients taken up by the PGPB-inoculated plants was simply owing to an increase in root surface area, the effect of PGPB would entirely depend on root development and not on the uptake function [23]. Furthermore, the PGPB inoculants can help plants acquire nutrients through greater mobilization. A range of diazotrophs (e.g. *Azospirillum, Azotobacter, Azorhizobium, Bacillus, Herbaspirillum* and *Klebsiella*) can supplement the use of urea in wheat production either by nitrogen fixation or growth promotion [121]. A number of studies have reported that bacterial inoculation can lead to an increase in nitrogen content in the biomass of peas and barley [110,111,121].

Both potassium and phosphorus are important macronutrients for plant growth. Since the availability of potassium and phosphorus in arid saline soil is limited, the most appropriate approach to address this situation is to use bioinoculants. These bioinoculants are able to enhance solubilization of applied phosphorus fertilizers and insoluble forms of potassium- and phosphorus-bearing minerals by excreting organic acids (e.g. citric acid) in soil [122–125]. For example, the *Bacillus edaphicus* NBT strain was found to increase potassium content of cotton and rape seed plants by 30% when the soil was treated with insoluble potassium sources [6]. On the contrary, *Bacillus megaterium* is considered to be the most effective among the phosphorus-solubilizing bacteria (PSB) [137]. A number of phosphate-solubilizing bacteria, such as NBRI0603, NBRI2601, NBRI3246 and NBRI4003 were isolated from the rhizosphere of alkaline soils used for growing chickpeas. All four strains have demonstrated diverse levels of phosphate solubilizing activity under *in vitro* conditions in the presence of various carbon and nitrogen sources [137]. The synergistic interactions of plants with coinoculation of PSB were identified to be *Azospirillum* species [38].

Application of rock phosphate along with PSB in phosphorus-deficient soil has reportedly improved phosphorus uptake by plants and enhanced crop yields, thus suggesting that PSB are effective for solubilizing phosphates and mobilizing available phosphorus to crops [126]. Phosphorus biofertilizers in the form of microorganisms can increase the solubility and availability of phosphates for plant growth [127,128]. It is reported that PSB-plant inoculations resulted in 10–15% increases in crop yields and phosphorus uptake in 10 out of 37 experiments [129]. Ectorhizospheric strains from *Pseudomonades* and *Bacilli* and endosymbiotic rhizobia have been described as effective phosphate solubilizers in soils [130,131]; however, their performance was influenced by environmental factors [132]. Highly saline soils may cause poor survival and growth of PSB. Weak organic acids, the end-products of microbial metabolism, often result in decrease of soil pH that probably plays a major role in solubilization of phosphorus-containing minerals or compounds [122,128]. In our previous work [133], the inoculation of cotton seeds with the salt-tolerant phosphate-solubilizing bacteria *R. meliloti* URM1 combined with phosphate had a significant effect on total dry matter, shoot and root production of plants. The yield of cotton was increased by 77%. The phosphorus content was significantly increased in cotton plants with treatment of PSB combined with phosphate (Table 14.10). The standard treatment with fertilizer did not affect phosphorus uptake by plants.

Table 14.10 Effects of PSB *R. meliloti* URM1 combined with phosphate on nitrogen and phosphorus uptake (%) of cotton grown in arid saline soils of Uzbekistan.

Treatments	Leaves		Stem		Bud case		Cotton fibers	
	N	P	N	P	N	P	N	P
NoPoKo	1.45b	0.51b	0.68a	0.21a	0.78a	0.19a	1.78ab	0.81a
NP$_{superp}$K	1.55a	0.75a	0.75a	0.24a	0.83a	0.22a	1.87a	0.84a
NP$_{phosphorit}$K	1.2c	0.2c	0.3b	0.1b	0.5b	0.1b	1.6b	0.4b
NP$_{PSB}$K	1.62a	0.8a	0.75a	0.24a	0.83a	0.25a	1.9a	0.89a

Mean values followed by the different lower case letters in the same column were significantly different at $P \leq 0.05$.

Other studies also reported that PSB increased phosphorus uptake with an increase in the yield of peas and barley [135]. The bacterial strains used were able to dissolve organophosphates. A number of studies reported that phosphorus uptake efficiency and crop yield increased with both phosphorus application and with inoculation [128,133,134]. Accordingly, *B. megaterium* is considered to be the most effective among the phosphorus-solubilizing bacteria. In short, phosphate-solubilizing *Pseudomonas* strains can enhance plant growth by producing plant growth regulators and vitamins, enhancing plant nutrient uptake and suppressing pathogenic or deleterious organisms [139].

14.7
Biomechanisms to Enhance Plant Growth

Plant growth promoting rhizobacteria may induce plant growth directly or indirectly. Direct influences include synthesis of phytohormones, siderophore production, nonsymbiotic nitrogen fixation and stimulation of disease resistance mechanisms [120,140,141]. Indirect effects include allelopathy, antibiotic production and competition with deleterious agents [142,143]. Some help increase the availability of essential nutrients through solubilization and chelation [144]. In addition, many PGPR produce ACC deaminase and metabolize ACC (precursors to plant ethylene) and thus reduce the inhibitive effects of stress-induced ethylene on plant root growth [21]. It is reported that a bacterium containing ACC deaminase can enhance the resistance of tomato seedlings to salt stress. However, the resistance mechanisms in plants are based on active and mutual association between bacteria and plant root surface [145].

One of the most commonly reported direct effects on plant growth by rhizobacteria is the production of plant growth substances. *B. amyloliquefaciens* and *B. subtilis* have been shown to synthesize plant growth promoting substances such as gibberellins and indole-acetic acid (IAA) in nature [146,147]. It is reported that *Azotobacter paspali* secreted IAA into culture medium and significantly increased the dry weight of leaves and roots of several plants after root treatment [148]. As mentioned earlier, the effects of PGPB on root morphogenesis is usually attributed to the release of phytohormones by rhizobacteria, auxin being the most cited of these plant growth factors. Organic substances capable of regulating plant growth by influencing physiological and morphological processes at very low concentrations were reported in a number of studies [149,150]. It is reported that the salt-tolerant rhizobacteria *A. globiformis* ArG1 produced 1.8 µg IAA 100 ml^{-1} filtrate, *A. simplex* ArS7 produced 7.3 µg IAA 100 ml^{-1} and *A. tumescens* ArT16 produced 7.4 µg IAA 100 ml^{-1} filtrate. The estimation of IAA in culture filtrates at different intervals showed a time-dependent increase in IAA production. Low amount of auxin production was also detected in cells grown in the absence of tryptophan. Production of 1.6–3.3 µg auxin ml^{-1} filtrate by rhizosphere bacteria isolated from wheat was reported in a recent study [151].

Biocontrol mechanisms are often classified based on their mode of action, such as root colonization, degradative parasitism and antibiosis or competitive antagonism [152]. Numerous studies have demonstrated the ability of several rhizobacteria to suppress diseases caused by fungus and parasitic nematodes [153,154]. The bacterial production of antifungal metabolites provides an important source of useful chemical compounds to enhance biocontrol mechanisms in nature. In general, the efficacy of the metabolites is related to their antagonistic activities and rhizosphere competence [155]. Mechanisms of bacterial antagonism toward plant pathogens include competition for food, nutrients and space, production of antibiotics and toxins or production of host cell wall degrading enzymes [156,157]. Cell wall degrading enzymes, such as β-1,3-glucanases, cellulases, proteases and chitinases, are involved in the antagonistic activity of some biological control agents against phytopathogenic fungi [158]. It is reported that β-1,3-glucanase produced by *B. amyloliquefaciens* MET0908 have shown strong activity against plant pathogens in watermelon and was found stable at high temperatures [159]. There were correlations between bacterial production of fungal cell wall degrading enzymes (e.g. chitinases, cellulases, etc.) and fungal antagonism [160,161]. Results of another study reported that *Arthrobacter* spp. that actively lyses *Fusarium roseum* was found to release protease and chitinase for biocontrol mechanisms [162]. Salt-tolerant *Arthrobacter globiformis* strain ArG1 and *M. phlei* MbP18 also produce enzymes such as lipase, collagenase, amylase and protease (Table 14.11) and were antagonistic to soil fungi (e.g. *Aspergillus flavus*, *A. insultus*, *A. ustus*, *Penicillum purpurogenum*, *P. sopii* and *Trichoderma lignorum*). These were also found antagonistic to several plant pathogenic fungi such as *Vertilcillum loteritum* and *Fusarium oxysporum*. The *P. alcaligenes* PsA15, *P. rathonis* PsR20, *B. polymyxa* BcP26 and *A. tumescens* ArT16 were found antagonist to *Fusarium oxysporum*, *F. culmorum*, *Rhizoctonia solani* and *Botrytis cinera*. In addition, isolates of *P. fluorescens* highly antagonistic to *R. solani* and *Pythium ultimum* have produced lytic enzymes with biocontrol properties [163]. Likewise, the strains of *Crinipellis perniciosa* isolated from cocoa (*Theobromae cacao*, L.) were found capable of producing extracellular enzymes to degrade cellulose, starch, lipids and lignin [164]. A number of studies have reported that the biocontrol effect of rhizobia is due to the secretion of secondary metabolites such as antibiotics and HCN [165], and the strains producing HCN were found antagonistic to fungal pathogens.

Table 14.11 Characteristics of bacterial strains isolated from sierozems soils of Uzbekistan.

Crops	Bacterial strains	HCN	Lypase	Amylase	Protease	Pectinase	Cellase
Maize	*P. alcaligenes* PsA15	+	+	+	−	+	−
Maize	*P. mendocins* PsM13	−	−	−	−	−	−
Alfalfa	*P. rathonis* PsR20	+	+	+	−	−	+
Melon	*B. amiloliquefaciens* BcA12	−	+	+	−	+	+
Cotton	*B. laevolacticus* BcL28	−	−	+	+	−	−
Melon	*B. latvianus* BcLt29	−	+	+	+	+	−
Tomato	*B. megaterium* BcMg33	−	−	−	+	+	−
Cotton	*B. polymyxa* BcP26	+	−	+	+	+	+
Wheat	*B. subtilis* TSAU180	−	+	+	+	+	−
Maize	*Mycobacterium phlei* MbP18	−	+	+	−	+	+
Wheat	*Mycoplana bullata* MpB46	−	+	+	−	−	−
Tomato	*A. globiformis* ArG1	+	+	+	+	−	−
Wheat	*A. simplex* ArS43	−	+	−	−	−	−
Alfalfa	*A. tumescens* ArT16	+	+	+	+	+	−

+ and − indicate the presence or absence of activity by a particular microorganism.

Recently, *Bacillus* has received much attention as a biocontrol agent because of its advantages over gram-negative bacteria and fungal biocontrol agents [12]. Strains of *Bacillus* have several advantages over other biocontrol bacteria owing to their production of several broad-spectrum antibiotics and their longer shelf lives as a result of their ability to form endospores [166]. They can be applied as spores on plant seeds or as inoculants to provide significant protection against several microbial pathogens as well as promote plant growth [167]. In our earlier studies, the salt-tolerant bacterial strain *B. subtilis* TSAU180, which has the ability to produce fungal cell wall degrading enzymes, significantly reduced diseases caused by *Fusarium oxysporum* and *Fusarium culmorum* in tomatoes and wheat in highly saline conditions. Although the presence of the pathogenic fungus caused diseases in 45% of the plants, using biological control organisms *B. subtilis* TSAU180 significantly reduced the diseases by 25–28%.

14.8
Conclusions

Salt-tolerant rhizobacteria isolated from arid saline soils have shown great potential to synthesize different biologically active compounds to control soilborne diseases and pathogens and improve plant establishment in ecologically disturbed environments. Study results show that seed inoculation with *B. japanicum* S2492, *B. polymyxa* BcP26 and *M. phlei* MbP18 enhanced nitrogen, phosphorus and potassium uptake by 40–50% and significantly increased cotton, soybean and wheat yields as compared to control. Salt-tolerant *A. globiformis* strain ArG1, *M. phlei* MbP18, *P. alcaligenes* PsA15, *P. rathonis* PsR20, *B. polymyxa* BcP26 *and A. tumescens* ArT16 were found very effective in controlling soilborne diseases and pathogens. These characteristics are useful in producing commercial inoculants for routine use to support plant production in stressed environments.

14.9
Future Directions

Although a number of rhizobacteria have been identified that are capable of tolerating extreme climates, their potential synergistic effects on plants or antagonistic effects on soilborne pests and diseases are still questionable. However, they have potential use to support plant establishment and growth under dry and saline environments. To further understand the highly complex nature of microbial adaptation and response to alterations in the biological, chemical and physical environment of the rhizosphere remains a significant challenge for plant biologists and soil microbiologists. Future research must address the following:

- Proper understanding of the behavior of rhizosphere microbes from a wide range of plants under extreme climatic conditions;
- Isolation, identification and comparison of rhizospheric microbes to develop a genetic data bank; and
- Development of more effective microbial strains, with longer shelf lives, as a 'biocontrol' to supplement and/or complement chemical fertilizers and pesticides in agriculture.

References

1 Flowers, T.J., Troke, P.F. and Yeo, A.R. (1977) *Annual Review of Plant Physiology and Plant Molecular Biology*, **28**, 89–121.

2 Requena, N., Perez-Solis, E., Azcón-Aguilar, C., Jeffries, P. and Barea, J.M. (2001) *Applied and Environmental Microbiology*, **67**, 495–498.

3 Soussi, M., Ocana, A. and Lluch, C. (1998) *Journal of Experimental Botany*, **49**, 1329–1337.

4 Cordowilla, M., Ligero, F. and Lluch, C. (1999) *Plant Science*, **140**, 127–136.

5 Hallman, J., Quadt-Hallman, A., Malhaffee, W.F. and Kloepper, J.W. (1997) *Canadian Journal of Microbiology*, **43**, 895–914.

6 Sheng, X.F. (2005) *Soil Biology & Biochemistry*, **37**, 1918–1922.

7 Hornby, D. (1990) *Biological Control of Soil-Born Plant Pathogens*, CAB International, UK.

8 Okon, Y., Bloemberg, G.V. and Lugtenberg, J.J. (1998) *Agricultural Biotechnology* (ed. A. Altman), Marcel Dekker, New York, pp. 327–349.

9 Rangarajan, S., Saleena, L.M. and Nair, S. (2002) *Microbial Ecology*, **43**, 280–289.

10 Zahran, H.H.J. (2001) *Biotechnology*, **91**, 143–153.

11 Mayak, S., Tirosh, T. and Glick, B.R. (2004) *Plant Physiology and Biochemistry*, **42**, 565–572.

12 Lugtenberg, B.J.J., Dekkers, L. and Bloemberg, G.V. (2001) *Annual Review of Phytopathology*, **39**, 461–490.

13 Handelsman, J. and Stabb, E.V. (1996) *Plant Cell*, **8**, 1855–1869.

14 Rainey, P.B. (1999) *Environmental Microbiology*, **1**, 243–257.

15 Egamberdiyeva, D. and Hoflich, G. (2003) *Archive of Agronomy and Soil Science*, **49(2)**, 203–213.

16 Pascual, A., Garcia, C., Hernandez, T., Moreno, J.L. and Ros, M. (2000) *Soil Biology & Biochemistry*, **32**, 1877–1883.

17 García, C., Hernandez, T., Albaladejo, J., Castillo, V. and Roldan, A. (1998) *Soil Science Society of America Journal*, **62**, 670–676.

18 Egamberdiyeva, D., Gafurova, L. and Islam, K.R. (2007) *Climate Change and Terrestrial C Sequestration in Central Asia* (eds R. Lal, M. Sulaimonov, B.A. Stewart, D. Hansen and P. Doraiswamy), Taylor and Francis, New York.

19 Atlas, R.M. (1984) *Advances in Microbial Ecology*, **7**, 1–47.

20 Lugtenberg, B.J.J., Chin-A-Woeng, T.F.C. and Bloemberg, G.V. (2002) *Antonie van Leeuwenhoek*, **81**, 373–383.

21 Glick, B.R., Jacobson, C.B., Schwarze, M.M.K. and Pasternak, J.J. (1994) *Canadian Journal of Microbiology*, **40**, 911–915.

22 Goldstein, A.H. and Liu, S.T. (1987) *Biotechnology*, **5**, 72–74.

23 Mantelin, S. and Touraine, B. (2004) *Journal of Experimental Botany*, **55**, 27–34.

24 Oren, A. (1993) *The Biology of Halophilic Bacteria* (eds Vreeland R.H. and Hochstein L.I.), CRC Press, Boca Raton, FL, pp. 26–53.

25 Tripathi, A.K., Verma, S.C. and Ron, E.Z. (2002) *Research in Microbiology*, **153**, 579–584.

26 Gupta, A., Saxena, A.K., Gopal, M. and Tilak, K.V.B.R. (2003) *Tropical Agriculture*, **80**, 28–35.

27 Miller, A.K.J. and Wood, J.M. (1996) *Annual Review of Microbiology*, **50**, 101–136.

28 Zahran, H.H. (1991) *Biology and Fertility of Soils*, **12**, 73–80.

29 Zahran, H.H., Moharram, A.M. and Mohammad, H.A. (1992) *Journal of Basic Microbiology*, **32**, 405–413.

30 Zahran, H.H., Ahmed, M.S. and Afkar, E.A. (1995) *Journal of Basic Microbiology*, **35**, 269–275.

31 Oren, A. (1994) *FEMS Microbiology Reviews*, **13**, 415–440.

32 Kamekura, M. (1998) *Extremophiles*, **2**, 289–295.

33 Loganathan, P. and Nair, S. (2004) *International Journal of Systematic and Evolutionary Microbiology*, **54**, 1185–1190.

34 Alexander, M. (1977) *Introduction to Soil Microbiology*, 2nd edn, John Wiley & Sons, Ltd, New York.

35 Cervantes, L.A. and Olivares, V.J. (1976) *Revista Latinoamericana de Microbiologia*, **18**, 73–76.

36 Logan, N.L. (2002) *Applications and Systematic of Bacillus and Relatives* (eds R. Berkeley, M. Hendricks, N. Logan and P. De Vos), Blackwell Science Ltd, Oxford, pp. 123–140.

37 Sunna, A., Tokajlan, S., Burgardt, J., Rainey, F., Autranikian, G. and Hashawa, F. (1997) *Systematic and Applied Microbiology*, **20**, 232–237.

38 Walker, R., Powell, A.A. and Seddon, B. (1998) *Journal of Applied Microbiology*, **84**, 791–901.

39 Sato, K. and Jiang, H.Y. (1996) *Biology and Fertility of Soils*, **23**, 121–125.

40 Zahran, H.H. (1999) *Microbiology and Molecular Biology Reviews*, **63**, 968–989.

41 Boncompagni, E., Østerå, S.M., Poggi, M. and le Rudulier, D. (1999) *Applied and Environmental Microbiology*, **65**, 2072–2077.

42 Delgado, M.J.F. and Lluch, L.C. (1994) *Soil Biology & Biochemistry*, **26**, 371–376.

43 Kassem, M., Capellano, A. and Gounot, M.A. (1995) *Journal of Applied Microbiology and Biotechnology*, **1**, 63–75.

44 Zahran, H.H. and Sprent, J.I. (1986) *Planta*, **167**, 303–309.

45 Tripathi, A.K., Mishra, B.M. and Tripathi, P. (1998) *Journal of Biosciences*, **23**, 463–471.

46 Khammas, K.M., Ageron, E., Grimont, P.A.D. and Kaiser, P. (1989) *Research in Microbiology*, **140**, 679–693.

47 Sanchez-Porro, C., Martin, S., Mellado, E. and Ventosa, A. (2003) *Journal of Applied Microbiology*, **94**, 295–300.

48 Bastos, A.E.R., Moon, D.H., Rossi, A., Trevors, J.T. and Tsai, S.M. (2004) *Archives of Microbiology*, **174**, 346–352.

49 Berg, G., Eberl, L. and Hartmann, A. (2005) *Environmental Microbiology*, **7**, 1673–1685.

50 Granowitz, E.V. and Keenholtz, S.L. (1998) *American Journal of Infection Control*, **26**, 146–148.

51 Roberts, D.P., Dery, P.D., Yucel, I. and Buyer, J.S. (2000) *Applied and Environmental Microbiology*, **66**, 87–91.

52 Morales, A., Garland, J.L. and Lim, D.V. (1996) *FEMS Microbiology Ecology*, **20**, 155–162.

53 Gilbert, G.S., Parke, J.L., Clayton, M.K. and Handelsman (1993) *Journal of Ecology*, **74**, 840–854.

54 Wilkinson, K.G., Sivasithamparam, K., Dixon, K.W., Fahy, P.C. and Bradley, J.K. (1994) *Soil Biology & Biochemistry*, **26**, 137–142.

55 Kloepper, J.W. and Beauchamp, C.J. (1992) *Canadian Journal of Microbiology*, **38**, 1219–1232.

56 Kloepper, J.W., McInroy, J.A. and Bowen, K.L. (1992) *Plant Soil*, **139**, 85–90.

57 Chadwick, M. (1981) Mycobacteria, Institute of Medical Laboratory Sciences Monographs, Wright, Bristol.

58 Tarnok, I. and Tarnok, Z. (1970) *Tubercle*, **51**, 305–312.

59 David, H.L. (1973) *The American Review of Respiratory Disease*, **108**, 1175–1185.

60 De Weger, L.A., Van der Bij, A.J., Dekkers, L.C., Simons, M., Wijffelmann, C.A. and Lugtenberg, B.J.J. (1995) *FEMS Microbiology Ecology*, **17**, 221–228.

61 Höflich, G., Wiehe, W. and Hecht-Buchholz, C. (1995) *Microbiological Research*, **150**, 39–47.

62 De Freitas, J.R. and Germida, J.J. (1992) *Soil Biology & Biochemistry*, **24**, 1127–1135.

63 Bloemberg, G.V. and Lugtenberg, B.J.J. (2001) *Current Opinion in Plant Biology*, **4**, 343–350.

64 McCully, M.E. (2001) *Australian Journal of Plant Physiology*, **28**, 983–990.

65 Van Overbeek, L.S. and van Elsas, J.D. (1997) *Chemosphere*, **39**, 665–682.

66 Williams, S.T., Shameenmullah, M., Watson, E.T. and Mayfield, C.I. (1972) *Soil Biology & Biochemistry*, **4**, 215–225.

67 Latour, X., Corberand, T., Laguerre, G., Allard, F. and Lemanceau, P. (1996) *Applied and Environmental Microbiology*, **62**, 2449–2456.

68 Rattray, E.A.S., Tyrrel, J.A., Prosser, J.L., Glover, L.A. and Killham, K. (1993) *European Journal of Soil Biology*, **29**, 73–82.

69 Bowers, J.H. and Parke, J.L. (1993) *Soil Biology & Biochemistry*, **25**, 1693–1701.

70 Egamberdiyeva, D. and Hoflich, G. (2003) *Soil Biology & Biochemistry*, **35**, 973–978.

71 Thompson, I.P., Cook, K.A., Lethbridge, G. and Burns, R.G. (1990) *Soil Biology & Biochemistry*, **22**, 1029–1037.

72 Loper, J.E., Haack, C. and Schroth, M.N. (1985) *Applied and Environmental Microbiology*, **49**, 416–422.

73 Van Veen, J.A., Van Overbeek, L.S. and van Elsas, J.D.((1997) *Microbiology and Molecular Biology Reviews*, **61**, 121–135.

74 Sinclair, J.L. and Alexander, M. (1984) *Applied and Environmental Microbiology*, **48**, 410–415.

75 Acea, M.J. and Alexander, M. (1988) *Soil Biology & Biochemistry*, **20**, 703–711.

76 Madkour, M.A., Smith, L.T. and Smith, G. M. (1990) *Applied and Environmental Microbiology*, **56**, 2876–2881.

77 Dommergues, Y.R., Belser, L.W. and Schmidt, E.L. (1978) *Advances in Microbial Ecology*, **2**, 49–104.

78 Schlegel, H.G. and Jannasch, H.W. (1981) *The Prokaryotes: A Handbook on Habitats Isolation, and Identification of Bacteria* (eds M.P. Starr, H. Stolp, H.G. Truper, A. Barlows and H.G. Schlegel), Springer-Verlag, New York, pp. 41–82.

79 Mohammad, R.M., Akhavan-Kharazian, M., Campbell, W.F. and Rumbaugh, M. D. (1991) *Plant Soil*, **134**, 271–276.

80 Gilmour, D. (1990) *Microbiology of Extreme Environments*, Open University Press, Milton Keynes, UK, pp. 147–177.

81 Kushner, D. and Kamekura, M. (1988) *Halophilic Bacteria* (ed. F. Rodriguez-Valera), CRC Press, Boca Raton, FL, pp. 109–138.

82 Welsh, D.T. (2000) *FEMS Microbiology Reviews*, **24**, 263–290.

83 Edwards, C. (1990) *Microbiology of Extreme Environments*, McGraw-Hill, New York.

84 Graham, P. (1992) *Canadian Journal of Microbiology*, **38**, 475–484.

85 Hungria, M., Franco, A.A. and Sprent, J.I. (1993) *Plant Soil*, **149**, 103–109.

86 Brock, T.D. (1978) *Thermophilic Microorganisms and Life at High Temperatures*, Springer-Verlag, New York.

87 Ratledge, C. (1982) *The Biology of the Mycobacteria* (eds C. Ratledge and J. Stanford), Academic Press, London, pp. 188–190.

88 Surange, S., Wollum, A.G., Kumar, N. and Nautiyal, C.S. (1997) *Canadian Journal of Microbiology*, **43**, 891–894.

89 Morita, R.Y. and Moyer, C.L. (1989) Proceedings of the 5th International Symposium on Microbial Ecology, Japan Scientific Societies Press, Tokyo, p. 8.

90 Poindexter, J.S. (1981) *Microbial Reviews*, **48**, 123–179.

91 Grey, T.R.G. (1976) *The Survival of Vegetative Microbes* (eds T.R.G. Gray and J. R. Postgate), Cambridge University Press, New York, pp. 327–364.

92 Keddie, R.M. and Jones, D. (1981) *The Prokaryotes: A Handbook on Habitats Isolation and Identification of Bacteria* (eds M.P. Starr, H. Stolp, H.G. Truper, A. Barlows and H.G. Schlegel), Springer Verlag, Berlin and New York, pp. 1838–1878.

93 Kuznetsov, S.I., Dubinina, G.A. and Lapteva, N.A. (1979) *Annual Review of Microbiology*, **33**, 377–387.

94 Chapman, S.J. and Gray, R.R.G. (1981) *Soil Biology & Biochemistry*, **13**, 11–18.

95 Nelson, L.M. and Parkinson, D. (1978) *Canadian Journal of Microbiology*, **24**, 1460–1467.

96 Dawes, E.A. (1976) *The Survival of Vegetative Microbes* (eds T.R.G. Gray and J. R. Postgate), Cambridge University Press, New York, 19–56.

97 Conn, H. (1948) *Journal of Bacteriological Reviews*, **12**, 257–273.

98 Tyler, B. (1978) *Annual Review of Biochemistry*, **47**, 1127–1162.

99 Poindexter, J.S. (1987) *Symposium of the Society for General Microbiology*, **41**, 283–317.

100 Egamberdiyeva, D. (2005) *The Scientific World Journal*, **5**, 501–509.

101 Olsen, R.A. and Bakken, L.R. (1987) *Microbial Ecology*, **13**, 59–74.

102 Paula, M.A., Urquiaga, S., Siqueira, I.O. and Döbereiner, J. (1992) *Biology and Fertility of Soils*, **14**, 61–66.

103 Wullstein, L.H. (1989) *Arid Soil Research and Rehabilitation*, **3**, 259–265.

104 Smith, R.S. (1992) *Canadian Journal of Microbiology*, **25**, 739–745.

105 Dashti, N., Zhang, F., Hynes, R. and Smith, D.L. (1997) *Plant Soil*, **188**, 33–41.

106 Molla, A.H., Shamsuddin, Z.H., Halimi, M.S., Morziah, M. and Puteh, A.B. (2001) *Soil Biology & Biochemistry*, **33**, 457–463.

107 Bai, Y., D'Aoust, F., Smith, D.L. and Driscoll, B.T. (2002) *Canadian Journal of Microbiology*, **48**, 230–238.

108 El-Maksoud, A., Moawad, H. and Saad, R. H. (1994) *Egyptian Journal of Microbiology*, **30** (3), 401–414.

109 Egamberdiyeva, D., Qarshieva, D. and Davranov, K. (2004) *Plant Growth Regulation*, **23**, 54–57.

110 Zhang, F., Dashti, N., Hynes, H. and Smith, D.L. (1996) *Annals of Botany*, **77**, 453–459.

111 Ping, S.Z., Lin, M., You, C.B. and Malik, K.A. (1996) *Journal of Agricultural Biotechnology*, **4**, 87–92.

112 Freiberg, C., Fellay, R., Bairoch, A., Broughton, W.J., Rosenthal, A. and Perret, X. (1997) *Nature*, **387**, 394–400.

113 Egamberdiyeva, D. and Hoflich, G. (2004) *Journal of Arid Environments*, **56**, 293–301.

114 Lazarovitz, G. and Norwak. (1997) *Journal of Horticultural Science*, **32**, 188–192.

115 Höfte, M., Boelens, J. and Vstrate, W. (1991) *Soil Biology & Biochemistry*, **23**, 407–410.

116 Javed, M. and Arshad, M. (1997) *Pakistan Journal of Botany*, **29**, 243–248.

117 Paau, M.A. (1989) *Applied and Environmental Microbiology*, **55**, 862–865.

118 Waldon, H.B., Jenkins, M.B., Virginia, R. A. and Harding, E.E. (1989) *Applied and Environmental Microbiology*, **55**, 3058–3064.

119 Höflich, G., Wolf, H.J., Chiefer, C., Kuhn, G., Hickisch, B. and Eich, D. (1987) *Agronomy and Soil Science*, **31**, 763–769.

120 Lifshitz, R., Kloepper, J.W., Kozlowski, M., Simonson, C., Carlson, J., Tipping, E.M. and Zaleska, I. (1987) *Canadian Journal of Microbiology*, **33**, 390–395.

121 Kennedy, I.R. and Islam, N. (2001) *Australian Journal of Experimental Agriculture*, **41**, 447–457.

122 Subba Rao, N.S. (1982) *Advances in Agricultural Microbiology* (ed. N.S. Subba Rao), Butterworth Science, London.

123 Subba Rao, N.S. (1982) *Biofertilizers in Agriculture*, 2nd edn, Oxford and IBH Publishing Co., New Delhi.

124 Groudev, S.N. (1987) *Acta Biotechnologica*, **7**, 299–306.

125 Ullman, W.J., Kirchman, D.L. and Welch, S.A. (1996) Laboratory evidence for microbially mediated silicate mineral dissolution in nature, *Chemical Geology*, **132**, 11–17.

126 Rogers, R.D. and Wolfram, J.H. (1993) *Phosphorus Sulfur, and Silicon Related Elements*, **77**, 137–140.

127 Goldstein, A.H. (1986) *American Journal of Alternative Agriculture*, **1**, 57–65.

128 Gyaneshwar, P. Naresh Kumar, G. and Parekh, L.J. (1998) *World Journal of Microbiology and Biotechnology*, **14**, 669–673.

129 Tandon, H.L. (1987) Fertilizer Development and Consultation Organization, New Delhi, India.

130 Igual, J.M., Valverde, A., Cervantes, E. and Velázquez, E. (2001) *Agronomy Journal*, **21**, 561–568.

131 Rodriguez, H. and Fraga, R. (1999) *Biotechnology Advances*, **17**, 319–339.

132 Yahya, A.I. and Al-Azawi, S.K. (1989) *Plant Soil*, **117**, 135–141.

133 Egamberdiyeva, D., Juraeva, D., Poberejskaya, S., Myachina, O., Teryuhova, P., Seydalieva, L. and Aliev, A. (2004) Proceedings of the 26th Annual Conservative Tillage Conference on Sustainable Agriculture Auburn, USA.

134 Shah, P., Kakar, K.M. and Zada, K. (2001) *Plant Nutrition – Food Security and Sustainability of Agroecosystems* (ed. W.J. Horst), Springer, The Netherlands, pp. 670–671.

135 Chaykovskaya, L.A., Patyka, V.P. and Melnychuk, T.M. (2001) *Plant Nutrition – Food Security and Sustainability of Agroecosystems* (ed. W.J. Horst), Springer, The Netherlands, pp. 68–669.

136 Asea, P.E.A., Kusey, R.M.N. and Stewart, J.W.B. (1988) *Soil Biology & Biochemistry*, **20**, 459–464.

137 Nautiyal, C.S. (2000) *FEMS Microbiology Letter*, **182**, 291–296.

138 Belimov, A.A., Kojemiakov, A.P. and Chuvarliyeva, C.V. (1995) *Plant Soil*, **173**, 29–37.

139 Glick, B.R. (1995) *Canadian Journal of Microbiology*, **41**, 109–117.

140 Bothe, H., Körsgen, H., Lehmacher, T. and Hundeshagen, B. (1992) *Symbiosis*, **13**, 167–179.

141 Bowen, G.D. and Rovira, A.D. (1999) *Advances in Agronomy*, **66**, 1–102.

142 Kloepper, J.W., Leong, J. and Shroth, M. N. (1980) *Current Microbiology*, **4**, 317–320.

143 Höflich, G., Wiehe, W. and Kühn, G. (1994) *Experientia*, **50**, 897–905.

144 Atlas, R.M. and Bartha, R. (1987) *Microbial Ecology: Fundamentals and*

Applications, 2nd edn,Benjamin-Cummings, Menlo Park, CA.

145 Kennedy, I.R., Choudhury, A.T. and Kecskes, M.L. (2004) *Soil Biology & Biochemistry*, **36**, 1229–1244.

146 Turner, J.T. and Backman, P.A. (1991) *Plant Disease*, **75**, 347–353.

147 Idriss, E.E., Makarewicz, O., Farouk, A., Rosner, K., Greiner, R., Bochow, H., Richter, T. and Borriss, R. (2002) *Microbiology*, **148**, 2097–2109.

148 Barea, J.M. and Brown, M.E. (1974) *The Journal of Applied Bacteriology*, **37**, 583–593.

149 Arshad, M. and Frankenberger, W.T.J. (1998) *Advances in Agronomy*, **62**, 145–151.

150 Chanway, C.P. (2002) *B. subtilis for Biocontrol in Variety of Plants* (eds R. Berkeley, M. Hendricks, N. Logan and P. De Vos), Blackwell Publishing, Malden, MA, pp. 219–235.

151 Leinhos, V. and Vasek, O. (1994) *Microbiological Research*, **149**, 31–35.

152 Kim, Y.S. and Kim, S.D. (1994) *Journal of Microbiology and Biotechnology*, **4**, 296–304.

153 Becker, J.O., Sabaleta-Mejia, E., Colbert, S.F., Schroth, M.N., Weinhold, A.R., Hancock, J.C. and Van Gundy, S.D. (1988) *Phytopathology*, **78**, 1466–1469.

154 Weller, D.M. (1988) *Annual Review of Phytopathology*, **73**, 463–469.

155 Cook, R.J., Thomashow, L.S., Weller, D. M., Fujimoto, D., Mazzola, M., Banger, G. and Kim, D.S. (1995) *Proceedings of the National Academy of Sciences of the United States of America*, **92**, 4197–4201.

156 Fravel, D.R. (1988) *Annual Review of Phytopathology*, **26**, 75–91.

157 Chet, I., Ordentlich, A., Shapira, R. and Oppenheim, A. (1990) *Plant Soil*, **129**, 85–92.

158 Patrick, G.M., Alice, G., Christopher, D. L., Michelle, M.C., Timothy, H., Angela, V.S., Antoni, P. and Maria, G.T. (2001) *Enzyme and Microbial Technology*, **29**, 90–98.

159 Kim, P. (2004) *FEMS Microbiology Letters*, **234**, 177–183.

160 Abd Rahman, R.N.Z., Geok, L.P., Basri, M. and Abd Rahmen, B.S. (2005) *Bioresource Technology*, **96**, 429–436.

161 Caballero, A.R., Moreau, J.M., Engel, L.S., Marquart, M.E., Hill, J.M. and O'Callaghan, R. (2001) *Journal of Analytical Biochemistry*, **290**, 2, 330–337.

162 Morrissey, R.F., Dugan, E.P. and Koths, J. S. (1976) *Soil Biology & Biochemistry*, **8**, 23–28.

163 Nielsen, M.N., Sorensen, J., Fels, J. and Pedersen, H.C. (1998) *Applied and Environmental Microbiology*, **64**, 3563–3569.

164 Cleber, N. (2005) Production of extracellular enzymes by *Crinipellis perniciosaFitopatologia Brasileira*, **30**, 286–288.

165 Deshwal, V.K., Dubey, R.C. and Maheshwari, D.K. (2003) *Current Science*, **84**, 443–444.

166 Emmert, E.A.B. and Handelsman, J. (1999) *FEMS Microbiology Letters*, **171**, 1–9.

167 Reva, O.N., Dixelius, C., Meijer, J. and Priest, F.G. (2004) *FEMS Microbiology Ecology*, **48**, 249–259.

15

The Use of Rhizospheric Bacteria to Enhance Metal Ion Uptake by Water Hyacinth, *Eichhornia crassipe* (Mart)

Lai M. So, Alex T. Chow, Kin H. Wong, and Po K. Wong

15.1
Introduction

Contamination of the aquatic environment by toxic metal ions is a serious problem worldwide [1–7]. Unlike organic pollutants, toxic metal ions cannot be degraded by chemical or biological process. To remediate the aquatic environment, toxic metal ions should therefore be concentrated in a form that can be extracted conveniently, possibly for reuse or at least for proper disposal. To achieve cost-effectiveness, toxic metal ions should be concentrated into a small mass. Conventional treatment methods, such as chemical precipitation, chemical oxidation or reduction, ion exchange, filtration, membrane technologies or evaporation process, are generally not effective at low metal concentration and are expensive [6,8]. Biological methods, such as using microbial biomass to remove metal ions, are well documented [9–18]; however, the performance of the microbe-based system is very sensitive to external physical and chemical factors and the operation costs incurred in cell culture and separation from effluent, cell immobilization and technical support [13] is quite high. Phytoremediation provides a viable alternative for metal ion removal with the merits of hyperaccumulating metal ions using low energy and at low cost.

In the plant rhizosphere, bacteria interact with roots to form mucigel, composed of plant mucilage and bacterial cells, at the root surface [19–25]. The interaction creates a 'rhizospheric effect' [19,22,25–27]. The microorganisms and their products interact with plant roots influencing plant growth and development, as well as change nutrient dynamics and susceptibility to disease and abiotic stress [25,28]. However, the role of rhizospheric bacteria in metal ion accumulation and tolerance of plants is not well known. Screening for metal-ion-resistant and/or accumulating bacteria from the rhizosphere and utilizing the natural colonizing ability of bacteria may increase the efficiency of plants to remove metal ion using the root surface for adhesion of metal ion removing bacteria. This principle is similar to application of biotechnology of immobilized cells on solid surface for metal ion removal [29,30]. This chapter attempts to combine these two technologies on water hyacinth, *Eichhornia crassipe* (Mart), a common aquatic plant used in phytoremediation.

Plant-Bacteria Interactions. Strategies and Techniques to Promote Plant Growth
Edited by Iqbal Ahmad, John Pichtel, and Shamsul Hayat
Copyright © 2008 WILEY-VCH Verlag GmbH & Co. KGaA, Weinheim
ISBN: 978-3-527-31901-5

15.2
Overview of Metal Ion Pollution

Heavy metals are generally defined as a group of approximately 65 metallic elements with density greater than $5\,g\,cm^{-3}$ [31]; for example, cadmium, chromium, copper, mercury, nickel, lead and zinc, plus the metalloids arsenic and selenium, which have long been contaminants in sewage [32]. These metal ions possess diverse physical, chemical and biological properties and have the ability to exert toxic effects on microbial and other life forms [33,34]. The ultimate cause for concern about metal ions in the environment is their toxicity toward humans [35,36] and other biota. Owing to their nondegradable nature, they will persist and cycle in different compartments of the environment. In addition to the accumulation of metals within food chains, elevated metal levels in the environment pose serious threats to human health [37]. Although some metal ions such as Cr^{6+}, Cu^{2+}, Ni^{2+} and Zn^{2+} are essential to life, they are required only in trace amounts and become toxic once they exceed the threshold bioavailable level [31].

Metal ion toxicity may be manifested from either an acute, single high exposure or chronic, long-term exposure to low concentrations [1]. Metals exert toxicity in man and animals in variable degrees by deactivating enzymes, replacing essential metal ions such as Ca^{2+} and Mg^{2+} or destabilizing biomolecules such as nucleic acids, which results in genotoxic or mutagenic effects producing heritable genetic disorders and cancers [1].

The rapid, unbridled industrialization without environmental controls has led to the serious contamination of our environment by metal ions [38]. The US Environmental Protection Agency (USEPA) estimates that there are 30 000 candidate sites for hazardous waste treatment services in the United States alone, including industrial sites that contain liquid and solid wastes contaminated with metals ions [39]. Current cleanup costs for all Superfund (National Priority) sites are estimated to be $16.5 billion [39–41]. Approximately 15% of these sites are contaminated by metals only, whereas 64% of the sites contain metals mixed with organic wastes [39,42].

Metals enter the aquatic environment by various sources that are divided into two main categories: anthropogenic and nonanthropogenic [1]. Natural geological processes release metal ions into the environment by weathering, erosion and runoff from the lithosphere. In the United States, it has been estimated that $1 million is spent every day to clean up 12 000 miles (31 080 km) of rivers and 180 000 acres (728 km^2) of lakes contaminated by mining wastes in which metal ions are the major pollutant, and the total cleanup costs of acid mine drainage have been estimated to be as high as $70 billion [6]. The major source of metal ions is human activities, which include mining, agriculture and industrial processes such as electroplating, oil refining, paper and textiles manufacturing [43]. For the major metal bearing industrial waste streams, the market for treatment of metals in the United States is estimated to be in the range of $1–2 billion per year [6,44]. To maintain a sustainable use of our water resources, an efficient and cost-effective technique must be developed to achieve a high effluent standard.

15.3
Treatment of Metal Ions in Wastewater

15.3.1
Conventional Methods

A variety of methods are available to remove metal ions from water, which include ion exchange, electrochemical reduction, adsorption, membrane filtration, chemical precipitation and reverse osmosis [6,8,42,45,46]. The most common method cleaning up the metal ion contaminated water is chemical treatment, by which metal ions are precipitated using basic reagents such as hydroxides (e.g. lime) to form sludge, which is then disposed off as a hazardous waste or, in some cases, reused or recycled. Chemical treatment remains the favored approach because it is reliable, efficient and reasonably inexpensive ($0.18–0.26/1000 l) [6]. However, the sludge thus created is difficult and costly to handle or dispose off. For some metal ions with special characteristics, large amounts of flocculating agents are required, resulting in the formation of excessive metal sludge that requires high disposal costs [45]. For instance, to remove 1 lb (0.45 kg) of copper from $1000\,\mathrm{mg\,l^{-1}}\,Cu^{2+}$ solution, costs average $76.99/1000 l of copper ion containing wastewater [39]. Over 80% of the cost is for disposal of the hazardous sludge. Moreover, this technology may prove very costly if large volumes of low metal concentration and high cleanup standards are involved.

15.3.2
Microbial Methods

Bacteria, algae and fungi can remove metal ions from the external environment by means of metabolism-dependent and metabolism-independent process to take up and accumulate metals on the cell surface and inside the cells [13]. Microbial cells can accumulate metal ions both in metabolism-dependent ways by precipitation, redox reactions and ion transport systems and in metabolism-independent ways by biosorption [13,47]. From toxicological perspective, microorganisms accumulate metal ions on the cell surface and within the cell by their metabolic activities, such as formation of metal sulfides by sulfate-reducing bacteria [48], oxidation of Fe^{2+} to Fe^{3+} by iron-oxidizing bacteria [49] and transportation of metal ions into cytoplasm by ion pumps; they are detoxified by complexing with siderophores or metallothionein in cytoplasm [18,50]. Biosorption involves nonactive uptake of metal ions by microbial biomass and includes physical adsorption, ion exchange, complexation, precipitation, crystallization and diffusion [12,18]. Metallic cations are attracted to negatively charged sites at the surface of the cells. The anionic ligands, which participate in metal binding, include phosphoryl, carboxyl, sulfuydryl and hydroxyl groups of membrane proteins [51].

The ability of microorganisms to remove metal has been utilized in metal recovery or metal-laden wastewater remediation. The process involves living or dead cells in a batch system or an immobilized cell system. Biosorption processes are more effective in metal removal than conventional methods when the metal concentration in wastewater is below $100\,mg\,l^{-1}$ and the effluent contains less than $1\,mg\,l^{-1}$ of metal ions [52]. However, microbe-based technologies have a number of disadvantages such as small particle size, low mechanical strength and low density, which make biomass-effluent separation difficult. The problem can be mediated using immobilized cells. The cost involved in immobilization is high because a well-adjusted chemical and physical environment is important to maintain the adhesion of cells and metal ion removal capacity of microbial cells. In addition to the cost of cell culture, the system should attain more than 99% metal ion removal efficiency with loading capacity greater than $150\,mg$ metal ions g^{-1} for a competitive niche [13].

As some metal ions are highly toxic in nature, the wastewater containing metal must be 'polished' after conventional treatment to meet discharge standards [53]. Thus, there is a growing interest in developing a reliable and inexpensive technology that can reduce toxic concentrations of metal ions to environmentally acceptable levels or to recover natural metal resources [53–56].

15.3.3
Phytoremediation

15.3.3.1 An Overview of Phytoremediation

Phytoremediation is defined as the use of green plants to remove pollutants such as metal ions from the environment [41,42,57–61]. This plant-based remediation technology mainly depends on the metal ion hyperaccumulating properties of certain plants [39,62,63]. The term 'hyperaccumulator' is used to describe a plant with a highly abnormal level of metal ion accumulation [64]. Baker and Brooks [65] have defined hyperaccumulators as plants that contain more than $1\,mg\,g^{-1}$ (0.1%) of Co^{2+}, Cu^{2+}, Cr^{6+}, Pb^{2+} or Ni^{2+} or $10\,mg\,g^{-1}$ (1%) of Mn^{2+} or Zn^{2+} in dry matter [60]. Studies on hyperaccumulator species by collecting plants in metal ion contaminated areas have been initiated. A detailed review of this project can be found in Reeves and Baker [64].

The goal of current phytoremediation efforts is to develop innovative, economical and environmentally compatible approaches to remove metal ions from the environment. Many studies have revealed the feasibility and importance of phytoremediation [40–42,60,61,66,67]. The most important features of phytoremediation include lower costs for treatment and the generation of a potentially recyclable metal ion rich plant residue [39]. Its cost, which includes the entire capital and operating expenses, is far below those of many competing technologies (Table 15.1). Phytoremediation also offers a cost advantage in wastewater treatment, because plants can remove up to 60% of their dry weight as metal ions, thus markedly reducing the disposal cost of the hazardous residue [39]. Ideally, one can use a good metal ion accumulator with a high accumulation rate, fast growth and high biomass production; however, there is no single species that fulfils all these requirements. Phytoremediation has its disadvantages (Table 15.2); the greatest limitation is that it needs a long time and large area. The time to remediate a site depends on the life cycle of plants and their growth requirements. Phytoremediation also needs a large surface area, which depends on the uptake rate of the plants, to accommodate the treatment

Table 15.1 Available technologies for remediation of metal ions
and radionuclides in wastewater [6].

Available technology	Cast ($1000 l^{-1} wastewater treated)
Ion exchange	0.03–0.38
Electrochemical reduction	0.13
Adsorption	0.22–4.40
Membrane technology: separation–filtration	0.26–1.28
Chemical precipitation	0.18–0.26
Reverse osmosis	0.66–1.36
Activated carbon adsorption	26.18–46.19
Biosorption	1.98–747.91
Phytoremediation	0.13–1.32

unit [6,40,42]. To overcome these difficulties, genetics or advanced molecular biological tools can be combined with microbial biotechnology such as colonizing of
metal ion removing bacteria on the root surface [59]. Using phytoremediation to
remediate metal ions in the environment is classified into phytoextraction, phytost-
abilization and rhizofiltration (Table 15.3) [39,42].

We are particularly interested in rhizofiltration that uses plants to remove metal
ions from water because it is widely used in wastewater treatment. The rhizofiltration
technology is divided into two categories. One involves the use of plant roots to
absorb, concentrate and precipitate metal ions from water [42]. This method is more
specifically defined as the use of hydroponically cultivated roots of terrestrial plants in
absorbing, concentrating or precipitating toxic metal ions from polluted effluents
[68,69]. Terrestrial plants are used instead of aquatic plants, because the terrestrial
plants develop much longer fibrous root systems covered with root hairs that have
large surface areas. Example of rhizofiltration include using Indian mustard (*Brassica
juncea*) and sunflower (*Helianthus annuus*) to concentrate Cd^{2+}, Cr^{6+}, Cu^{2+}, Ni^{2+}, Pb^{2+}
and Zn^{2+} in root systems and to translocate only a small part of the metal ion absorbed
to the shoots [42,68]. The second mode of rhizofiltration, which is more fully developed, involves using aquatic or/and wetland plants for treatment of wastewater.
Kadlec and Knight [70] provided the classification of the major plant groups used
for wastewater treatment in wetlands. The aquatic macrophytes commonly known for
removing metal ions and organic pollutants include *Eichhornia crassipes* (water hyacinth), *Pistia stratiotes* (water lettuce), *Elodea nuttallii* (western waterweed), *Typha
latifolia* (cattail), *Myriophyllum* (water milfoil), *Heterophyllum* (water milfoil), *Juncus
roemerian* (marsh rush), *Lemna* species (duckweed) and *Azolla* species (fern) [71,72].

15.3.3.2 Using Water Hyacinth for Wastewater Treatment

The significance of using macrophytes for wastewater treatment and nutrient absorption was recognized as early as 1938, when species such as water hyacinth were
found to have the capacity to remove nitrogen and phosphorus from secondary
sewage [73]. Many studies, in laboratory as well as in the field, have shown the ability
of water hyacinth to remove a variety of waste constituents such as suspended
particulates, organic matter, metal ions, nitrogen, phosphorus, phenol, dyes, cyanide

Table 15.2 Advantages and disadvantages of phytoremediation [6].

Advantages	Disadvantages
Cost	Time and space
• Low capital and operating costs • Metal ion recycling provides further economic advantages	• Slower than some alternatives; seasonally dependent • Many natural metal ion hyperaccumulators are slow growers • Application to treat groundwater or wastewater may require large land area
Performance	Performance
• Permanent treatment solution • *In situ* application avoids excavation • Capable of mineralizing organics • Applicable to a variety of contaminants, including some recalcitrant compounds • Trees are capable of high hydraulic pumping pressures	• Biological methods are not capable of 100% reduction • May not be applicable to all mixed wastes • High metal ion concentrations or other contaminants may be toxic • Phytoremediation applicable only to surface soils
Others	Others
• Public acceptance; aesthetically pleasing • Compatible with risk-based remediation, brownfields • Can be used during site investigation or after closure	• Need to displace existing facilities (e.g. wastewater treatment) • Regulators may be unfamiliar with the technology and its capabilities • Lack of recognized economic performance data

Table 15.3 Types of metal ion phytoremediation technologies [39,42].

Type	Description	Applicability
Phytoextraction	• Use plant to transport metal ions from soil and concentrate into roots and aboveground shoots that can be harvested	• Potential for transport of metal ions to surface. Plant residue can be isolated as hazardous waste or recycled as metal ore
Phytostabilization	• Use plants to limit the mobility and bioavailability of metal ions in soils by sorption, precipitation, complexation or reduction of metal valences	• Stabilizing the metal ions present in the soil and the soil matrix to minimize erosion and migration of sediment • Is containment rather than removal • Can be used in the sites at which removal of metal ions is not economically feasible
Rhizofiltration	• Use roots to absorb, concentrate and precipitate metal ions from wastewater • Use wetlands or reed beds for treatment of contaminated wastewater	• Applicable for treatment of water only and is cost-effective for the treatment of large volumes of wastewater with low concentrations of metals

and fecal coliform bacteria [27,74–84]. A number of industrial wastes such as dairy waste, sugar industry waste, tannery waste, electroplating industry waste, textile industry waste, paper and pulp mill waste and domestic waste have been treated by water hyacinth [55,72,85,86]. Treatment efficiency cannot be summarized as it varies with wastewaters having different chemical properties; however, the method can attain high treatment efficiencies. For example, water hyacinth reduced 87% of chemical oxygen demand (COD), 95% of total suspended solids (TTS) and 98.6% of fecal bacteria in 7 days in municipal wastewater treatment [79].

Water hyacinth has attracted considerable attention because of its ability to grow in heavily polluted water together with its capacity for metal ion accumulation [87–91]. It removes nutrients and pollutants in wastewater through a complex array of physicochemical processes, including absorption, flocculation, precipitation and sedimentation, and biological mechanisms including plant and bacterial processes [53,44,92,93]. Water hyacinth can grow in heavily polluted water and has a high capacity of metal ion accumulation, even with relatively toxic metal ions such as Hg^{2+} and Pb^{2+} [87–91,94]. Sutton and Blackburn [95] studied Cu^{2+} uptake by water hyacinth from Hoagland nutrient solution. Many data have been accumulated on the uptake of metal ions including As^{3+}, Cu^{2+}, Cd^{2+}, Cr^{6+}, Hg^{2+}, Ni^{2+}, Pb^{2+}, Se^{2+}, Eu^{2+} and Zn^{2+} by this species [92,96–106]. The highest levels of Cd^{2+}, Cr^{2+} and Cu^{2+} found in roots were 6103, 3951 and 2655 $mg\,l^{-1}$, respectively [90]. Water hyacinth can also extract metal ions at low concentrations. The bioconcentration factors (BCFs) of the roots of water hyacinth for Cu^{2+}, Ni^{2+} and Zn^{2+} were 2.5×10^3, 1.6×10^3 and 6.1×10^3, respectively, and were comparatively higher than those of other biological systems [104]. Thus, water hyacinth can be used as an efficient biofilter for metal ions and has been widely used in plant-based wastewater treatment systems [55].

15.4
Biology of Water Hyacinth

Water hyacinth is a free-floating aquatic plant, which has spread throughout the world by human activity. It is a successful invader of freshwater and eutrophic environments having the property of rapid vegetative growth and multiplication, wide ecological amplitude and great phenotypic variation [87]. The dramatic growth of water hyacinth has deteriorated the utilization of water resources [55,87,107] and ranks first among the aquatic plants as a nuisance weed [87]. Nowadays, however, scientists and engineers have turned this nuisance to merit and are using water hyacinth to combat water pollution as it removes nutrients and inorganic pollutants from wastewater [55].

15.4.1
Scientific Classification

Water hyacinth belongs to the family Pontederiaceae. The scientific name is *Eichhornia crassipes*. The genus *Eichhornia* was named in 1843 by Kunth in the honor of

John Albert Friedrich Eichhorn [87]; and the name hyacinth is derived from Latin word 'hyacinthus', which means 'a bulbous plant bearing bell shaped purplish flowers'.

15.4.2
Morphology

Water hyacinth is an attractive free-floating aquatic plant with beautiful lilac violet flowers (Figure 15.1). It consists of a shoot with a rosette of petiolate leaves and numerous roots hanging in water (Figure 15.1). The shoot consists of a sympodially branched, stoloniferous rhizome, with several short internodes [55]. Each node bears a leaf and roots. Being a stoloniferous herb, vegetative reproduction occurs with the help of stolons (Figure 15.1). The stolons, with the auxiliary buds or the elongation of the internode, develop by growing out at an angle of about 60° from the rhizome. Stolons are purplish violet and vary in length, extending up to 50 cm. A leaf consists of a petiole, isthmus (thin part between petiole and blade) and a blade. The petioles may be elongated, swollen or may form a bulbous float, which varies with

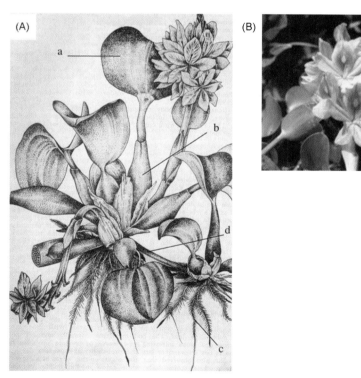

Figure 15.1 Morphology of water hyacinth. (A) Whole plant shows (a) leaf, (b) petiole, (c) fibrous root and (d) stolon [87]. (B) Inflorescence of violet flowers.

Figure 15.2 Elastoplasticity of morphology of water hyacinth. (a) Swollen petioles and (b) elongated petioles.

growing conditions (Figure 15.2). In crowded conditions, the float is not produced but a long petiole develops [55,87]. The leaf blade is orbicular to ovoid in shape and the size also varies with growing conditions [55,87]. The roots are adventitious, fibrous and unbranched and have a conspicuous root cap. The roots produce many laterals but are usually small in size and vary with growing conditions. This gives them a fine feathery appearance. Root length varies from 10 to 300 cm. They are usually whitish in color but turn pinkish violet because of the presence of anthocyanin when exposed to light [87]. The inflorescence of the water hyacinth is a spike with 4–25 yellow-spotted lavender-colored flowers. The fruit is a many-seeded capsule.

15.4.3
Ecology

Water hyacinth is a perennial, tropical, aquatic weed. It grows in a variety of freshwater habitats ranging from shallow temporary ponds, marshes and sluggish flowing waters to large lakes, reservoirs and rivers [87]. The plants can survive on wet mud for prolonged periods or perennate in the form of seeds [55,87]. The plants exist within a wide range of temperatures from as low as 1 °C during winter in northern latitudes to over 40 °C during summer in dry tropics [87]. They can also survive in soils poor in nutrients, even in water highly polluted with a variety of organic and inorganic industrial effluents containing metal ions [55,76,81,82,87]. Being highly adaptive to severe conditions, free-floating and capable of rapid vegetative reproduction and surviving in diverse habitats, water hyacinth is well distributed throughout the world. With a maximum productivity as high as 54.4 g of dry weight m^{-2} day^{-1} [55,87], plants grow rapidly during warm season and can cover 15% more surface area every day. Optimum conditions for growth are temperatures ranging between 26 and 35 °C, pH 6–7 and 240 000 lx h [87,107,108].

15.4.4
Environmental Impact

Water hyacinth grows so abundantly in rivers and other navigable waters that it obstructs the passage of ships and water flow in irrigation channels and hinders hydroelectric power generation. Water hyacinth has devastating impacts also on fresh water ecosystems in various ways [55,87,109]. The extensive growth of water hyacinth depletes oxygen completely, as the rate of organic matter production is so high that large amounts of dead organic matter accumulates in the water. The decomposing organic matter depletes oxygen and hence kills biota and generates obnoxious odors [55,109]. Decomposition also releases free CO_2 that reacts with water to produce H_2CO_3, which decreases the pH of water [55,87]. Excessive growth of water hyacinth causes the reduction of light penetration into water bodies, which leads to the reduction of water temperature and affects the growth of phytoplanton [55].

15.4.5
Management of Water Hyacinth

Because of its high reproductive rate, ability to adapt to adverse environments and free-floating nature, it is difficult to restrict the growth of water hyacinth [55,87,109,110]. Numerous studies have been conducted to develop suitable management techniques to control its prolific growth [87]. In a conference organized by the Common Wealth Council, various control measures involving chemical, biological and mechanical devices including using pesticides, herbivorous aquatic mammals and fish, insects and microbial plant pathogens have been suggested and practiced [111–113]. Mechanical removal has been reported as the most complete and effective method to control water hyacinth, but the process is relatively slow, expensive and labor-intensive [87,110]. Another method to control growth of water hyacinth is economic utilization. Studies have reported that water hyacinth has potential as an effective absorber of organic and inorganic water pollutants [55,87]. Extensive studies have been conducted on the possibility of using water hyacinth for secondary and tertiary treatment of sewage and various industrial effluents [55].

Water hyacinth contains about 26% crude protein, 26% fiber, 17% ash and 8% available carbohydrate on a dry weight basis [87]. Because of its high protein content, water hyacinth can be considered as a protein source for nonruminant animals and humans [87]. However, numerous purification steps are needed to prepare water hyacinth for human consumption. Water hyacinth is also a good source for the production of biogas. One kilogram of dried water hyacinth can yield 374 l of biogas containing 60–80% methane with a fuel value of $21\,000\,BTU\,m^{-3}$, as shown by experiments at the National Space Technology Laboratory, USA [87]. The sludge remaining after the production of biogas can also be used as a fertilizer and soil conditioner [55,87]. The fiber composition of water hyacinth is chemically and physically similar to sugarcane bagasse [87]. The plant can be utilized for manufacturing paper pulp and board [55,87].

15.5
Microbial Enhancement of Metal Ion Removal Capacity of Water Hyacinth

15.5.1
Biology of the Rhizosphere

The rhizosphere is defined as the environment influenced biologically and biochemically by the living root [23]. It is an ideal environment for many organisms and communities because it provides water, oxygen, organic substrates and physical protection. The surface of the plant root is coated with a layer of mucigel composed of plant mucilage, bacterial cells, metabolic products, organic colloids and mineral materials. The plant mucilage is released from the roots as exudates and secretions [114]. The exudates contain amino acids, sugars, organic acids, proteins, polysaccharides and growth-promoting and growth-inhibiting substances [114]. The organic carbon source and the binding surface provide a dynamic force for the colonization of microbes in the rhizosphere and create the so-called rhizospheric effect [19,26,22,27].

The microbe–plant interaction in the rhizosphere is dynamic and complicated. Some microbes contribute to plant health by mobilizing nutrients, some are detrimental to plant health as they compete with the plant for nutrients or cause disease and some stimulate plant growth by producing hormones or suppressing pathogens [19,21,23,25].

The rhizosphere of floating aquatic vegetation, which constitutes a different biotope of bacteria from those in waters without vegetation and the sediment of the same aquatic system, is densely populated by many specialized organisms [115]. These organisms can control the biochemical environments in the rhizosphere and protect plants against pests through the release of bioactive chemicals. Polprasert and Khatiwada [93] reported that the film of bacteria attached to the roots of water hyacinth is involved in the reduction of BOD_5. Bioactive chemicals such as anaerobic metabolites, alkaloids, phenolics, terpenoids and steroids are found in abundance in roots and rhizospheres in wetlands. Bioactivities include allelopathy, growth regulation, extraorganismal enzymatic activities, metal manipulation by phytosiderophores and phytochelatins, various pest-control effects and poisoning [116]. Alteration of the biological and chemical composition of the rhizosphere can, therefore, be expected to modify the microbial ecology of rhizosphere [19–23,117]. This can be attained by various methods such as addition of organic amendments, microbial inoculation, manipulation of plant genotypes and exposure to environmental stresses such as heavy metal contamination.

15.5.2
Mechanisms of Metal Ion Removal by Plant Roots

Possible mechanisms of toxic metal ion removal by plant roots include extracellular precipitation, cell wall precipitation and adsorption, intracellular uptake followed by cytoplasm compartmentalization or vacuolar deposition (Figure 15.1) [130]. The metal ion uptake and detoxification mechanisms will vary with different plant species. Most metal ions enter plant cells by an energy-dependent, saturable process through specific or generic metal ion carriers or channels [42]. Toxic metal ions may employ the same mechanisms that are responsible for the uptake of essential ions. Paganetto *et al.* [118] used the 'patch-clamp' techniques to study the transport properties of vacuolar ion channels from the roots of water hyacinth, *Eichhornia crassipes* (Mart. Solms, Pontederiaceae). The vacuolar currents for the transport of Ni^{2+} and Zn^{2+} were found to be different from other common ions such as Na^+, Ca^{2+} and NH_4^+. However, membrane transport systems such as aqueous pores, ion efflux pumps, ion selective channels and proton–anion contraports may fail to discriminate among different metal ions that have similar ionic radii and the same ionic charge [42,87].

Metabolically important cations from the external solution can accumulate in a nonmetabolic step (Figure 15.3) [87]. Entry and association of metal ions with plant cells may occur by a number of physical processes, including diffusion, ion exchange, mass flow and adsorption. The cation-exchange occurs within the 'free space' of the roots where ions can penetrate without passage through a living membrane. The exchangeable cations are electrostatically bound to negatively charged functional groups, probably the free carboxyl groups of pectic cell wall matrix substances [119–121]. Precipitation and exchangeable sorption remove metal ions by forming insoluble compounds in the free space (Figure 15.3) [87]. Some natural hyperaccumulators extend and proliferate their roots positively into patches of high metal ion availability. In contrast, nonaccumulators actively avoid these areas; this is one of the mechanisms by which hyperaccumulators absorb more metal ions when grown in the same soil [122]. Also, phytochelatins play an important role in the accumulation and detoxification of excess metal ions such as Cd^{2+} in plant cells [58,87,123–127]. This mechanism further assists phytofiltration technology, as it increases the specificity of metal ions to binding domains in the plant root. Some species of plants also have multiple binding capacities for metal ions owing to induced formation of phytochelatins by different metal ions such as Cu^{2+}, Zn^{2+}, Pb^{2+} and Cd^{2+} [87]. Indeed, the application of chelates has been shown to induce substrate accumulation of metals such as Pb^{2+}, U^{2+} and Au^{3+} in the shoots of nonhyperaccumulators by increasing metal solubility and root-to-shoot translocation [122].

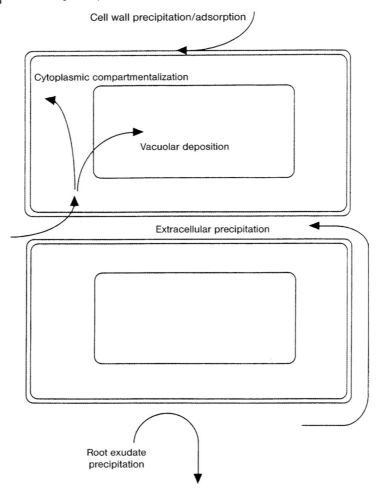

Cell wall precipitation/adsorption

Cytoplasmic compartmentalization

Vacuolar deposition

Extracellular precipitation

Root exudate
precipitation

Figure 15.3 Mechanisms of metal ion removal from the solution
by plant root [130].

15.5.3
Effects of Rhizospheric Bacteria on Metal Uptake and Plant Growth

It is known that rhizospheric microorganisms facilitate the uptake of essential metal
ions such as Fe^{3+} and Mn^{2+} by plants [128,129]. Several strains of *Pseudomonas* and
Bacillus were shown to increase the total amount of Cd^{2+} accumulated from hydroponic solution by 2-week-old *B. juncea* seedlings [42].

The purpose of rhizospheric bacteria inoculation in phytoremediation is to optimize the rate and the yields of absorption of pollutants within plants. In addition to
metal ion uptake, colonization of the rhizosphere is also supposed to be beneficial to

plant functions, especially their ability to produce plant growth promoting (PGP) compounds, such as indole acetic acid, or to protect the plant from pathogens. Some bacteria can reduce the toxicity of metal ions toward plants [131,132]. To enhance adsorption processes, metal ion resistant bacterial strains with the capability to produce PGP compounds are targets for bacterial isolation from the rhizosphere of some spontaneous plants in contaminated soil, such as *Brassica* sp., *Trifolium repens*, *Trifolium pratense* and *Chenopodium album* [133].

In plants, the toxic effect of heavy metals is partially due to the metal ion induced iron deficiency [131]. Siderophores produced by bacteria can eliminate iron deficiency and hence the metal ion toxicity. Burd *et al.* [131] showed that the plant growth promoting bacterium, *Kluyvera ascorbata* SUD 165, could produce an enzyme, 1-aminocyclopropane-1-carboxylic acid (ACC) deaminase, and decrease Ni^{2+} toxicity toward the plant. Metal ions can also induce ethylene production by plants. Excess ethylene can inhibit plant development. The bacterial ACC deaminase can hydrolyze and decrease the amount of ACC, an ethylene precursor, in plants and hence eliminate the inhibitory effect of metal ions on the plant without affecting the amount of Ni^{2+} accumulated by the plant.

In another study conducted by Śouza *et al.* [134], inoculation of rhizospheric bacteria into the plant rhizosphere could facilitate Se^{3+} and Hg^{2+} accumulation in two wetland plants, saltmarsh bulrush (*Scirpus robustus* Pursh) and rabbit-foot grass (*Polypogon monspeliensis* (L) Desf.). The selenate uptake into roots was enhanced by these bacteria, which interacted with the plant to produce a heat-labile proteinaceous compound that stimulates selenate uptake, possibly by a novel pathway [134].

Pischik *et al.* [135] showed the mutually beneficial interactions between plants and bacteria under cadmium stress. They found that under cadmium stress, the bacteria synthesize phytohormones (IAA and ethylene), root excretory activity increases, the number of bacteria in the rhizoplane grows, the flux of bacteria migrating to rhizosphere increases, the number of bacteria binding Cd^{2+} in the rhizosphere increases and the number of free ions entering the plant decreases [135]. A metal ion tolerant PGP rhizobacterium (PGPR) was isolated from cadmium-polluted soil. The resistance mechanism was studied to evaluate the potential influence of the bacterial inoculation on phytoremediation [133]. Their results find that bacteria isolated from the root zone of polluted site remove about 50% of cadmium in the culture medium after 3 days.

Amico *et al.* [136] isolated several heavy metal ion resistant bacterial strains belonging to different genera, such as *Pseudomonas*, *Mycobacterium*, *Agrobacterium* and *Arthrobacter*, from the rhizoplane of perennial Graminaceae from polluted water meadow soil; some of them were found to possess the plant growth promoting characteristics including possessing ACC deaminase and/or the ability to produce IAA and siderophores. The strains resistant to Cd^{2+} and Zn^{2+}, *Pseudomonas tolaasii* RP23 and *Pseudomonas fluorescens* RS9, possessed all three PGP characteristics [136]. Several studies have documented that metal ion resistant Proteobacteria can protect plants from the toxic effects of metal ions or even enhance metal ion uptake by hyperaccumulator plants (reviewed by Amico *et al.* [136]). Shiley *et al.* [137] found that *Pseudomonas fluorescens* had a positive influence on sunflower plant growth and As^{3+} accumulation and that higher growth rate of the plants was observed when they were inoculated with *P. fluorescens* [137].

There are some studies on the interaction of rhizospheric bacteria with root exudates of water hyacinth [138] and the role of rhizospheric bacteria on the degradation of phenol [27]. These studies indicate that the presence of phenol increased the rhizospheric effect and the inoculation of rhizospheric bacteria into the rhizosphere can enhance the degradation of phenol by increasing the amount of polyphenoloxidase and peroxidase. So *et al.* [139] had isolated the Cu^{2+}-resistant bacteria (Strain CU-1) from water hyacinth with higher metal ion removal capacity compared to a Ni^{2+}- or Zn^{2+}-resistant strain. Results showed that Strain CU-1 colonized the plant root and led to the increase in the capacity of water hyacinth roots to remove Cu^{2+} [139].

Kamnev *et al.* [28] reviewed the role of soil microorganisms in phytoremediation. Soil microorganisms interacted with plants in many different ways to reduce metal ion toxicity and enhance metal ion absorption by plants; for example, by acidifying the local environment, producing plant growth regulators, enzymatic reduction of metal ions to less toxic forms, chelating the metal ion into a phytoavailable form, physical binding on to the cell wall or precipitation of the metal ions [28]. To develop a plant–bacterium remediation system for metal ions, more basic studies on the interactions of rhizospheric bacteria and metal ion uptake by plants are needed.

15.6
Summary

Urbanization and industrialization have increased the quantity of metal-polluted wastewater discharged to the environment. As metal ions are nonbiodegradable, their elevated concentrations in the environment are an increasing threat to human health and pose the problem of depletion of water resources and damage of ecosystems [37]. To encourage sustainable development, a simple, inexpensive and effective treatment of metal ions in wastewater is needed.

Water hyacinth has been proposed as a cheap biological treatment system for the removal and recovery of metal ions from wastewater [53,74,89,98,99,140]. Water hyacinth has a high metal ion accumulating ability and can remove metal ions from wastewater by concentrating metals in the roots [80,141].

The microbial biotechnology that can immobilize microbial cells onto solid surfaces to remove metal ions is also well developed nowadays [29,30]. We suggest that such microbial biotechnology can increase the efficiency of the biological systems using water hyacinth. Thus, we propose to use the roots of water hyacinth as the binding surface for forming a biofilm with metal ion removing bacteria. However, the ecology of the rhizosphere is dynamic and complex; to develop a microbial–plant remediation system for metal pollution in water, following points should be considered:

1. A better understanding of the community and population dynamics of rhizospheric bacteria of water hyacinth in metal-polluted environments.
2. The roles of rhizospheric bacteria and the plant root in metal removal processes and metal resistance of water hyacinth.
3. The mechanism(s) of the mutualistic relationship between rhizospheric bacteria and plants and characterization of the plant growth promoting properties of metal-resistant rhizobacteria.

References

1 Rudd, T. (1987) *Heavy Metals in Wastewater and Sludge Treatment Processes, vol. I: Source, Analysis, and Legislation* (ed. J.N. Lester), CRC Press, Boca Raton, pp. 1–30.
2 Blackmore, G. (1998) *Science of the Total Environment*, **214**, 21–48.
3 Ansari, A.A., Singh, I.B. and Tobschall, H.J. (1999) *India Environmental Geology*, **38**, 25–33.
4 Moiseenko, T.I. (1999) *Science of the Total Environment*, **236**, 19–39.
5 Schintu, M. and Degetto, S. (1999) *Science of the Total Environment*, **241**, 129–141.
6 Glass, D.J. (2000) *Phytoremediation of Toxic Metals: Using Plants to Clean Up the Environment* (ed. I. Raskin), John Wiley & Sons, Inc., New York, pp. 15–32.
7 Khamar, M.D., Bouya, D. and Ronneau, C. (2000) *Water Quality Research Journal of Canada*, **35**, 147–161.
8 Hamby, D.M. (1996) *Science of the Total Environment*, **191**, 203–224.
9 Berveridge, T. and Koval, S.F. (1981) *Applied and Environmental Microbiology*, **2**, 325–335.
10 de Rome, L. and Gadd, G.M. (1987) *Applied Microbiology and Biotechnology*, **26**, 84–90.
11 Wong, P.K. and Choi, C.Y. (1988) *Recent Advances in Biotechnology and Applied Biology* (eds S.T. Chang, K.Y. Chan and N.Y.S. Woo),The Chinese University Press, Hong Kong, pp. 271–280.
12 Mullen, M.D., Wolf, D.C., Ferris, F.G., Berveridge, T.J., Flemming, C.A. and Bailey, G.W. (1989) *Applied and Environmental Microbiology*, **55**, 3143–3149.
13 Gadd, G.M. (1990) *Experientia*, **46**, 835–840.
14 Harris, P.O. and Ramelow, G.J. (1990) *Environmental Science & Technology*, **24**, 220–228.
15 de Rome, L. and Gadd, G.M. (1991) *Journal of Industrial Microbiology*, **7**, 97–104.
16 Shuttleworth, K.L. and Unz, R.F. (1993) *Applied and Environmental Microbiology*, **59**, 1274–1282.
17 Mago, R. and Srivastava, S. (1994) *Applied and Environmental Microbiology*, **60**, 2367–2370.
18 Gadd, G.M. (1998) *Biotechnology*, vol. 6b (eds H.J. Rehm and G. Reed),Verlag Cheme, Weinheim, pp. 401–433.
19 Bazin, M.J., Markham, P., Scott, E.M. and Lynch, J.M. (1990) *The Rhizosphere* (ed. J. M. Lynch), John Wiley & Sons, Ltd, New York, pp. 99–128.
20 Parke, J.L. (1991) *The Rhizosphere and Plant Growth* (eds D.L. Keister and P.B. Cregon), Kluwer Academic Publishers, Boston, pp. 33–42.
21 Anderson, T.A., Gutherie, E.A. and Walton, B.T. (1993) *Environmental Science & Technology*, **27**, 2630–2636.
22 Bolton, J.H., Fredickson, J.K. and Elliott, L.F. (1993) *Soil Microbial Ecology*, (eds F. Blaine and J. Metting),Dekker, New York, pp. 27–63.
23 O'Connell, K.P., Goodman, R.M. and Handelsman, J. (1996) *Trends in Biotechnology*, **14**, 83–88.

24 Westover, K.M., Kennedy, A. and Kelley, S. (1997) *Journal of Ecology*, **85**, 863–873.

25 Morgan, J.A.W., Bending, G.D. and White, P.J. (2005) *Journal of Experimental Botany*, **56**, 1279–1739.

26 Elliot, L.F., Gilmour, G.M., Lynch, J.M. and Tittemore, D. (1984) *Microbial–Plant Interactions* (eds R.L. Todd and J.E. Giddens), Soil Science Society of America, Madison, pp. 1–16.

27 Wang, M. and Zheng, S.Z. (1994) *Chinese Journal of Applied Ecology*, **5**, 309–313.

28 Kamnev, A.A. and Lelie, D. (2000) *Bioscience Reports*, **20**, 239–258.

29 Subramanian, V.V., Sivasubramanian, V. and Gowrinathan, K.P. (1994) *Journal of Environmental Science and Health Part A*, **29**, 1723–1733.

30 Chang, J.S. and Huang, J.C. (1998) *Biotechnology Progress*, **14**, 735–741.

31 Gadd, G.M. (1992) *FEMS Microbiology Letters*, **100**, 197–204.

32 Stephenson, T. (1987) *Heavy Metals in Wastewater and Sludge Treatment Processes, vol. I: Source, Analysis, and Legislation* (ed. J.N. Lester), CRC Press, Boca Raton, pp. 31–64.

33 Tyler, G., Pahlsson, M.B., Benntsson, G., Baath, E. and Tranvik, L. (1989) *Water Air and Soil Pollution*, **47**, 189–215.

34 Khan, K.S. and Huang, C.Y. (1999) *Journal of Environmental Sciences*, **11**, 10–47.

35 Wada, O., Matsui, H., Arakawa, Y. and Nagahashi, M.P. (1984) *Journal of Hygiene Chemistry*, **30**, 1–4.

36 Chan, H.M., Kim, C., Khoday, K., Receveur, O. and Kuhnlein, H.V. (1995) *Environmental Health Perspectives*, **103**, 740–746.

37 Newman, M.C. and McIntosh, A.W. (1991) *Metal Ecotoxicology: Concepts and Applications*, Lewis Publisher, Inc., Detroit.

38 Benin, A.L., Sargent, J.D., Dalton, M. and Roda, S. (1999) *Environmental Health Perspectives*, **107**, 279–284.

39 Ensley, B.D. (2000) *Phytoremediation of Toxic Metals: Using Plants to Clean Up the Environment* (ed. I. Raskin), John Wiley & Sons, Inc., New York, pp. 3–11.

40 Black, H. (1995) *Environmental Health Perspectives*, **103**, 1106–1108.

41 Boyajian, G.E. and Carreira, L.H. (1997) *Nature Biotechnology*, **15**, 127–128.

42 Salt, D.E., Blaylock, M., Kumar, N.P.B.A., Dushenkov, V., Ensley, B.D., Chet, I. and Raskin, I. (1995) *Biotechnology*, **13**, 468–474.

43 Chiu, H., Tsang, K.L. and Lee, R.M.L. (1987) *Hong Kong Engineering*, 43–49.

44 Lantz, W.L. (1992) *Water Environmental Technology*, **4**, 12–15.

45 Christensen, E.R. and Delwiche, J.T. (1982) *Water Research*, **16**, 729–737.

46 Volesky, B. and Holan, Z.R. (1995) *Biotechnology Progress*, **11**, 235–250.

47 Cheremisinoff, P.N. (1995) *Handbook of Water and Wastewater Treatment Technology*, Dekker, New York.

48 Aiking, H., Kok, K., van Heerikhuizen, H. and van't Riet, J.((1982) *Journal of Applied and Environmental Microbiology*, **44**, 938–944.

49 Viswanathan, M.N. and Boettcher, B. (1991) *Water Science and Technology*, **23** (2), 1437–1446.

50 Ohtake, H., Cervantes, C. and Silver, S. (1987) *Journal of Bacteriology*, **169**, 3853–3856.

51 Volesky, B. (1990) *Biosorption of Heavy Metals*, CRC Press, Florida.

52 Shumate, S.E. II, Strandberg, G.W. and Parrott, J.R. Jr (1978) *Biotechnology & Bioengineering Symposium*, **8**, 13–20.

53 Jamil, K. (1998) *Advances in Wastewater Treatment Technologies*, vol. 1 (ed. R.K. Trivedy), Global Science, India, pp. 151–154.

54 Lovely, D.R. and Coates, J.D. (1997) *Current Opinion in Biotechnology*, **8**, 285–289.

55 Trivedy, R.K. (1998) *Advances in Wastewater Treatment Technologies*, vol. 1 (ed. R.K. Trivedy),Global Science, Aligarh, pp. 463–486.

56 Qian, J.H., Zayed, A., Zhu, Y.L., Yu, M. and Terry, N. (1999) *Journal of Environmental Quality*, **28**, 1448–1455.

57 Moffat, A.S. (1995) *Science*, **269**, 302–303.

58 Cunningham, S.D. and Ow, D.W. (1996) *Plant Physiology*, **110**, 715–719.

59 Ow, D.W. (1996) *Resources Conservation and Recycling*, **18**, 135–149.

60 Wantanabe, M.E. (1997) *Environmental Science & Technology*, **31**, 182A–186A.

61 Harrigan, K. (1999) *Pollution Engineering*, **12**, 24–26.

62 Salt, D.E., Smith, R.D. and Raskin, I. (1998) *Annual Review of Plant Physiology and Plant Molecular Biology*, **49**, 643–668.

63 Salt, D.E. and Krämer, U. (2000) *Phytoremediation of Toxic Metals: Using Plants to Clean Up the Environment* (ed. I. Raskin), John Wiley & Sons, Inc., New York, pp. 231–245.

64 Reeves, R.D. and Baker, A.J.M. (2000) *Phytoremediation of Toxic Metals: Using Plants to Clean Up the Environment* (ed. I. Raskin), John Wiley & Sons, Inc., New York, pp. 193–229.

65 Baker, A.J.M. and Brooks, R.R. (1989) *Biorecovery*, **1**, 81–126.

66 Raskin, I., Kumar, P.B.A.N., Dushenkov, S. and Salt, D.E. (1994) *Current Opinion in Biotechnology*, **5**, 285–290.

67 Dini, J.W. (1998) *Plating and Surface Finishing*, **30**, 42–44.

68 Dushenkov, V., Kumar, N.P.B.A., Motto, H. and Raskin, I. (1995) *Environmental Science & Technology*, **29**, 1239–1245.

69 Brooks, R.R. and Robinson, B.H. (1998) *Plants that Hyperaccumulate Heavy Metals* (ed. R.R. Brooks), CAB International, New York, pp. 203–226.

70 Kadlec, R.H. and Knight, R.L. (1996) *Treatment Wetlands*. Lewis Publishers, New York.

71 Chandra, P., Sinha, S. and Rai, U.N. (1997) *Phytoremediation of Soil and Water Contaminants* (eds E.L. Kruger, T.A. Anderson and J.R. Coats), American Chemical Society, Washington, DC, pp. 275–282.

72 Lee, C.L., Wang, T.C., Hsu, C.H. and Chiou, A.A. (1998) *Bulletin of Environmental Contamination and Toxicology*, **61**, 497–504.

73 Clock, R.M. (1938) *Nitrogen and phosphorus removal from a secondary sewage treatment effluent*, PhD Dissertation, University of Florida, Gainville.

74 Mischra, P.C., Partri, M. and Panda, M.J. (1991) *Ecotoxicology and Environmental Monitoring*, **1**, 218–224.

75 Tripathi, B.D. and Shukla, S.C. (1991) *Environmental Pollution*, **69**, 69–78.

76 Granato, M. (1993) *Biotechnology Letters*, **15**, 1085–1090.

77 Fett, J.P., Cambraia, J., Oliva, M.A. and Jordao, C.P. (1994) *Journal of Plant Nutrition*, **17**, 1219–1230.

78 Fett, J.P., Cambraia, J., Oliva, M.A. and Jordao, C.P. (1994) *Journal of Plant Nutrition*, **17**.

79 Mandi, L. (1994) *Water Science and Technology*, **29** (4), 283–287.

80 Low, K.S., Lee, C.K. and Tai, C.H.J. (1994) *Journal of Environmental Science and Health Part A*, **29**, 171–188.

81 Ćasabianca, M.L., Laugier, T. and Posada, F. (1995) *Waste Management*, **15**, 651–655.

82 Ćasabiance, M.L. and Laugier, T. (1995) *Bioresource Technology*, **54**, 39–43.

83 Wang, M. and Zheng, S.Z. (1995) *Acta Phytophysiologica Sinica*, **21**, 254–258.

84 Ghosh, C. (1998) *Advances in Wastewater Treatment Technologies*, vol. 1 (ed. R.K. Trivedy), Gobal Science, India, pp. 267–284.

85 Verma, V.K., Gupta, R.K. and Rai, J.P.N. (2005) *Journal of Scientific & Industrial Research*, **64**, 778–781.

86 Maine, M.A., Sune, N., Hadad, H., Sanchez, G. and Bonetto, C. (2006) *Ecological Engineering*, **26**, 341–347.

87 Gopal, B. (1987) *Water Hyacinth*, Elsevier, New York.

88 Ismail, A.S., Abdel-Sabour, M.F. and Radwan, R.M. (1996) *Egyptian Journal of Soil Science*, **36**, 43–354.

89 Jamil, K., Salifakala, G. and Mahesh, K.B. (1998) *Advances in Wastewater Treatment Technologies*, vol.1 (ed. R.K. Trivedy), Gobal Science, India, pp. 7–11.

90 Zhu, Y.L., Zayed, A.M., Qian, J.H., de Souza, M. and Terry, N. (1999) *Journal of Environmental Quality*, **28**, 339–344.

91 Skinner, K., Wright, N. and Porter-Goff, E. (2007) *Environmental Pollution*, **145**, 234–237.

92 Muramoto, S. and Oki, Y. (1983) *Bulletin of Environmental Contamination and Toxicology*, **30**, 170–177.

93 Polprasert, C. and Khatiwada, N.R. (1998) *Water Research*, **32**, 179–185.

94 Wang, Q., Cui, Y. and Dong, Y. (2002) *Acta Biotechnologica*, **22**, 199–208.

95 Sutton, D.L. and Blackburn, R.D. (1971) *Hyacinth Control Journal*, **9**, 18–20.

96 Chigbo, F.F., Smith, R.W. and Shore, F.L. (1982) *Environmental Pollution*, **33**, 31–36.

97 Turnquist, T.D. (1990) *Journal of Environmental Science and Health, Part A: Environmental Science & Engineering & Toxic & Hazardous Substance Control*, **25**, 897–912.

98 Akcin, G., Guldede, N. and Saltabas, O. (1993) *Journal of Environmental Science and Health Part A: Environmental Science & Engineering & Toxic & Hazardous Substance Control*, **28**, 1725–1727.

99 Delgado, M., Bigeriego, M. and Guardiola, E. (1993) *Water Research*, **27**, 269–272.

100 Rai, S., Hasan, S.H., Joshi, V.C., Narayanaswami, M.S. and Rupainwar, D. C. (1993) *Indian Journal of Environmental Health*, **35**, 178–184.

101 Jenatte, P.F., Cambraia, J., Oliver, M.A. and Jordao, C.P. (1994) *Journal of Plant Nutrition*, **17**, 1219–1230.

102 Singaram, P. (1994) *Indian Journal of Environmental Health*, **36**, 197–199.

103 Zaranyika, M.F., Mutoko, F. and Murahwa, H. (1994) *Science of the Total Environment*, **153**, 117–121.

104 Zaranyika, M.F. and Ndapwadza, T. (1995) *Journal of Environmental Science and Health Part A*, **30**, 157–169.

105 Panda, A.K. (1996) *Indian Journal of Environmental Health*, **38**, 51–53.

106 Kelly, C., Mielke, R.E., Diamaquibo, D., Curtis, A.J. and Dewitt, J.G. (1999) *Environmental Science & Technology*, **33**, 1439–1443.

107 Mehrotra, R. and Aowal, A.F.S.A. (1982) *Journal of Industrial Engineering*, **62**, 43–46.

108 Urbanc-Berčič, O. and Gaberščik, A. (1989) *Aquatic Botany*, **35**, 409–413.

109 Aneija, K.R. and Singh, K. (1992) *Proceedings of the Indian National Science Academy*, **B58**, 357–364.

110 Smith, G. (1998) *LakeLine*, **18**, 20–21.

111 Commonwealth Science Council (1979) Review Meeting on Management of Water Hyacinth, Report of the First Review Meeting on Management of Water Hyacinth, Commonwealth Science Council, London.

112 Commonwealth Science Council (1981) Review Meeting on Management of Water Hyacinth, Report of the Second Review Meeting on Management of Water Hyacinth, Commonwealth Science Council, London.

113 Commonwealth Science Council (1980) Interim Project Review Meeting on Management of Water Hyacinth, First Interim Project Review Meeting on Management of Water Hyacinth, Commonwealth Science Council, London.

114 Klein, D.A., Salzwedel, J.L. and Dazzo, F.B. (1990) *Biotechnology of Plant–Microbe Interactions* (eds J.P. Nakas and C. Hagedorn), McGraw-Hill Publishing Company, New York, pp. 190–225.

115 Emiliani, F., Lajmanovich, R. and Gonzalez, S.M. (2001) *Revista Argentina de Microbiologia*, **33**, 65–74.

116 Noeori, A. and Reddy, K.R. (2000) *The Botanical Review*, **66**, 350–378.

117 Zhao, D. and Zheng, S. (1996) *Chinese Journal of Applied Ecology*, **7**, 435–438.

118 Paganetto, A., Carpaneto, A. and Gambale, F. (2001) *Plant Cell & Environment*, **24**, 1329–1336.

119 Pitman, M.G. (1965) *Australian Journal of Biological Sciences*, **18**, 547–553.

120 Lauchli, A. (1976) *Encyclopedia of Plant Physiology, Transport in Plants. II. Part B. Tissues and Organs* (eds U. Luttge and A.M.G. Pitman), Springer-Verlag, Berlin, pp. 3–34.

121 Haynes, R.J. (1980) *The Botanical Review*, **46**, 76–99.

122 McGrath, S.P., Zhao, F.J. and Lombi, E. (2000) *Plant Soil*, **232**, 207–214.

123 Ding, X., Jiang, J., Wang, Y.Y., Wang, W.Q. and Ru, B.G. (1994) *Environmental Pollution*, **84**, 93–96.

124 Ding, X., Wang, W.Q., Jiang, J., Ru, B.G. and Wang, Y.Y. (1994) *Science in China. Series B Chemistry, Life Sciences & Earth Sciences*, **37**, 303–309.

125 Enany, A.E. and Mazen, A.M.A. (1996) *Water Air, and Soil Pollution*, **87**, 357–362.

126 Cobbett, C.S. and Goldsbrough, P.B. (2000) *Phytoremediation of Toxic Metals: Using Plants to Clean Up the Environment* (ed. I. Raskin), John Wiley & Sons, Inc., New York, pp. 247–269.

127 Khan, A.G., Kuek, C., Chaudhry, T.M., Khoo, C.S. and Hayes, W.J. (2000) *Chemosphere*, **41**, 197–207.

128 Barber, D.A. and Lee, R.B. (1974) *New Phytologist*, **43**, 97–106.

129 Crowely, D.E., Wang, Y.C., Reid, C.P.P. and Szaniszlo, P.J. (1991) *Plant Soil*, **130**, 179–198.

130 Dushenkov, S. and Kapulnik, Y. (2000) *Phytoremediation of Toxic Metals: Using Plants to Clean Up the Environment* (ed. I. Raskin), John Wiley & Sons, Inc., New York pp. 89–105.

131 Burd, G.I., Dixon, D.G. and Glick, B.R. (1998) *Applied and Environmental Microbiology*, **64**, 3663–3668.

132 Burd, G.I., Dixon, D.G. and Glick, B.R. (2000) *Canadian Journal of Microbiology*, **46**, 237–245.

133 Carlot, M., Giacomini, A. and Casella, S. (2002) *Acta Biotechnologica*, **22**, 13–20.

134 Śouza, M.P., Oilot-Snits, M.A.H. and Terry, N. (1999) *Phytoremediation of Toxic Metals: Using Plants to Clean Up the Environment* (ed. I. Raskin), John Wiley & Sons, Inc., New York, pp. 171–190.

135 Pishchik, V.N., Vorob'ev, N.I. and Provorov, N.A. (2005) *Microbiology*, **74** (6), 735–740.

136 Amico, E.D., Cavalca, L. and Andreoni, V. (2005) *FEMS Microbiology Ecology*, **52**, 153–162.

137 Shirley, S., Fernandezi, A., Benlloch, M. and Sancho, E.D. (2006) *Ohytoremediation of Metal-Contaminated Soils* (ed. J.L. Morel), Springer, Netherlands, pp. 315–318.

138 Zheng, S.Z. and He, M. (1990) *Journal of Ecology*, **9**, 56–57.

139 So, L.M., Chu, L.M. and Wong, P.K. (2003) *Chemosphere*, **52**, 1499–1503.

140 Akcin, G., Guldede, N. and Saltabas, O. (1994) *Journal of Environmental Science and Health Part A*, **29**, 2177–2183.

141 Vesk, P.A., Nockolds, C.E. and Allaway, W.G. (1999) *Plant Cell, & Environment*, **22**, 149–158.

Index

Plant-Bacteria Interactions. Strategies and Techniques to Promote Plant Growth
Edited by Iqbal Ahmad, John Pichtel, and Shamsul Hayat
Copyright © 2008 WILEY-VCH Verlag GmbH & Co. KGaA, Weinheim
ISBN: 978-3-527-31901-5